Structure Correlation

Volume 1

Edited by
H.-B. Bürgi and J. D. Dunitz

©VCH Verlagsgesellschaft mbH, D-69451 Weinheim (Federal Republic of Germany), 1994

Distribution:

VCH, P. O. Box 101161, D-69451 Weinheim, Federal Republic of Germany

Switzerland: VCH, P. O. Box, CH-4020 Basel, Switzerland

United Kingdom and Ireland: VCH, 8 Wellington Court, Cambridge CB1 1HZ, United Kingdom

USA and Canada: VCH, 220 East 23rd Street, New York, NY 10010–4606, USA

Japan: VCH, Eikow Building, 10-9 Hongo 1-chome, Bunkyo-ku, Tokyo 113, Japan

ISBN 3-527-29042-7 (VCH, Weinheim)   ISBN 1-56081-752-6 (VCH, New York)

Structure Correlation

Volume 1

Edited by
Hans-Beat Bürgi and
Jack D. Dunitz

VCH

Weinheim · New York
Basel · Cambridge · Tokyo

CHEMISTRY

Prof. Dr. H.-B. Bürgi
Laboratorium für Kristallographie
der Universität
Freiestr. 3
CH-3012 Bern
Switzerland

Prof. Dr. J. D. Dunitz
Laboratorium für Organische Chemie
ETH Zentrum
Universitätsstr. 16
CH-8092 Zürich
Switzerland

Published jointly by
VCH Verlagsgesellschaft mbH, Weinheim (Federal Republic of Germany)
VCH Publishers, Inc., New York, NY (USA)

Editorial Director: Dr. Thomas Mager
Production Manager: Elke Littmann

Library of Congress Card No. applied for.

A catalogue record for this book is available from the British Library.

Deutsche Bibliothek Cataloguing-in-Publication Data:
Structure correlation / ed. by Hans-Beat Bürgi and Jack D. Dunitz. –
Weinheim ; New York ; Basel ; Cambridge ; Tokio :
VCH.
ISBN 3-527-29042-7 (Weinheim...)
ISBN 1-56081-752-6 (New York...)
NE: Bürgi, Hans-Beat [Hrsg.]
Volume 1 (1994)

©VCH Verlagsgesellschaft mbH, D-69451 Weinheim (Federal Republic of Germany), 1994

Printed on acid-free and chlorine-free paper.

Composition: K + V Fotosatz GmbH, D-64743 Beerfelden. Printing: betz-druck gmbh, D-64291 Darmstadt. Bookbinding: Industrie- und Verlagsbuchbinderei Heppenheim GmbH, D-64630 Heppenheim.
Printed in the Federal Republic of Germany

Preface

Since the middle of the 19th century, the molecular hypothesis has established itself as one of the most far reaching products of the human intellect and certainly as the foundation of modern chemistry. Especially since the Braggs started to investigate the structure of matter at the atomic scale, about 80 years ago, the molecular paradigm (in the sense proposed by T. S. Kuhn [1]) has been at the origin of important developments in physics, chemistry, and biology, and has been instrumental in achieving the coalescence of large areas of these sciences. The preoccupation with molecules has led, among other achievements, to the invention and development of ever better, more efficient and diverse methods of molecular structure determination in the gaseous, liquid, and solid states of matter. Molecular models, portrayed as formulas, perspective drawings, collections of balls and sticks, or as elaborate computer-drawn stereoviews, encode an enormous amount of information and provide problems as well as their solutions to the community of physicists, chemists, biochemists and molecular biologists — or, as Hoffmann has written [2]:

> "There is no more basic enterprise in chemistry than the determination of the geometrical structure of a molecule. Such a determination, when it is well done, ends all speculation as to the structure and provides us with the starting point for the understanding of every physical, chemical and biological property of the molecule."

For many years the standard approach to many problems in the study of matter was to determine a single relevant molecular structure and relate it to the material properties of interest. It was soon discovered that, irrespective of the problem at hand, the metrical information obtained in this way showed a high degree of regularity. Pauling expressed this state of affairs in his book "The Nature of the Chemical Bond" [3]:

> "It has been found that the values of interatomic distances corresponding to covalent bonds can be correlated in a simple way in terms of a set of values of covalent bond radii of atoms."

Since then this statement has been confirmed by structure determinations of about 100,000 different molecules. Pauling's finding has been expanded and put to practical use in tabulations of standard bond lengths and angles and of standard dimensions for structural fragments that are important in macromolecular crystallography, such as the peptide moiety and the nucleic acid bases.

Pauling and his co-workers also noticed that standard distances based on covalent radii sums needed to be slightly adjusted for certain types of bond (e.g. those involving large electronegativity differences between the atoms) and also depended somewhat on the number and type of other atoms surrounding the bond in question. More recently, such correlations involving bond types and the environments in which they occur have been exploited systematically in the tables of standard bond lengths published by the Cambridge Crystallographic Data Centre (CCDC) [4].

This book deals with the extension and generalization of these two ideas – standardized fragments and their correlation with environment. The underlying concept is to specify a chemical fragment, to analyze all available structural data for this fragment, and hence to obtain not only one or more prototypical structures of the fragment but also a detailed description of the distribution of fragment structures. Such a description shows – more often than not – correlation among some of the structural parameters. This correlation is characteristic of the fragment, as characteristic as the prototypical structures themselves. It is this finding that has led us to the title of this book, "Structure Correlation".

The results of such a correlation analysis would be of limited interest if restricted to mere statistical relationships; they call to be interpreted in terms of their chemical content. It seems to us that the concept of energy surfaces provides a natural, adequate, and fertile basis for such interpretation. Over the last decade or so, this approach has led to a better understanding of fragment flexibility, conformational analysis, reaction pathways or mechanisms, and of the intricacies in molecular design, especially when combined with other physico-chemical quantities. Of course, many ideas to be presented in this book have been invented and developed under headings quite other than "Structure Correlation". However, our main goal here has been to bring together experts willing to cooperate in producing an overview of modern methods for interpreting structural data, in the spirit of Hoffmann's statement. The following synopsis is intended to illustrate the thoughts that have guided our selection of topics.

Part 1 covers basics. Chapter 1 summarizes different ways of describing the geometric structure of molecules. Chapter 2 addresses the unavoidable topic of symmetry, which has proved to be of great conceptual and practical importance in this type of work. Chapter 3 deals with the central problem of getting at the necessary data and putting them into a form suitable for analysis. Chapter 4 introduces the statistical tools for analyzing distributions, and Chapter 5 adds the physical concepts necessary for the interpretation in chemical terms.

Part 2 points at one possible escape route of small-molecule crystallography from the confines of a narrow, static viewpoint into the wider realm of chemical dynamics. It collects contributions in the broad area of reaction pathways, ranging from purely structural aspects of chemical transformations (structure-structure correlations) to the relations between structure and energy (structure-energy correlation). Chapter 6 deals with nucleophilic addition to carbonyl groups and the reverse decay of ketal fragments. Chapter 7 illustrates the widespread occurrence of three-center four-electron systems and their relevance to organic and inorganic substitution reactions. Chapter 8 focuses on other substitution reactions and on fluxionality in molecular inorganic chemistry. Chapter 9 takes up the problem of molecular flexibility in the context of conformational analysis of organic molecules. In some of these chapters results of quantum chemical calculations are discussed. Increase in computing power and improvements in software are continuing to make such calculations an important source of molecular structures and energies, which are only beginning to be analyzed with the methods of structure-structure and structure-energy correlations.

The boundary between bonding and non-bonding interactions, between molecular and extended crystal structures, is not always very clear cut. Part 3 gives a discussion of crystal packing from several points of view. Chapter 10 reviews the extension of local correlations to non-molecular, extended inorganic solids. Chapter 11 takes up the important topic of hydrogen bonding and its role in molecular packing. Chapter 12 is concerned with the problem of deriving more precise descriptors of molecular shape and size, as well as suitable non-bonded potential energy functions. The problem of predicting the crystal structure of a given molecule of known structure is still unsolved, but the insights emerging from systematic analysis of the factors governing crystal packing are very useful in the burgeoning areas of crystal design and supramolecular chemistry.

Finally, Part 4 aims at a demonstration of the use that can be made of the concepts and methods developed for small molecules for the description and understanding of the structure of proteins, where the degree of structural detail common for smaller molecules may be neither necessary nor experimentally available. Chapter 13 attempts such an amalgamation quite directly by discussing the mode of action of enzymes with the help of reaction path data and structure correlations derived from the small molecule area. Chapter 14 tackles the problem of mapping the active sites of steroid hormone receptors using structural data on more than a thousand steroid molecules; this is an area where crystal-structure analyses of the protein receptors are just beginning to provide the necessary checks and insights. Chapter 15 summarizes currently recognized patterns in the secondary and tertiary structure of globular proteins. It emphasizes the recurring motives, rather than their variability, in a way that is reminiscent of Pauling's early systematization of interatomic distances. Chapter 16 takes up the search for correlations among amino-acid variations (primary structure) and the conservation of tertiary structure in families of homologous proteins. Chapter 17 addresses the problem of predicting secondary and tertiary structure from local hexapeptide sequence patterns culled from the Protein Data Bank [5]. The scale of this problem can be appreciated by the consideration that of the $20^6 = 64 \cdot 10^9$ possible hexapeptide fragments only a tiny fraction (about $5 \cdot 10^4$) occur in proteins of known tertiary structure. The final Chapter 18 is concerned with structural patterns in nucleic acids and in short oligonucleotide fragments. Whereas fiber diffraction diagrams of DNA could yield only averaged dimensions for the well known helical structures, we now have more detailed information for several oligonucleotide duplexes that reveal the structural variability associated with individual base sequences.

In Appendix A published tables of standard interatomic distances [4] are reproduced in slightly abbreviated form. We have limited ourselves to a selection of the more reliable standards, those based on at least four observed distances for the type of interatomic contact in question; for the others the reader is referred to the original tables [4]. Appendix B gives database identification codes (refcodes) to crystal structure analyses mentioned in the book and will be useful in locating the original literature sources for the primary structural data. Appendix C lists the one- and three-letter codes for the amino acids found in proteins and may be helpful to readers who have not learned this by heart.

A certain degree of overlap among the various chapters is unavoidable − and perhaps even useful. Our authors were given a fairly free hand, and although they may share a broad common viewpoint, they certainly do not agree about everything. Topics discussed in different chapters will reflect the views, insights, and prejudices of the individual authors. Where there is overlap we have the privilege of listening to the different voices of these authors discussing the same material. Any attempt at complete avoidance of overlap would have meant intolerable restrictions in the terms of reference of the various authors and would have led to an undesirable rigidity in the subject matter of the chapters and to endless cross referencing. Most of the individual chapters can be read on their own and do not depend excessively on an understanding of material in other parts of the book.

We take particular satisfaction from the consideration that this book is a fulfilment of a credo stated by one of us more than 25 years ago in the preface to a short-lived series of review articles on aspects of structural chemistry [6]:

"One of the most important tasks in structural chemistry today is, we believe, that of making critical assessments of the mass of already published material in the search for unifying ideas. To mention only two examples: we know almost nothing about the weak interactions that control preferred packing arrangements of molecules in crystals, and we are only beginning to understand the conformational complexities of organic molecules, although a vast amount of information of these and other untouched matters is stored away in the literature. Much of this potentially valuable structural information implicit in the results of diffraction analyses is often passed over by the authors themselves (who are presumably happy enough to have completed the analysis successfully) and has to be actively sought and recovered from the authors' description of their results." ...

One difficulty, it was realized at that time, was that of finding a sufficient number of knowledgeable, suitably qualified, and willing authors. We are very glad to have assembled the team of knowledgeable, suitably qualified, and willing authors whose names appear in the list of contributors. It is time to express our thanks to them not only for the thoroughness and clarity of their contributions, but also for the patience and composure with which they endured our editorial carping criticisms.

Hans-Beat Bürgi
Jack D. Dunitz

References

[1] Kuhn, T.S., *The Structure of Scientific Revolutions*, University of Chicago Press, **1970**
[2] Hoffmann, R., Foreword to Vilkov, L.V., Mastryukov, V.S., Sadova, N.I., *Determination of the Geometrical Structure of Free Molecules*, MIR Publishers, Moscow, **1983**
[3] Pauling, L., *The Nature of the Chemical Bond*, 3rd Edn., Cornell University Press, **1960**, p. 221
[4] Allen, F.H., Kennard, O., Watson, D.G., Brammer, L., Orpen, A.G., Taylor, R., *J. Chem. Soc. Perkin II*, **1987**, S1−S19; Orpen, A.G., Brammer, L., Allen, F.H., Kennard, O., Watson, D.G., Taylor, R., *J. Chem. Soc. Dalton* **1989**, S1−S83
[5] Protein Data Bank, Chemistry Department, Brookhaven National Laboratory, Upton, NY 11973, USA
[6] Dunitz, J.D., Ibers, J.A. (Eds.) *Perspectives in Structural Chemistry*, Wiley, New York. Volume 1, **1967**; Volume 2, **1968**; Volume 3, **1970**; Volume 4, **1971**

Contents

Part II Molecular Structure and Reactivity

Part III Crystal Packing

Part IV Proteins and Nucleic Acids

XXIV *Contents*

List of Contributors

Frank H. Allen
Cambridge Crystallographic
Data Centre
Union Road
Cambridge CB2 1EZ
England

Thomas Auf der Heyde
Department of Chemistry
University of the Western Cape
Bellville 7535
South Africa

Joel Bernstein
Department of Chemistry
Ben-Gurion University
of the Negev
Beer Sheva 84105
Israel

Tom Blundell
Department of Crystallography
Birkbeck College
London University
Malet Street
London WC1E 7HX, England

L. Brammer
Department of Chemistry
University of New Orleans
New Orleans, LA 70148
USA

Clemens Broger
F. Hoffmann-La Roche Ltd.
Pharma Research − New Technologies
CH-4002 Basel
Switzerland

I. D. Brown
Institute for Materials Research
McMaster University
Hamilton, Ontario L8S 4MI
Canada

Hans-Beat Bürgi
Laboratory of Crystallography
University of Bern
Freiestraße 3
CH-3012 Bern
Switzerland

Andrzej Stanislaw Cieplak
Department of Chemistry
Yale University[†]
New Haven, CT 06517
USA

William L. Duax
Medical Foundation
of Buffalo, Inc.
73 High Street
Buffalo, NY 14203-1196
USA

[†] Visiting scientist

Jack D. Dunitz
Organic Chemistry Laboratory
ETH-Zentrum
CH-8092 Zürich
Switzerland

Martin Egli
Organic Chemistry Laboratory
ETH-Zentrum
CH-8092 Zürich
Switzerland

Margaret C. Etter[††]
Department of Chemistry
University of Minnesota
Minneapolis, MN 55455
USA

A. Gavezzotti
Dipartimento di Chimica Fisica
ed Elettrochimica e Centro CNR
University of Milan
Milano, Italy

Debashis Ghosh
Medical Foundation
of Buffalo, Inc.
73 High Street
Buffalo, NY 14203-1196
USA

Jane F. Griffin
Medical Foundation
of Buffalo, Inc.
73 High Street
Buffalo, NY 14203-1196
USA

E. Gail Hutchinson
Department of Biochemistry
and Molecular Biology
University College
Gower Street
London WC1E 6BT
England

Olga Kennard
Cambridge Crystallographic
Data Centre
Union Road
Cambridge, CB2 1EZ
England

Gerhard Klebe
Main Laboratory of BASF-AG
Carl-Bosch-Straße
D-67056 Ludwigshafen
Germany

Leslie Leiserowitz
Department of Materials
and Interfaces
Weizmann Institute of Science
Rehovoth 76100
Israel

A. Louise Morris
Biomolecular Structure
and Modeling Unit
Department of Biochemistry
and Molecular Biology
University College
Gower Street
London WC1E 6BT
England

Klaus Müller
F. Hoffmann-La Roche Ltd.
Pharma Research – New Technologies
CH-4002 Basel
Switzerland

A. G. Orpen
School of Chemistry
University of Bristol
Bristol BS8 1TS
England

W. Bernd Schweizer
Organic Chemistry Laboratory
ETH Zentrum
CH-8092 Zürich
Switzerland

[††] deceased

Valery Shklover
Institute of Crystallography
and Petrography
ETH-Zentrum
CH-8092 Zürich
Switzerland

Robin Taylor
Zeneca Agrochemicals
Jealott's Hill Research Station
Bracknell, Berkshire, RG12 6EY
England

Janet M. Thornton
Department of Biochemistry
and Molecular Biology
University College
Gower Street
London WC1E 6BT
England

David G. Watson
Cambridge Crystallographic
Data Centre
Union Road
Cambridge CB2 1EZ
England

Part I
Basics

1 Molecular Structure and Coordinate Systems

Jack D. Dunitz and Hans-Beat Bürgi

1.1 Molecules and Molecular Fragments

Chemists think about molecular structure in many different ways and are helped by many kinds of symbols – from squiggles on the backs of old envelopes to mechanical models or, nowadays, to elaborate pictorial representations obtained with computer graphics. Most molecular models are based on a few simple rules with respect to the number of nearest-neighbor interactions between atoms and the magnitude of corresponding interatomic distances, leading to the familiar line formulae or the more abstract incidence matrices representing connectivity. Such a matrix contains on the main diagonal the types of atomic nuclei in the molecule, while the off-diagonal elements denote neighborhood relationships, usually coded in terms of some model of chemical bonding (single, double, triple, fractional bond, non-bond, hydrogen bond, etc., Figure 1.1, left).

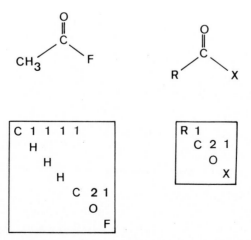

Fig. 1.1. Line formulae and incidence matrix representations; left: acetyl fluoride; right: the family of alkyl substituted carbonyl compounds with any substituent X

For structure correlation purposes the notion of a molecule is unnecessarily restrictive. It has long been recognized that the geometry of certain fragments within molecules remains fairly constant and depends only slightly on their environment. To obtain a suitable definition of a molecular or structural fragment, we rely on the widespread chemical practice of dissecting molecules into a constant backbone, frame or fragment, and variable functional groups, substituents or ligands. We can denote a family of molecules incorporating a common fragment in terms of a modified incidence matrix. It contains the atoms and bonds that are the same for all members of the family and a number of variable substituents, R or X as in Figure 1.1, right.

If these symbols and models are to have any metrical significance, if they are to portray the relative atomic positions in a reasonably realistic way, they need to be based on numerical data. To define the atomic positions we need numbers − a set of coordinates. These may be external coordinates, referred to some set of coordinate axes, or they may be internal coordinates, defining the various interatomic distances and angles in the molecule of interest.

If one wishes to compare a family of related molecules differing only in peripheral substituents, one runs into the question of how to define the "position" of a variable substituent. One possibility is to use the position of the substituent atom directly linked to the constant fragment but without assigning an elemental symbol to it. Another possibility is to consider interatomic vectors; for a vector between an atom of the constant fragment and the first atom of the substituent, only the direction is important, not the length.

In this chapter we discuss different ways of describing molecular structure and the relationships among them. For our purpose, we consider the structure of a molecule to be given by a set of atomic (nuclear) positions and associated interatomic vectors. In different contexts this set of atomic positions could correspond to a single arrangement of atomic nuclei or to an average over some probability distribution connected with the nature of the experiment (usually crystal structure analysis); similarly it could represent an actual energy minimum or an unstable nuclear configuration. Such matters will be taken up in later chapters.

1.2 Positional Coordinates

1.2.1 Crystal Coordinates

The results of a crystal structure are usually expressed in the primary literature as a numerical table of positional coordinates and vibrational parameters for the atoms contained in an asymmetric unit of structure. The asymmetric unit is repeated by the appropriate combination of space-group symmetry operations and lattice translations to give the crystal structure.

The asymmetric unit does not necessarily correspond to a single molecule and may contain more than one molecule. For a molecule that occupies a site of non-trivial crystallographic symmetry, the asymmetric unit corresponds to only part of the complete molecule, or sometimes even to parts of more than one molecule. The symmetry operations necessary to build up complete molecules are sometimes provided in the paper. More often, they have to be deduced from the space-group information given in International Tables for X-Ray Crystallography [1]. The choice of any particular asymmetric unit in a symmetric, repeating pattern is, of course, arbitrary. Most authors will take care that atoms in the same molecule belong to the same asymmetric unit. But there are many ways of choosing the asymmetric unit, and it can happen that the unit chosen in some particular case comprises nonequivalent atoms belonging to different molecules.

Atomic positions are usually expressed as fractional coordinates x_i, y_i, z_i, fractional scalar components along the a, b, c crystal axes, respectively. The scalar quantities a, b, c are the lengths of a, b, c and the interaxial angles are usually denoted by the Greek letters $\alpha(bc)$, $\beta(ca)$ and $\gamma(ab)$. In some space groups, International Tables allows a choice of origins (i.e. inversion center or intersection of rotation axes). Ideally, there should be no ambiguity about the origin of the crystal coordinates, but this ideal is not always attained. Unless it is unambiguously defined, the location of the origin should always be checked.

Fractional scalar components can be converted to distance components by multiplying them by the axial lengths, but the resulting system is Cartesian only if the crystal axes are mutually orthogonal. This is the case only for the orthorhombic, tetragonal, and cubic crystal systems. For triclinic, monoclinic, rhombohedral and hexagonal lattices, the crystal axes are oblique and the simple formulas for interatomic distances and angles are no longer valid. In an oblique system the length of a vector $v(x,y,z)$ with scalar components x, y, z along a, b, c respectively is

$$|v| = (v \cdot v)^{1/2} = [(xa+yb+zc) \cdot (xa+yb+zc)]^{1/2}$$

$$= (x^2a^2+y^2b^2+z^2c^2+2xyab \cos \gamma+2xzac \cos \beta+2yzbc \cos \alpha)^{1/2} \ .$$

The expression inside the brackets can be written as a matrix product:

$$[xyz] \begin{bmatrix} a^2 & ab \cos \gamma & ac \cos \beta \\ ab \cos \gamma & b^2 & bc \cos \alpha \\ ac \cos \beta & bc \cos \alpha & c^2 \end{bmatrix} \begin{bmatrix} x \\ y \\ z \end{bmatrix}$$

$$= [xyz] \begin{bmatrix} a \cdot a & a \cdot b & a \cdot c \\ a \cdot b & b \cdot b & b \cdot c \\ a \cdot c & b \cdot c & c \cdot c \end{bmatrix} \begin{bmatrix} x \\ y \\ z \end{bmatrix} = v^{\mathrm{T}} G v$$

so that

$$|v| = (v^{\mathrm{T}} G v)^{1/2} \ .$$

Here v^{T} represents the components (x,y,z) of the vector along the three axes a, b, c as a row vector while v contains the same information expressed as a column vector.

(The distinction between row and column vectors is a formality required by the rules of matrix multiplication.) The matrix G is known as the metric matrix of the three basis vectors. Its determinant is equal to the volume squared of the parallelepiped formed by the three basis vectors, i.e. of the unit cell:

$$|G| = V^2 = a^2 b^2 c^2 (1 - \cos^2 \alpha - \cos^2 \beta - \cos^2 \gamma + 2 \cos \alpha \cos \beta \cos \gamma)$$

as can be checked by multiplying out.

The angle between two vectors v_1 and v_2 with components along oblique axes is given by the dot product formula

$$\cos \theta_{12} = [(x_1 a + y_1 b + z_1 c) \cdot (x_2 a + y_2 b + z_2 c)] / |v_1 v_2|$$

while the cross product is given by

$$v_1(x_1, y_1, z_1) \times v_2(x_2, y_2, z_2)$$

$$= (x_1 a + y_1 b + z_1 c) \times (x_2 a + y_2 b + z_2 c)$$

$$= (x_1 y_2 - x_2 y_1)(a \times b) + (y_1 z_2 - y_2 z_1)(b \times c) + (z_1 x_2 - z_2 x_1)(c \times a) .$$

Many of the complications of working in oblique coordinate systems can be simplified by the use of reciprocal basis vectors, as described in any crystallographic textbook. And, of course, they can be avoided by working, where possible, in Cartesian coordinates, based on unit vectors along mutually orthogonal directions. In this coordinate system the usual expressions for dot and cross product in terms of vector components apply:

$$v_1(x_1, y_1, z_1) \cdot v_2(x_2, y_2, z_2) = x_1 x_2 + y_1 y_2 + z_1 z_2$$

$$v_1(x_1, y_1, z_1) \times v_2(x_2, y_2, z_2) = \begin{bmatrix} i & j & k \\ x_1 & y_1 & z_1 \\ x_2 & y_2 & z_2 \end{bmatrix} .$$

The expressions for calculating distances and angles become correspondingly simpler:

$$|v| = (x^2 + y^2 + z^2)^{1/2}$$

$$\cos \theta_{12} = (x_1 x_2 + y_1 y_2 + z_1 z_2) / |v_1 v_2| .$$

1.2.2 Linear Transformations

It is often desirable to change from one coordinate system to another. For example, we may want to compare the results of two crystal structure analyses of the same compound, each based on a different choice of axes; or we may want to change from an oblique coordinate system to a Cartesian one, or vice versa. Suppose that one coordinate system is based on a particular set of three non-coplanar basis vectors a_1, a_2, a_3, and that the second set b_1, b_2, b_3 is related to the first by the equations:

$$b_1 = A_{11}a_1 + A_{12}a_2 + A_{13}a_3$$

$$b_2 = A_{21}a_1 + A_{22}a_2 + A_{23}a_3$$

$$b_3 = A_{31}a_1 + A_{32}a_2 + A_{33}a_3$$

where the coefficients are any real numbers. In matrix notation this is

$$\begin{bmatrix} b_1 \\ b_2 \\ b_3 \end{bmatrix} = \begin{bmatrix} A_{11} & A_{12} & A_{13} \\ A_{21} & A_{22} & A_{33} \\ A_{31} & A_{32} & A_{33} \end{bmatrix} \begin{bmatrix} a_1 \\ a_2 \\ a_3 \end{bmatrix}$$

which can be condensed to

$$b = A a \quad \text{or} \quad b_i = A_{ij}a_j$$

where summation over the repeated index is implied.

The reverse transformation is

$$a = B b \quad \text{or} \quad a_i = B_{ij}b_j$$

where $A^{-1} = B$ is the inverse matrix of A. The elements of the inverse matrix B are:

$$B_{ij} = (-1)^{i+j} (\text{minor of } A_{ji})/|\det A| \ .$$

A vector is unaffected by a change of axes, but its components are altered. Let $v(x_1, x_2, x_3)$ be the vector in coordinate system a_i, and let $v(y_1, y_2, y_3)$ be the same vector in coordinate system b_i:

$$v = y_1 b_1 + y_2 b_2 + y_3 b_3 = x_1 a_1 + x_2 a_2 + x_3 a_3$$

$$= \quad x_1 (B_{11}b_1 + B_{12}b_2 + B_{13}b_3)$$

$$+ x_2 (B_{21}b_1 + B_{22}b_2 + B_{23}b_3)$$

$$+ x_3 (B_{31}b_1 + B_{32}b_2 + B_{33}b_3)$$

$$= \quad (B_{11}x_1 + B_{21}x_2 + B_{31}x_3) b_1$$

$$+ (B_{12}x_1 + B_{22}x_2 + B_{32}x_3) b_2$$

$$+ (B_{13}x_1 + B_{23}x_2 + B_{33}x_3) b_3 \ .$$

Equating coefficients, we obtain

$$y_1 = B_{11}x_1 + B_{21}x_2 + B_{31}x_3$$

$$y_2 = B_{12}x_1 + B_{22}x_2 + B_{32}x_3$$

$$y_3 = B_{13}x_1 + B_{23}x_2 + B_{33}x_3$$

or

$$y_i = B_{ji}x_j \ .$$

Thus the transformation matrix for vector components is the transpose of the inverse of A, the transformation matrix for basis vectors. For transformations among Cartesian coordinate systems, we have the special relationships,

$$\boldsymbol{b}_i \cdot \boldsymbol{b}_i = 1 \qquad \boldsymbol{b}_i \cdot \boldsymbol{b}_j = 0$$

and similarly for the $\boldsymbol{a}_i$'s. Under these conditions, the matrix elements are direction cosines of one set of vectors referred to the other set; sums of squares along rows or columns equal unity, and sums of products of different rows or columns equal zero. For such transformations (orthonormal transformations)

$$B = A^{-1} = A^{\mathrm{T}} \ ,$$

i.e. the transpose of a matrix is the same as its inverse. Thus, among Cartesian coordinate systems, vector components transform in the same way as the basis vectors.

To avoid the complications of working in oblique coordinate systems, it is often convenient to transform immediately from crystal coordinates to Cartesians. The most general transformation of this kind is from triclinic crystal coordinates. We assume that the triclinic axes form a right-handed coordinate system and choose the Cartesian basis vectors $\boldsymbol{e}_1$, $\boldsymbol{e}_2$, $\boldsymbol{e}_3$ such that $\boldsymbol{e}_1$ is a unit vector along $\boldsymbol{a}$, $\boldsymbol{e}_2$ is a unit vector in the $\boldsymbol{ab}$ plane, and $\boldsymbol{e}_3$ is $\boldsymbol{e}_1 \times \boldsymbol{e}_2$, completing a right-handed Cartesian system. The calculation is straightforward but laborious, and we merely give the result [2]

$$
\begin{bmatrix} \boldsymbol{e}_1 \\ \boldsymbol{e}_2 \\ \boldsymbol{e}_3 \end{bmatrix}
=
\begin{bmatrix}
\dfrac{1}{a} & 0 & 0 \\[2ex]
\dfrac{-\cos\gamma}{a\sin\gamma} & \dfrac{1}{b\sin\gamma} & 0 \\[2ex]
\dfrac{\cos\gamma\cos\alpha-\cos\beta}{au\sin\gamma} & \dfrac{\cos\gamma\cos\beta-\cos\alpha}{bu\sin\gamma} & \dfrac{\sin\gamma}{cu}
\end{bmatrix}
\begin{bmatrix} a \\ b \\ c \end{bmatrix}
$$

where u is the volume of the parallelepiped formed by the unit vectors $\boldsymbol{a}/a$, $\boldsymbol{b}/b$, $\boldsymbol{c}/c$. The transformation we usually want is from the fractional crystal coordinates x, y, z to the Cartesian coordinates s_1, s_2, s_3. It is

$$\begin{bmatrix} s_1 \\ s_2 \\ s_3 \end{bmatrix} = \begin{bmatrix} a & b\cos\gamma & c\cos\beta \\ 0 & b\sin\gamma & \dfrac{c\,(\cos\alpha - \cos\beta\,\cos\gamma)}{\sin\gamma} \\ 0 & 0 & \dfrac{cu}{\sin\gamma} \end{bmatrix} \begin{bmatrix} x \\ y \\ z \end{bmatrix}$$

or in short:

$$s = Pv \ .$$

1.2.3 Symmetry Transformations

A symmetry transformation is a special kind of linear transformation (see Chapter 2). In crystals, symmetry transformations generally have two parts: a rotation (or rotation plus reflection or inversion) S_i and a translation t_i. Thus,

$$v_i = S_i v_1 + t_i \ .$$

For each space group there is a group of operations S_i that leave at least one point in space unchanged – point group operations. The set of operations (S_i, t_i) then converts a given point (vector) v_1 into all symmetry-equivalent points in the crystal. The matrices S_i and the translation vectors t_i are expressed most simply and concisely in crystal coordinates, but compromises may sometimes be called for. For example, in the study of molecular packing in crystals (see Chapter 12) it may be desirable to combine the necessity of using symmetry relationships with the advantages of working in an orthogonal coordinate system. The required transformations are then (see Section 1.2.2):

$$s_i = Pv_i = P(S_i v_1 + t_i) = PS_i P^{-1} s_1 + Pt_i \ .$$

1.2.4 Molecule or Fragment Centered Coordinate Systems

When comparing molecules or molecular fragments from different environments, e.g. from different crystal structures with different unit cells and space groups, it is usually convenient to use a coordinate system related to the molecule or molecular fragment itself rather than to the structure in which it is embedded. For this purpose we need agreement on the choice of the origin and orientation of such a coordinate system. The origin s_0 is usually chosen as some weighted average of the positional coordinates s_i of the set of atoms involved:

$$s_0 = \sum (w_i s_i) / \sum w_i \ .$$

Various choices of w_i are in common use:

- if $w_i = 1$, unit weights, s_0 is the geometric center of the collection of points (atoms);
- if $w_i = m_i$, the mass (or atomic weight) of atom i, s_0 is the center of mass of the molecule or fragment;
- if $w_i = q_i$, the electric charge of atom i, s_0 is the center of charge;
- if $w_i = q_i$, and the sum is taken only over the positive (negative) charges, s_0^+ (s_0^-) is the center of positive (negative) charge in the molecule or fragment, which may be needed for the calculation of the electric dipole moment, $d = s_0^- - s_0^+$. Note that the dipole moment d is independent of the choice of origin only if the molecule is electrically neutral, $\sum q_i = 0$.

The orientation of the coordinate system is obtained from a second moment matrix M that is constructed as follows:

$$r_i = s_i - s_0$$

$$M = \sum w_i r_i r_i^{\mathrm{T}}$$

where r^{T} is the transpose of r, or written out in full:

$$M = \begin{bmatrix} \sum w_i r_{i1}^2 & \sum w_i r_{i1} r_{i2} & \sum w_i r_{i1} r_{i3} \\ \sum w_i r_{i1} r_{i2} & \sum w_i r_{i2}^2 & \sum w_i r_{i2} r_{i3} \\ \sum w_i r_{i1} r_{i3} & \sum w_i r_{i2} r_{i3} & \sum w_i r_{i3}^2 \end{bmatrix}$$

where $r_{ij} = s_{ij} - s_{0j}$ are the components of r_i.

With $w_i = 1$, M is a coordinate second-moment matrix with eigenvectors e_1, e_2, e_3. The corresponding eigenvalues M_1, M_2, M_3 (in decreasing order) are the sums of squares of deviations of the atoms from the planes normal to these vectors. Thus e_3 is normal to the "best" plane, e_1 is normal to the "worst" plane, and e_2 is perpendicular to e_1 and e_3.

With $w_i = m_i$, the mass of atom i, e_1, e_2, e_3 are the principal inertial axes of the molecule. The matrix $M(m_i)$ is closely related to the inertial matrix I, namely,

$$I = \mathrm{tr} \, [M(m_i)] \, E - M(m_i)$$

where tr $[M(m_i)]$ is the sum of the diagonal elements of $M(m_i)$ and E is the identity matrix. Written out in full,

$$I = \begin{bmatrix} \sum m_i(r_{i2}^2 + r_{i3}^2) & \sum -m_i r_{i1} r_{i2} & \sum -m_i r_{i1} r_{i3} \\ \sum -m_i r_{i1} r_{i2} & \sum m_i(r_{i3}^2 + r_{i1}^2) & \sum -m_i r_{i2} r_{i3} \\ \sum -m_i r_{i1} r_{i3} & \sum -m_i r_{i2} r_{i3} & \sum m_i(r_{i1}^2 + r_{i2}^2) \end{bmatrix} .$$

Note that $M(m_i)$ and I have the same eigenvectors. The eigenvalues I_1, I_2, I_3 are the principal moments of inertia of the molecule. From the relationship between $M(m_i)$ and I, $I_1 = M_2 + M_3$, $I_2 = M_3 + M_1$, $I_3 = M_1 + M_2$.

Finally, with $w_i = q_i$, the charge of atom i, $I(q_i)$ is the quadrupole moment matrix of the molecule and I_1, I_2, I_3 are its principal values.

The transformation of the vectors r_i to the coordinate system defined by the eigenvectors e_1, e_2, e_3 is:

$$q_i = R^T r_i$$

where R^T is a matrix whose rows are the transposed vectors e_1, e_2, e_3. For symmetric molecules with rotation axes of order greater than two, at least two of these three eigenvalues become equal (or nearly equal if the symmetry is inexact). The directions of the corresponding principal axes are then poorly determined. Note also that most methods for determining eigenvectors of matrices leave the sense of polarity of these vectors more or less to chance; after all, if e_1 is an eigenvector corresponding to a given eigenvalue then $-e_1$ is also an eigenvector with the same eigenvalue.

1.3 Invariants of Molecular or Fragment Structure

The descriptions of molecular structure that we have used so far all depend on the choice of a coordinate system. The question then arises: which geometric aspects of structure are coordinate system independent? It is clear that, for many purposes, we may regard the position and orientation of the molecule as irrelevant, in which case only the relative atomic positions are required. For a system of N atoms, these can be specified by $3N-6$ parameters. In terms of a Cartesian coordinate system (see below), we need $3N$ parameters in general, but the origin can always be chosen to coincide with one of the atomic positions, say atom 1; then the x axis can always be chosen along the direction of the vector from atom 1 to 2 and the y axis in the plane of atoms 1, 2, and 3. Thus, if atom 1 is assigned the coordinates $(0,0,0)$, atom 2 $(x_2,0,0)$, and atom 3 $(x_3,y_3,0)$ the list of coordinates contains six zeros (fixed, non-adjustable coordinates). The relative positions of the atoms are then defined by the remaining $3N-6$ parameters (degrees of freedom). To define the structure of a planar molecule, only $2N-3$ independent parameters are needed.

The relative positions of the atoms may be specified by interatomic vectors. Their lengths and the angles between them are coordinate-system independent. For N atoms, there are $N(N-1)/2$ such vectors. Clearly, if $N>4$ this number is larger than $3N-6$. For describing molecules we usually choose a subset of these, those corresponding to the chemical "bonds". Their lengths are then the bond lengths, and the angles between vectors emanating from the same atom are the bond angles. The angle between the two planes defined by three vectors along a chain of four atoms is known as a dihedral angle, or, more commonly in a chemical context, as a torsion angle. These quantities and others derived from the interatomic vectors — such as the geometric center, the centers of mass and charge, the dipole moment (if the molecule is uncharged), the principal values of the inertial and quadrupolar tensors and the orientation of their eigenvectors to the interatomic vectors — can be regarded as the internal coordinates of a molecule.

In a deep sense, the description of a structure in terms of its internal coordinates is more fundamental than the description in terms of its external coordinates with respect to some set of basis vectors. The external coordinates depend on the arbitrary choice of axes and are thus not *invariants* of the structure, whereas the internal coordinates characterize the structure completely — if there are enough of them and if they are properly chosen. As we shall see in the next section, the problem of calculating internal coordinates from external ones is perfectly straightforward. The reverse problem, that of calculating Cartesian coordinates from distance and angle information, can be solved provided that the molecule is described in terms of $N-1$ distances, $N-2$ bond angles, and $N-3$ torsion angles in such a way as to form a main chain with branches. The work is straightforward but laborious. When more than $3N-6$ parameters are given, there are problems of self-consistency and the calculation becomes much more difficult.

1.3.1 Internal Coordinates

The calculation of internal coordinates from Cartesian coordinates is a straightforward matter. Bond distances are just lengths of vectors and bond angles are angles between vectors, and we have already seen in Section 1.2.1 how these may be calculated. The calculation of torsion angles needs a more detailed discussion.

For a sequence of four atoms, A, B, C, D (Figure 1.2) the torsion angle ω(ABCD) is defined as the angle between the directions BA and CD in projection down BC, or, what amounts to the same, as the angle between the normals to the planes ABC and BCD. For this reason, the term dihedral angle is often used instead of torsion angle. (Chemists usually prefer the latter because they think in terms of rotation around the bonds in a molecule.) By convention [3] ω is defined as positive if the sense of rotation from BA to CD, viewed down BC, is clockwise; it is negative if this

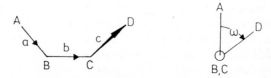

Fig. 1.2. Definition of torsion angle ω(ABCD) for a sequence of four atoms A, B, C, D

sense is counterclockwise. A positive value of ω means that the sequence of atoms A, B, C, D forms a right-handed screw. Note that if ω(ABCD) is positive, then so is ω(DCBA) – a right-handed screw remains right-handed when it is turned back to front. From the above definition of a torsion angle

$$\cos \omega = \frac{(\overline{AB} \times \overline{BC}) \cdot (\overline{BC} \times \overline{CD})}{AB\,(BC)^2\,CD\,\sin\theta_{ABC}\,\sin\theta_{BCD}}$$

$$\frac{(\overline{BC})}{BC}\sin \omega = \frac{(\overline{AB} \times \overline{BC}) \times (\overline{BC} \times \overline{CD})}{AB\,(BC)^2\,CD\,\sin\theta_{ABC}\,\sin\theta_{BCD}}\,.$$

These vector formulae hold in any coordinate system, but the torsion angle is undefined when one of the bond angles, or both, is 0° or 180°. As we have seen, the sign of a torsion angle is unaltered by rotation. It is reversed by reflection or inversion. For many years, it has been customary to give results of crystal structure analyses in a right-handed coordinate system. For chiral crystals, the sense of chirality, if known, is then specified by the atomic coordinates. The expressions given in Section 1.2.2 for transforming coordinates from crystal axes to a Cartesian system ensure that the sense of chirality is preserved. Generally, in transforming from one set of basis vectors to another, the sense of chirality of the axial triple is preserved if the determinant of the transformation matrix is positive. If all vectors are reversed in direction (as on going from a right-handed to a left-handed coordinate system or by reversing the signs of all coordinates without changing the coordinate system) the signs of all torsion angles are reversed. This can be seen by inspection of the cross product formula for the sine of the torsion angle. On the right-hand side of the equation, all cross products are unchanged, but the direction of the vector BC, and hence the sign of the left-hand side, is reversed.

The magnitude of a torsion angle (but not the sign) can also be calculated from interatomic distances and angles. In the four atom sequence A, B, C, D described above let $AB = a$, $BC = b$, $CD = c$, $ABC = \theta_1$, $BDC = \theta_2$. Then the distance AD is related to the torsion angle about the central bond (Figure 1.2) by

$$(AD)^2 = (a)^2 + (b)^2 + (c)^2 - 2ab \cos\theta_1 - 2bc \cos\theta_2$$

$$+ 2ab\,(\cos\theta_1 \cos\theta_2 - \sin\theta_1 \sin\theta_2 \cos\omega)$$

so that only $\cos \omega$ but not $\sin \omega$ is determined in this way. Since interatomic distances and bond angles are the same for a pair of enantiomeric structures, it is clear that the sign of the torsion angle cannot be determined from these quantities alone.

As everyone knows who has handled mechanical molecular models, such as Dreiding models, bond distances and angles alone do not, in general, completely define the three-dimensional structure of a molecule, i.e. models constructed with fixed values of of bond distances and angles are generally not rigid. In order to make them so, we need to fix some of the torsion angles as well. How many?

We have already seen that the relative positions of N atoms are completely specified by $3N-6$ adjustable parameters. If the N atoms are arranged in a chain, their relative positions are fixed by assigning definite values to the $N-1$ bond distances, the $N-2$ bond angles and the $N-3$ torsion angles. These are mutually independent, and there are $3N-6$ in all. For a branched chain, we can still choose the $3N-6$ parameters in the same way, but there are now several ways to choose the bond angles and torsion angles, and care is necessary to ensure that the ones chosen are truly independent. For example, at each non-terminal atom of ligancy m, there are $m(m-1)/2$ distinct bond angles, but only $2m-3$ of these are independent:

Ligancy of central atom	2	3	4	5	6
Distinct bond angles		1	3	6	10	15
Independent bond angles	1	3	5	7	9 .

Similarly, each non-terminal bond between atoms of ligancy p and q can be associated with $(p-1)(q-1)$ distinct torsion angles, of which only one is independent once the bond angles have been specified. Additional torsional degrees of freedom may be introduced, but only at the cost of reducing the number of independent bond angles.

Relationships between dependent and independent parameters are generally quite complicated. Even for apparently simple geometrical systems, we soon become involved in a quadratic mess of trigonometric expressions. Consider a four-atom fragment consisting of a central atom linked to three other atoms, as shown in Figure 1.3. With $N = 4$, $3N-6 = 6$. We can choose the six independent parameters as the three distances, d_1, d_2, d_3 and the three angles θ_{12}, θ_{23}, θ_{31}. The distance Δ of the central atom from the plane defined by the three others is a complicated function of these six independent parameters:

$$\Delta = \left\{ \left(\frac{\sin \theta_{12}}{d_3} \right)^2 + \ldots + \frac{2 (\cos \theta_{12} \cos \theta_{31} - \cos \theta_{23})}{d_2 d_3} + \ldots \right\}^{-1/2} u$$

where the missing terms in the summation are obtained by permuting the indices and where u is the volume of the parallelepiped formed by unit vectors along the three bond directions (see Section 1.2.1) [4]. For systems with approximate C_{3v} symmetry where the pyramidality of the central atom is not too large, the distance Δ is given approximately by

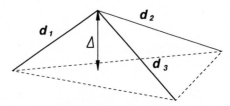

Fig. 1.3. A central atom linked to three other atoms

$$\Delta = \left\{ \frac{\pi(360-S)}{180\sqrt{3}} \right\}^{1/2} d$$

where S is the sum of the three bond angles (in degrees) and d is the mean of the three bond distances. If the Cartesian coordinates of the atoms are available, it is generally simpler to calculate the equation of the plane through the three bonded atoms.

Another frequently occurring fragment is the tetrahedral carbon atom, a central atom linked to four other atoms, i.e. $N = 5$. Having fixed the four bond distances, we still have six bond angles (or, equivalently, the six distances defining the edges of the bounding tetrahedron), ten parameters in all. But only nine of them are independent, that is, once we assign values to nine of them, the value of the tenth is fixed. The relationship among the six bond angles at a tetrahedral center is given by the determinantal equation:

$$\det \begin{bmatrix} 1 & C_{12} & C_{13} & C_{14} \\ C_{12} & 1 & C_{23} & C_{24} \\ C_{13} & C_{23} & 1 & C_{34} \\ C_{14} & C_{24} & C_{34} & 1 \end{bmatrix} = 0$$

where $C_{12} = \cos\theta_{12}$, etc. This is in fact the metric matrix for the set of four unit vectors, taken as basis vectors. The value of the determinant can be regarded as the square of the volume of the four-dimensional hyperparallelepiped formed by these four unit vectors emanating from the tetrahedral center. (It should be clear that the determinant is merely a four-dimensional analog of the expression given in Section 1.2.1 for the volume squared of a parallelepiped formed by three basis vectors in three-dimensional space). Just as the volume of a parallelepiped is zero if the three basis vectors lie in a common plane, so the volume of the hyperparallelepiped is zero if the four basis vectors lie in a common three-dimensional space. Thus, given five of the six bond angles, the value of the sixth is fixed, although its accurate evaluation involves some fairly laborious arithmetic. We say that of the six bond angles at the tetrahedral center, one of them is *redundant*. Note that if the six angles are assigned arbitrary values, the resulting figure cannot be constructed in three-dimensional space.

This problem may arise if one attempts to obtain an "averaged" structure of a tetrahedral fragment by averaging each of the six bond angles over several examples. In general, the average values will not satisfy the determinantal equation (nor will they do so if the cosines of the angles are averaged).

In principle, a five-membered ring ($N = 5$) could be treated in exactly the same way; choose one atom as a "central" atom and the other four as vertices of a tetrahedron. For a chemist, however, it would be more natural to choose the nine independent parameters as the five bond distances and four of the five bond angles. The remaining bond angle is then redundant – its possible values, as well as those of the ring torsion angles, are determined from the values of the nine given parameters.

The six-membered ring is interesting. With $N = 6$, one might think that the figure should be fixed when the six bond distances and the six bond angles have been specified. In general, this is the case; a Dreiding model of the chair form of cyclohexane is rigid, it has no remaining degrees of freedom. However, if the ring possesses a non-intersecting twofold rotation axis (an axis that does not pass through any atoms or bonds) then it is flexible — it still has a degree of freedom [5]. Whereas the torsion angles in the chair form of cyclohexane are fixed by the values of the bond distances and angles, this is not the case for the "flexible" family comprising the boat and twist forms. In Chapter 2 some relationships among torsion angles in cyclic molecules will be discussed; in general, for cyclic molecules it is not always obvious which parameters can be chosen as independent.

The description of a molecule in terms of its bond lengths, bond angles, and torsion angles is often convenient because it is usually possible to guess the approximate values of these quantities (or at least of some of them) on the basis of prior knowledge. Thus, given the constitution of a molecule, the bond lengths can usually be regarded as fixed within narrow limits at standard values that are characteristic of the various bond types (see Appendix A). Likewise, bond angles do not vary much from characteristic values unless forced to do so by ring constraints (e.g. angles in small rings). This is the reason why useful information about the possible shapes of molecules can often be obtained by examining mechanical models in which the bond lengths and angles are held fixed. However, it is also possible to describe the relative atomic positions in a molecule by specifying only interatomic distances. Clearly, two bond lengths and the included bond angle define a triangle, the third side of which is fixed by the information given; and conversely, from the lengths of the three sides of the triangle, the two bond lengths and the third nonbonded distance, the bond angle can easily be found. In a similar way, information about torsion angles can always be translated into information about the interatomic distances in sets of four atoms, or in other words, about relationships among the edges of tetrahedra. Thus, just as the solution of triangles is fundamental to plane geometry, the solution of tetrahedra is needed for describing the relative arrangement of points in three-dimensional space.

1.3.2 Distance Geometry

The general problem of deriving relative positions of atoms from distance information alone is the subject of Distance Geometry [6]. The main problem can be described as follows: Whereas any set of $3N$ numbers regarded as Cartesian coordinates will describe some arrangement of points in three-dimensional space, this is not the case for a set of $3N$ numbers regarded as interatomic distances. Even though there are sufficient numbers to do the job in principle, i.e. at least $3N-6$ for a system of N points, the numbers will not in general describe any feasible arrangement of points because they are not mutually consistent. More generally, in three-dimensional space, the total number of distances, $N(N-1)/2$, exceeds the number of independent distances $3N-6$ for $N>4$. From the examples given in the previous section, it is clear that relationships between dependent and independent parameters are complex, even for simple frameworks. For more complicated ones, the problems of self-consistency of the given parameters and the evaluation of the unknown distances and angles from the given parameters become immeasurably more complicated. Moreover, in any practical case, the information about distances in complex molecules has to be obtained experimentally; it is therefore at best inexact and some of it may even be wrong, which makes the problem even more difficult.

One problem is that since the relationships among the parameters involve quadratic and higher equations, they do not generally yield a single-valued solution for the dependent parameters. For example, for $N = 5$, given nine ($3N-6$) distances, there are still two possible values for the tenth. The nine can always be chosen as the distances of two points (1 and 2) from three reference points (3, 4 and 5; Figure 1.4). Clearly points 1 and 2 may be on the same side of the plane through 3, 4 and 5 or on opposite sides. The tenth distance, d_{12}, is needed to resolve this ambiguity. If, however, we know the approximate value of the tenth distance, we can determine it exactly from the values of the nine given distances. A similar ambiguity occurs for each set of 5 points.

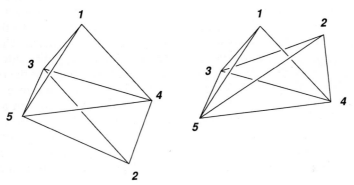

Fig. 1.4. A set of five points in space; given the lengths of the nine indicated lines, points 1 and 2 may be on the same side of the plane through 3, 4, 5 or on opposite sides

If all $N(N-1)/2$ distances in a figure are given, it is possible to find a set of Cartesian coordinates for the N vertices. The method makes use of the properties of the metric matrix. First, the connectivity matrix C, the symmetric $N \times N$ matrix of distances squared is formed. From it we form an $(N-1) \times (N-1)$ symmetric matrix G by subtracting the first row and column from all other rows and columns, respectively. The diagonal elements are then of the form $-2(d_{1n})^2$ and the off-diagonal ones are $(d_{nm})^2 - (d_{1n})^2 - (d_{1m})^2$, which is the same as $-2d_{1n} \cdot d_{1m}$ from the properties of the triangle formed by the three points 1, n, m. The new matrix divided by -2 is thus recognized to be the metric matrix of the vectors from point 1 to all the other vertices:

$$
G = \begin{bmatrix}
d_{12} \cdot d_{12} & d_{12} \cdot d_{13} & d_{12} \cdot d_{14} & \cdots & d_{12} \cdot d_{1N} \\
d_{12} \cdot d_{13} & d_{13} \cdot d_{13} & d_{13} \cdot d_{14} & \cdots & d_{13} \cdot d_{1N} \\
\cdots & \cdots & \cdots & \cdots & \cdots \\
d_{12} \cdot d_{1N} & d_{13} \cdot d_{1N} & d_{14} \cdot d_{1N} & \cdots & d_{1N} \cdot d_{1N}
\end{bmatrix} .
$$

Since this is a symmetric matrix it can be factorized into the product of a lower triangular matrix and its transpose. The former gives the x, y, z Cartesian coordinates of the points with point 1 at the origin, d_{12} along the x axis and d_{13} in the x, y plane. The tetrahedron formed by the points 1, 2, 3, 4 is thus the basis for the construction of the coordinate system used to describe the figure. Since the figure is three-dimensional, the coordinates relating to higher dimensions are zero. However, if the $N(N-1)/2$ distances are not mutually consistent then the figure may require higher dimensions for its description (compare the discussion in the previous section of the interdependence of the six angles at a tetrahedral center). The remaining coordinates will then not be exactly zero; they give a measure of the non-consistency in the distances.

As an example, we take an anonymous cyclohexane ring from the thousands in the Cambridge Structural Database (CSD). From the crystal coordinates, the squares of the interatomic distances were computed (in Å^2) and rounded off to give the 6×6 connectivity matrix:

$$
C = \begin{bmatrix}
0 & 2.39 & 6.45 & 8.47 & 6.05 & 2.25 \\
2.39 & 0 & 2.40 & 6.35 & 8.70 & 6.40 \\
6.45 & 2.40 & 0 & 2.29 & 6.45 & 8.88 \\
8.47 & 6.35 & 2.29 & 0 & 2.35 & 6.35 \\
6.05 & 8.70 & 6.45 & 2.35 & 0 & 2.36 \\
2.25 & 6.40 & 8.88 & 6.35 & 2.36 & 0
\end{bmatrix} .
$$

The 5×5 metric matrix is obtained by subtracting the first row and column from all other rows and columns and dividing by -2:

$$
G = \begin{bmatrix}
2.39 & 3.22 & 2.26 & -0.13 & -0.88 \\
3.22 & 6.45 & 6.32 & 3.03 & -0.09 \\
2.26 & 6.32 & 8.47 & 6.09 & 2.19 \\
-0.13 & 3.03 & 6.09 & 6.05 & 2.97 \\
-0.88 & -0.09 & 2.19 & 2.97 & 2.25
\end{bmatrix}
$$

which is factorized into the lower triangular matrix L and its transpose;

$$
L = \begin{bmatrix}
1.546 & 0 & 0 & & \\
2.083 & 1.453 & 0 & & \\
1.459 & 2.255 & 1.121 & & \\
-0.084 & 2.202 & 1.107 & 0.182 & \\
-0.569 & 0.754 & 1.173 & 0.204 & 0.153
\end{bmatrix} .
$$

The first three columns of L give the Cartesian coordinates of the ring atoms, referred to atom 1 at the origin. The fourth and fifth columns would contain only zeroes if the elements of the connectivity matrix C were perfectly self-consistent and not rounded off, as in the example.

Distance geometry techniques have become especially important in the structure analysis of macromolecules, where pairs of hydrogen atoms separated by distances of less than about 5 Å can be identified by NMR spectroscopy using the nuclear Overhauser effect [7]. Additional distance constraints are imposed by the known properties of the covalent polypeptide or polynucleotide structure. The details of such calculations are outside the scope of this Chapter [8, 9]; it is enough to say that for the large number of atoms involved the computer time and memory requirements are formidable.

1.4 External or Internal Coordinates?

Molecular structures may be described and compared in terms of external or internal coordinates. The question of which is to be preferred depends on the type of problem that is to be solved. For example, one problem that is much easier to solve in a Cartesian system is that of finding the principal inertial axes of a molecule; indeed, if only internal coordinates are given then, in general, the first step is to convert them to Cartesian ones and then proceed as described in Section 1.2.4. Similarly, the optimal superposition of two or more similar molecules or molecular fragments, i.e. with the condition of least-squared sums of distances between all pairs of corresponding atoms, is best done in a Cartesian system. On the other hand, systematic trends in a collection of molecular structures and correlations among their structural parameters are more readily detectable in internal coordinates.

1.4.1 Superposition of Molecules

If we want to compare several specimens of the same molecule or fragment, we can begin by transforming the origin for each molecule to its center of mass and orienting the axial directions along the principal inertial axes. The next step is to superimpose the atomic coordinates of the various fragments and determine the rotations of the individual axial frames that minimize the weighted sum of mutual squared deviations [10]. The matching is first done in pairs with respect to a single selected target molecule, but in a second stage the sum of deviations over all molecules is minimized. For any pair of molecules, we can define a "distance" between them in terms of the sum of squares of deviations between corresponding atoms.

   A related method is derived from the description of molecular motions in rotational and vibrational spectroscopy. The basic idea is to relate a distorted structure with coordinates s_i^d to a reference structure with coordinates s_i^r. The conditions imposed for superposition are:

(1) that the centers of mass are the same for both, i.e.

$$\sum w_i s_i^{\text{d}} = \sum w_i s_i^{\text{r}} = 0 .$$

(2) that the distortion does not contain a rotational component, or, more exactly, that the coordinates s_i^{r} of the reference molecule and the coordinates s_i^{d} of the distorted molecule obey the condition

$$\sum w_i s_i^{\text{r}} \times (s_i^{\text{d}} - s_i^{\text{r}}) = 0 .$$

In general, the inertial coordinates q_i^{r} described in Section 1.2.4 do not fulfill the second condition, and the necessary rotation from these to coordinates that do satisfy condition (2) has been described by Eckart for molecules that rotate and vibrate simultaneously [11]. The procedure has recently been applied to the analysis of static molecular distortions [12]. Once a reference geometry is fixed, the Eckart procedure has the advantage of not being iterative. Whereas for molecules in motion w_i is usually chosen as m_i, the alternative choice $w_i = 1$ emphasizes the purely geometric aspects of the comparison between structures.

1.4.2 Configuration Space

Another quite different way of comparing several specimens of molecules or molecular fragments is to think of them not as objects in three-dimensional space but, more abstractly, as a set of representative points (termini of vectors) in an n-dimensional hyperspace, one dimension for each internal coordinate or rather one for each internal coordinate of interest. We choose some set of geometric parameters of interest and associate with each a basis vector p_i of unit length. The actual value of the parameter is then defined by a scalar displacement d_i along the corresponding direction. If the parameters are independently variable, then the basis vectors are orthogonal. Different specimens of the same molecule or molecular fragment are then associated with different representative points (RP's). Again, a "distance" between different structures can be defined in terms of the metric of the hyperspace. (To avoid scaling problems, the basis vectors should be chosen to have the same dimensional units, that is, all distances or all angles.) For closely similar molecules the RP's will cluster in the same region of the hyperspace. Distributions of RP's may be analyzed with the help of statistical methods such as cluster and principal component analysis (see Chapter 4).

If we wish to compare not complete molecules in different crystal environments but rather a particular structural fragment in different molecules, we can restrict the configuration space to those coordinates necessary to describe the fragment; the other parameters can be assumed to show only small and insignificant variation. Although the definition of a structural fragment is largely arbitrary and depends on the particular problem under investigation, we can usually specify it as an invariant part of the structural formulae (or incidence matrices) for a collection of related molecules (see Section 1.1).

One advantage of configuration space is that it provides an immediate connection with the concept of the molecular potential energy hypersurface, a concept almost indispensable to any modern discussion of chemical reaction dynamics. Very briefly, each point of the n-dimensional space is associated with a potential energy V; the minima of V correspond to stable chemical species, the passes to transition states, and the curves following the energy valleys and connecting minima via the lowest passes between them correspond to deformation of the molecule along chemical reaction paths.

It may be helpful to consider a simple two-dimensional illustration – a linear triatomic molecule, such as the triiodide anion, which has been observed in very many different crystal environments. Assuming linearity (which holds in practice quite closely for triiodide ions in crystals), we completely describe the structure of any given example of a triiodide ion by the two interatomic distances, d_1 and d_2. A collection of triiodide ions is then described by a collection of points in the plane spanned by the two corresponding basis vectors, p_1 and p_2 (Figure 1.5). It is evident that the points corresponding to observed structures lie close to a rather well defined curve. We can interpret this curve in different ways, some of which are described in later Chapters (see especially Chapter 5). For example, we can regard it as the reaction path for the exchange reaction: $I_2 + I^- \rightleftharpoons I^- + I_2$.

The main point at present is that the distribution is far from random. Note too that the distribution shown is Figure 1.5 is symmetrical about the diagonal. This symmetry occurs because if the point (d_1, d_2) corresponds to an observed structure, so does the point (d_2, d_1). Each of the two points corresponds to a different way of assigning the two interatomic distances to the two coordinate axes. To avoid ar-

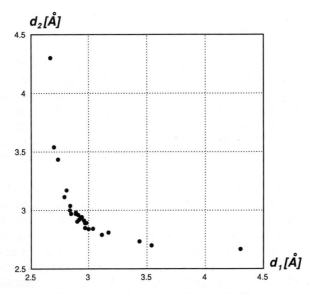

Fig. 1.5. Correlation plot of interatomic distances in linear triiodide anions

bitrariness, we must use both ways. In Chapter 2 the question of symmetry in configuration space is taken up in a more general way.

1.4.3 Deformation Coordinates and Reference Structures

Often, our interest will lie not so much in the actual structure of a particular molecule or molecular fragment in a particular environment as in the details of how this molecule or fragment deviates from some reference structure with the same atomic connectedness or constitution. Insofar as we can usually ignore the absolute position and orientation of our molecule, comparisons of this kind are most conveniently made in terms of internal coordinates. The distortion can be expressed in terms of a total displacement vector $D = \sum d_j p_j$, where the d_j's are displacements along some set of basis vectors p_j. The only difference to the internal coordinates described in the previous section is that for deformation coordinates the displacements d_j are defined to be zero for the reference structure. This could be an observed structure, or a calculated one, or an idealized, more symmetric version of the structure we are interested in.

1.4.4 Linear Transformations in Configuration Space

Linear transformations, including symmetry transformations in configuration space are analogous to those in Cartesian space (see Sections 1.2.2 and 1.2.3). For symmetrical reference structures, it is usually better to use not the internal coordinates themselves but to choose a new coordinate system in which the basis vectors are symmetry adapted linear combinations of the internal displacement coordinates with the special property that they transform according to the irreducible representations of the point group of the idealized, reference molecule (symmetry coordinates, see Chapter 2).

Thus, in the triiodide example, there are obvious advantages in choosing new basis vectors (and corresponding displacements) as sums and differences of the original ones:

$$q_1 = (p_1 + p_2)/\sqrt{2} , \quad q_2 = (p_1 - p_2)/\sqrt{2}$$

$$f_1 = (d_1 + d_2)/\sqrt{2} , \quad f_2 = (d_1 - d_2)/\sqrt{2} .$$

The division by $\sqrt{2}$ is necessary for normalization, i.e. to assign unit length to the new basis vectors. These new coordinates are symmetry coordinates. Displacement f_1 is left unchanged by interchanging d_1 and d_2 whereas f_2 changes sign. Structures for which $f_2 = 0$ have equal interatomic distances; they have the maximum possible symmetry of the structure in question. We say that f_1 is symmetric under the permutation whereas f_2 is antisymmetric. From Figure 1.5 we see that the first displacement is along the diagonal of the figure, while the second is perpendicular to this direction. We shall have more to say about symmetry coordinates in the following chapter.

References

[1] *International Tables for X-Ray Crystallography*, published for the International Union of Crystallography, is an indispensable reference for every aspect of the subject. Of particular interest for our purposes are: *Vol. 1, Symmetry Groups*. 3rd edn., **1969**; *Vol. A, Space-Group Symmetry*, **1983**

[2] A detailed derivation can be found in Dunitz, J.D., *X-Ray Analysis and the Structure of Organic Molecules*, Cornell University Press, **1979**, pp. 235–238

[3] Klyne, W., Prelog, V., *Experientia* **1960**, *16*, 521

[4] Crystallographers will recognize this as the same problem as calculating the distance between the (111) planes of a triclinic crystal, given the lengths of the cell edges and the interaxial angles. It is best handled using the reciprocal set of basis vectors

[5] Dunitz, J.D., Waser, J., *J. Am. Chem. Soc.* **1972**, *94*, 5645–5650

[6] The central problem of Distance Geometry was explicitly stated by the French mathematician Lazare Carnot (also famous as military engineer, statesman, and father of Sadi Carnot) in his book *Géometrie de Position*, published in Paris in **1803**. Compare Carnot's statement of the problem: *"Dans un systême quelconque de lignes droites, tracées ou non dans un même plan, quelques-unes d'elles, ou les angles qui résultent de leur assemblage, soit entre elles-mêmes, soit entre les plans qui les contiennent, étant données en nombre suffisant pour que toute la figure soit déterminée, trouver tout le reste"*. Translated: "In any system of straight lines, drawn in the same plane or not, given some of them, or the angles that result from their assembly, whether between the lines themselves or between the planes that contain them, in sufficient number to determine the complete figure, find the remainder."

[7] Wüthrich, K., *NMR of Proteins and Nucleic Acids*, Wiley Interscience, New York, **1986**

[8] Crippen, G.M., *Distance Geometry and Conformational Calculations*, Research Studies Press, Wiley, New York, **1981**

[9] Crippen, G.M., Havel, T.F., *Distance Geometry and Molecular Conformation*, Research Studies Press, Wiley, New York, **1988**

[10] Gerber, P.R., Müller, K., *Acta Cryst.* **1987**, *A43*, 426–428; Kearsley, S.K., *J. Comp. Chem.* **1990**, *11*, 1187–1192

[11] Eckart, C., *Phys. Rev.* **1935**, *47*, 552–558

[12] Cammi, R., Cavalli, E., *Acta Cryst.* **1992**, *B48*, 245–252

2 Symmetry Aspects of Structure Correlation

Jack D. Dunitz and Hans-Beat Bürgi

2.1 Introduction

The concepts of molecular and fragment structure were defined in the previous chapter, Section 1.1, in terms of incidence matrices. In structure correlation we compare molecules with the same incidence matrix coming from different crystal structures, or fragments with the same incidence matrix coming from different molecules. More often than not, the molecules or fragments in which we are interested show little or no symmetry. Why then should a book on structure correlation contain a chapter dealing with symmetry aspects? What symmetry aspects?

There are several reasons for including a chapter on symmetry: The first has to do with the different equivalent ways we label the individual atoms in molecules or molecular fragments and hence with the concept of isometric structures. Consider a linear triatomic molecule, say the triiodide anion introduced in Section 1.4.2 to illustrate the idea of configuration space. We may arbitrarily label the atoms as I(1)-I(2)-I(3) with interatomic distances d_{12} and d_{23}, of length a and b, respectively. Each observed structure can then be represented in configuration space by a point in a plane with coordinates (a, b). But we can also interchange the labels of the two outer atoms to obtain a representative point with the coordinates (b, a). The two points correspond to the same molecule, differently labeled. But this is usually just what we do in structure correlation: we detach a molecule from its environment and assign some set of labels to its constituent atoms. As long as there is nothing to choose between equivalent labeling schemes we have to treat them all on an equal footing. To avoid any preference for one or another way of labeling the atoms we have to take them *all* into account. For the simple case of the linear triiodide there are just two of these equivalent labeling schemes; we have two equivalent descriptions of the same molecule.

An equivalent alternative viewpoint is to consider the two points as two molecules I(1)-I(2)-I(3) that differ only by a permutation of the two distances. We say that the two molecules described by (a, b) and by (b, a) are isometric, since they are characterized by the same set of interatomic distances between the same types of atoms [1]. The structure invariants (bond distances, angles, principal moments) are the same for both molecules. Indeed, there is obviously nothing to choose between the two

$$
\begin{array}{ccccc}
\text{I} & - & \text{I} & - & \text{I} \\
a & & b & & \text{distances}
\end{array}
\qquad
\begin{array}{ccccc}
\text{I} & - & \text{I} & - & \text{I} \\
1 & & 2 & & 3 \quad \text{labeling}
\end{array}
$$

1	2	3	labeling
3	2	1	

a	b	distances
b	a	

Scheme 2.1. Two equivalent descriptions of a triiodide molecule (left), contrasted with two triiodide molecules that differ only by permutation of the interatomic distances *a* and *b* (right)

molecules, at least as long as we disregard any difference between the environments of the two outer atoms.

These two viewpoints are summarized in Scheme 2.1 and both will be used in this chapter, the choice being determined by what seems most suitable in the particular context.

For the triiodides, more generally, we might also choose to remove the restriction of colinearity of the three atoms, in which case there are no "outer" atoms and no "inner" one. We would then need to specify three interatomic distances to describe the molecules: say d_{12}, d_{23}, d_{31} of lengths a, b, and c, respectively. There are now six permutations, i.e. six ways of choosing the atomic labels or the corresponding interatomic distances that lead to isometric structures.

In technical language, these sets of two and six permutations are examples of groups, namely the symmetric groups of 2 and 3 elements (of order 2 and 6), respectively. Thus, even for this very simple example of a homonuclear triatomic molecule we find ourselves in the midst of group theory. Note too that the two groups involved here are not the same as the geometric point groups of the molecules concerned ($C_{\infty v}$ for the linear case and C_s for the angular one).

As mentioned in Section 1.4.4, it is often advantageous to use symmetry coordinates instead of the usual internal coordinates in such analyses. Thus, in the first example, instead of the two interatomic distances d_{12} and d_{23} we could use the sum (or average) and the difference of these two quantities. The first combination describes the structure of a hypothetical symmetrized molecule, the second expresses the deformation of the actual molecule from this averaged structure.

Symmetry coordinates are linear combinations of the internal coordinates that remain invariant, apart from a possible change of sign, under the action of a group of symmetry operations. They have been used for many years to describe molecular vibrations [2] but they are also useful for describing static distortions of molecules. In this type of description the total distortion is regarded as a sum of displacements along some set of symmetry coordinates of the reference structure. For the reference structure itself, the displacements are all zero. Thus, by summing the displacements (or rather their squares) over all the symmetry coordinates, we may be able to quantify deviations of actual structures from symmetric reference structures and to give quantitative expression to statements like "the molecule has approximate symmetry so-and-so". Also, by fixing some but not all of the displacements to zero, the total distortion can be broken down into various components, each of which preserves some but not all of the symmetry elements of the reference structure.

As we hope to explain in Chapter 5, symmetry arguments can be used not only to classify molecular distortions and simplify their description, but also to draw certain conclusions about the form of molecular potential energy functions, in particular about their low-lying regions. This extension introduces energy as an additional variable; the molecular energy (of the electronic ground state) can be considered to be another structure invariant, analogous to the principal moments of inertia and the electric multipole moments discussed in Chapter 1.

Of the two ways of looking at the triiodide anion, as linear or as bent, the second is more general and correspondingly more complicated than the first. One might therefore wonder which of the two is to be preferred in any given case. The answer will depend on what one hopes to achieve from the analysis of the experimental data. If one is merely interested in the variability of the distances in a collection of more or less linear triiodide anions in a variety of different environments, then the simpler description is sufficient. If, however, one were planning to investigate tri-atomic fragments in which the choice of a central atom and of terminal atoms is not so obvious, then the second, more complicated description would clearly be called for (see Section 2.5).

In summary, symmetry is a useful tool in structure correlation for at least three purposes: (1) to enumerate all equivalent ways of labeling atoms in a given molecule or molecular fragment and hence to eliminate the element of arbitrariness inherent in any particular labeling scheme; (2) to quantify the notion of "approximate symmetry" with the help of symmetry coordinates; (3) to analyze the symmetry properties of the molecular potential energy surface.

2.2 Permutation Groups and Point Group Symmetries

Group theory is an abstract branch of mathematics that brings out the close connections between permutation groups, symmetry groups, and other kinds of groups. It provides in essence a systematic way of classifying and analyzing symmetry. In this and the following section we shall provide certain definitions and results from group theory, but we do not have space to provide the necessary formal background within the limitations of the present book [3]. In order to make use of symmetry arguments in structure correlation it is important to specify right at the outset of an analysis which permutations of atomic labels, associated interatomic distances, and other structural parameters, are of interest for the particular problem at hand. It may therefore be helpful to include here a short introduction to the subject of permutation groups in general.

A permutation is a rearrangement of a finite number of objects. For example, one permutation of the six letters a, b, c, d, e, f is to change a into b, b into c, c into a, to transpose d and e, and to leave f alone; there are several ways of denoting this permutation. One way is to list the letters twice, once in the initial, natural order and again in the revised order:

$$abcdef$$

$$bcaedf$$

A second way is in terms of cycles; in each cycle the first element is replaced by the second, the second by the third, etc., and the last by the first. Each cycle is enclosed in brackets, so that the above permutation is denoted by:

$$(abc)(de)(f)$$

where the cycle (f) consisting of a single element can be omitted since it merely denotes replacement of f by itself. A cycle consisting of n elements is said to be of period n, since by repeating the rearrangement n times the original order of the elements is restored. For $n = 2$ the cycle is often referred to as a transposition. Clearly, any rearrangement can be achieved by a sequence of transpositions. A set of permutations that consists of a single cycle is said to be cyclic.

A third way of denoting permutations is in terms of matrices, where in an obvious notation the above permutation can be described as:

$$
\begin{vmatrix}
0 & 1 & 0 & 0 & 0 & 0 \\
0 & 0 & 1 & 0 & 0 & 0 \\
1 & 0 & 0 & 0 & 0 & 0 \\
0 & 0 & 0 & 0 & 1 & 0 \\
0 & 0 & 0 & 1 & 0 & 0 \\
0 & 0 & 0 & 0 & 0 & 1
\end{vmatrix}
$$

A permutation matrix contains only one non-zero element, a "1" in each row and column. The determinant of the matrix is then ± 1; if positive the permutation is called an *even* permutation, if negative an *odd* permutation. To avoid multiplying out, the parity of the permutation can be ascertained simply by counting the number of transpositions or the number of cycles of even period; if this number is even, then the permutation is even; if it is odd, then the permutation is odd. The number of cycles of odd period is irrelevant for determining the parity.

The set of permutations of atomic labels in a molecule forms a group: (1) there is an identity permutation, one that leaves the order of the elements unchanged; (2) to each permutation there is an inverse permutation, one that undoes its effect; (3) for any sequence of successive permutations there is always a single permutation from the set that has the same effect; (4) in performing successive permutations the associative law holds. The complete set of permutations of n elements (letters, numbers, atomic labels) forms the complete permutation group $\mathbb{P}_n$ of order $n!$. The set of even permutations also forms a group, of half this order, known as the alternating group of n elements.

In the triiodide example we considered the rearrangements of 2 and 3 elements, the atomic labels, corresponding to the permutation groups $\mathbb{P}_2$ of order 2 and $\mathbb{P}_3$ of order 6. Readers should convince themselves that the permutations of the four labels associated with the vertices of a tetrahedron (or the ligands X of a tetrahedral MX_4 molecule) form the permutation group $\mathbb{P}_4$ of order 24.

The two terminal atoms of the linear triiodide anion define a line, the basic one-dimensional figure; the three atoms of the bent anion define a triangle, the basic two-dimensional figure; and the four X atoms of an MX_4 molecule define a tetrahedron, the basic three-dimensional figure. These basic figures are called the simplexes in one-, two- and three-dimensional space. The various permutations of the groups $\mathbb{P}_2$, $\mathbb{P}_3$ and $\mathbb{P}_4$ can be associated with the geometric symmetry operations that transform regular simplexes into themselves. Thus $\mathbb{P}_2$ has the same formal structure as the geometric point-symmetry groups of order 2, namely C_i, C_s and C_2. Likewise, it is easily seen that there is a direct one-to-one correspondence between the various rearrangements of $\mathbb{P}_3$ and the symmetry operations of either of the two point groups D_3 or C_{3v} of order 6 (but not with the symmetry operations of the other three point groups of order 6, C_6, S_6, and C_{3h}, which have a different structure). Again, readers are invited to convince themselves of the one-to-one correspondence between the 24 rearrangements of the permutation group $\mathbb{P}_4$ and the symmetry operations of the point group T_d; it will be seen that for a tetrahedron with vertices labeled *1,2,3,4* the permutation *(12)(34)* corresponds to a C_2 operation, *(234)* to a C_3 operation, *(12)* to a reflection operation and *(1234)* to an S_4 operation. The various sets of symmetry operations that bring geometric figures into self-coincidence therefore also form examples of groups. Groups whose elements are in one-to-one correspondence with each other are said to be isomorphous.

When we come to the complete groups of more than four objects, this identification of permutations with symmetry operations of a three-dimensional figure is no longer possible; the complete group is no longer isomorphous to any point group. Thus, for example, there are 5! = 120 ways of permuting five objects, but there is no three-dimensional point group with this number of operations. In other words, there is no symmetrical arrangement of five X atoms around a central M atom which is a representation of the complete permutation group of order 120 − nor for that matter, of the corresponding alternating group of order 60. There is, however, an isomorphism if we are prepared to deal with figures in four dimensions, here the four-dimensional simplex consisting of five equivalent points − an MX_5 figure with 5 equal M−X distances and 10 equal X−X distances or X−M−X angles. In fact, all we need are 5 unit vectors making mutual angles of 104.48 ° ($\cos^{-1}(-1/4)$), which can be done in four-dimensional space but not in three. Nevertheless, there will still be certain sub-groups of permutations that are isomorphous with normal point groups, and for these the various rearrangements can be made to correspond to symmetry operations of symmetrical figures. This will allow the use of symmetry coordinates in such cases, although the analysis of the relevant permutations becomes much more complicated.

Notice that, for the one-, two- and three-dimensional simplexes, an even permutation of labels corresponds to a cyclic rearrangement, i.e. the permuted arrangement of labels has the same sense of chirality as in the initial figure. An odd permutation corresponds to a non-cyclic rearrangement, i.e. the permuted arrangement of labels has the opposite sense of chirality as in the initial figure. Correspondingly, an even permutation of distances leads to a properly congruent isometric figure, superimposable on the initial one, whereas an odd permutation of distances leads to an improperly congruent isometric figure, enantiomeric to the initial one. Thus the two structures of I_3^- shown on the right of Scheme 2.1 are of opposite polarity, that is to say, they are one-dimensionally chiral and are not superimposable in the line; two scalene triangles with distances (abc) and (cba) are two-dimensionally chiral figures and are not superimposable in the plane, and two tetrahedra with $M-X$ distances $(abcd)$ and $(cbad)$ are not superimposable in three dimensions.

An alternative representation of the permutation groups can be given in configuration space. Here we need one dimension for each internal coordinate of interest, and this number may be greater than the number of elements (atomic labels) to be permuted. As we have seen, this is not the case for the simplest examples we have studied. For the linear triiodide anion we need a two-dimensional coordinate system, the interatomic distances a and b, and there are just two atomic labels of the terminal atoms; for the triangular triiodide anion we need a three-dimensional space and there are three atomic labels. But for a tetrahedral MX_4 molecule there are 10 distances ($4\,M-X$ and $6\,X-X$) that need to be considered and only four atomic labels. For this example, the permutations of the labels produce 24 equivalent points in the ten-dimensional space of the internal coordinates (which can be reduced to 9 dimensions if the redundancy among the $X-X$ distances is taken into account; see Section 1.3.1).

2.3 Symmetry Coordinates, a Simple Example and some Generalizations Related to Point Group Symmetry

To recall some basic concepts, it may be useful to discuss a simple example: the geometry of the idealized ketal fragment shown in Figure 2.1. The relative positions of the five nuclei are described by 9 ($= 3 \times 5 - 6$) *internal coordinates*. Taking these as bond distances and angles, we see immediately that the 4 bond distances are independent quantities – any one can be varied independently of the others. It follows

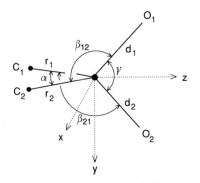

Fig. 2.1. Diagram of idealized ketal fragment showing atom labeling

that of the 6 bond angles involving the central atom, only 5 ($= 9-4$) can be in-dependently varied, and, indeed, the 6 angles are related by the determinantal equa-tion given in Section 1.3.1. The relevant *permutations* are those involving interchange of the two $C-C$ distances and of the two $C-O$ distances, either separately or both together. They form a group isomorphous to the *point group* C_{2v}, which describes the most symmetric version of the fragment, whereby one mirror plane $\sigma(xz)$ coin-cides with the CCC plane and cuts the OCO plane, the other mirror plane $\sigma(yz)$ coincides with the OCO plane and cuts the CCC plane, and the twofold rotation axis $C_2(z)$ bisects both the $C-C-C$ angle α and the $O-C-O$ angle γ.

Thus, in this most symmetric version of the fragment, the two $C-O$ distances are equal and so are the two $C-C$ distances; likewise, the four $C-C-O$ angles are equal. A *standard geometry* of the fragment can be determined in one way or another: for example, as the average of many experimental structure determinations of different ketals, or from a single, especially accurate determination, or from a reliable calculation for a suitable molecule containing the fragment, or in any other way. In general, a structure observed for such a fragment will be found to have no symmetry. Any particular example of such a fragment can now be treated as a distorted version of the standard structure with C_{2v} symmetry.

The *symmetry coordinates* are linear combinations of the initial coordinates with special properties when transformed according to the symmetry operations of the reference structure. To see what this means, let us look at what is called the *character table* of the point group C_{2v} [4] (Table 2.1).

Table 2.1 shows, for example, how the symmetry operations of the point group transform the Cartesian coordinates (x, y, z) of a representative point, or combina-tions thereof; the C_2 operation changes the sign of the x and y coordinates but leaves the z coordinate unaltered; and similarly, reading downwards, for the other symmetry operations. For this symmetry group there are just four essentially dif-ferent classes of function of x, y, z. When each of the symmetry transformations is applied to x, y, z, these functions are either left unchanged (symmetric, $+1$) or altered in sign (anti-symmetric, -1). The four classes of function are associated with dif-ferent kinds of transformation behavior, each of which can be summarized by a horizontal sequence of $+1$ and -1 entries (Table 2.1), its *character*. A function showing one of these kinds of transformation behavior is called an *irreducible representation* of the symmetry group. A function that is left unchanged by all the symmetry operations of the group is said to belong to the *totally symmetric* irreduci-ble representation, called A_1 in the present example. Note that the B_1 and B_2

Table 2.1. Character table for the point group C_{2v}

C_{2v}	I	C_2	$\sigma(xz)$	$\sigma(yz)$		Kernel
A_1	$+1$	$+1$	$+1$	$+1$	z, x^2, y^2, z^2	C_{2v}
A_2	$+1$	$+1$	-1	-1	xy	C_2
B_1	$+1$	-1	$+1$	-1	x, xz	$C_s(xz)$
B_2	$+1$	-1	-1	$+1$	y, yz	$C_s(yz)$

representations are distinguished merely by their symmetry or anti-symmetry with respect to reflection across the two mirror planes, and this depends only on the way the axes are labeled. For point groups with rotation axes of order higher than two, there are complications; for then it is no longer true that all the group operations leave a function unchanged or alter its sign. We shall deal with some of these complications as they arise (see Section 2.4.1).

The internal coordinates shown in Figure 2.1 do not transform according to the symmetry operations of the C_{2v} point group. They are neither left unchanged, nor are they simply reversed in sign; instead, they interchange their values. However, it is not difficult to construct linear combinations of these coordinates that do have the desired property [5]. From the 10 internal coordinates, we obtain 10 linearly independent combinations called *symmetry coordinates* (Table 2.2), each having the property we seek.

Inspection of Table 2.2 will show that S_1, S_3, S_5, S_6 and S_{10} transform as the A_1 representation; they are left unaltered by all the symmetry operations of the group. The S_2 and S_7 combinations transform as B_2; they are left unaltered by $\sigma(yz)$ but change sign under the $\sigma(xz)$ and C_2 operations. Similarly, the S_4 and S_8 combinations transform as B_1, and S_9 transforms as A_2. Although these ten combinations are linearly independent, one of the angle combinations depends on the values assigned to the other five, because, as mentioned above, the six bond angles themselves are not independent. One of the angle combinations is said to be "redundant"; its value is fixed by the values assigned to the other five.

As we have seen in Chapter 1, we can regard the structure of a molecule or molecular fragment in an abstract way as a representative point in many-dimensional configuration space, with basis vectors corresponding to the various kinds of geometrical parameter used to describe the structure. The list of numerical values of bond lengths, bond angles, etc., can be considered as a list of displacements along these various basis vectors. The representative point defining the structure is then given by the resulting vector sum of displacements.

For a symmetrical fragment such as the ketal shown in Figure 2.1, there are several equivalent ways of choosing the representative point. Imagine we happen to find the

Table 2.2. Symmetry coordinates of a ketal fragment, based on the internal coordinates shown in Figure 2.1, and their irreducible representations

$S_1 = (d_1 + d_2)/\sqrt{2}$	A_1
$S_2 = (d_1 - d_2)/\sqrt{2}$	B_2
$S_3 = (r_1 + r_2)/\sqrt{2}$	A_1
$S_4 = (r_1 - r_2)/\sqrt{2}$	B_1
$S_5 = \alpha$	A_1
$S_6 = \gamma$	A_1
$S_7 = (\beta_{11} + \beta_{12} - \beta_{21} - \beta_{22})/2$	B_2
$S_8 = (\beta_{11} - \beta_{12} + \beta_{21} - \beta_{22})/2$	B_1
$S_9 = (\beta_{11} - \beta_{12} - \beta_{21} + \beta_{22})/2$	A_2
$S_{10} = (\beta_{11} + \beta_{12} + \beta_{21} + \beta_{22})/2$	A_1

structure of 1,4,9,12-tetraoxadispiro[4.2.4.2]tetradecane in the literature [6]. From this work, the $C-O$ bond lengths are 1.424 and 1.431 Å, and the $C-C$ bond lengths are 1.500 and 1.513 Å. We could choose the shorter $C-O$ length as d_1 and the longer as d_2, the shorter $C-C$ length as r_1 and the longer as r_2, leading to a point with coordinates (1.424, 1.431, 1.500, 1.513) in the four-dimensional space spanned by the basis vectors d_1, d_2, r_1, r_2. But we could equally well have interchanged the labels of the $C-O$ pair of bonds, or of the $C-C$ pair, or of both pairs, leading to four possible representative points that all define the same structure. The symmetry of the reference structure is reflected in the symmetry of the abstract space describing it.

We can express the observed distortion of a molecule from a reference structure of the same or of higher symmetry in terms of a displacement vector,

$$D = d_i p_i = [d_i(\text{obs}) - d_i(\text{ref})] p_i$$

where the d_i's are components along some set of displacement vectors p_i (summation over the repeated index implied). The reference structure then corresponds to the origin of the coordinate system, and any collection of distorted structures will yield a collection of points grouped round the origin and distributed according to the symmetry of the reference structure. Internal coordinates (bond lengths, bond angles, etc.) can always be transformed to a new coordinate system in which the basis vectors are particular linear combinations of these internal coordinates that transform according to the irreducible representations of the point group of the reference molecule. A set of basis vectors with this property is called a set of symmetry displacement coordinates $S_i = T_{ij} p_j$. When the matrix T_{ij} that transforms the initial basis vectors p_j into the new basis vectors S_i also transforms the initial displacements d_j into the new symmetry displacements D_i, the distinction between the symmetry coordinates S_i and the symmetry displacements D_i along these coordinates can be ignored (always the case when the transformation T_{ij} is orthonormal).

$$D_i = T_{ij} [d_j(\text{obs}) - d_j(\text{ref})] \ .$$

Moreover, in transforming d_j to symmetry displacements the actual values of $d_j(\text{ref})$ are important only for the totally symmetric displacements since they cancel out for all the others.

The operation of the total displacement vector on the reference structure leads to the observed structure. This could also be reconstructed by distorting the reference structure along each S_i in turn and summing the corresponding displacements. Or we can choose a subset of symmetry displacements containing only part of the total distortion of the observed molecule and corresponding to the structure that would be produced if displacements along the remaining symmetry coordinates were set to zero. In this way certain symmetry elements of the reference structure can be maintained. These form sub-groups of the point group of the reference molecule known as kernel or co-kernel symmetries; the corresponding structures are kernel or co-kernel configurations. The kernel symmetry of an irreducible representation is simply the sub-group formed by those operations whose character equals the character of the identity operation. For example, in Table 2.1, the kernel of the A_2 irreducible representation is C_2, and any displacement along an A_2 symmetry coordinate (e.g. S_9 in Table 2.2) will preserve C_2 symmetry. For degenerate representations, the situation is more complicated; certain displacements can preserve a higher symmetry than the kernel symmetry – the co-kernel symmetry. A fuller discussion of kernel and co-kernel symmetries is provided elsewhere [7].

The number of internal coordinates generally exceeds the number of degrees of freedom. The displacement along some coordinates is fixed by the displacements along others. The choice of independent and dependent (redundant) coordinates is not always unique and may be left to the discretion of the user in any given case. Moreover, care is necessary with symmetrized structures obtained by averaging over appropriate internal coordinates of less symmetric structures, because the averaged structures may be geometrically infeasible. An example is provided by the six bond angles in the spiro-ketal fragment mentioned above (OCO: 105.9°; CCC: 111.0°; CCO: 108.5, 109.6, 111.3, 110.3°) [6]. Their C_{2v}-symmetric averages (⟨CCO⟩: 109.925°) do not correspond, strictly speaking, to a structure that is feasible in three-dimensional space; the value of the determinant relating the six angles (see Section 1.3.1) is 0.002 instead of zero. The average of all six angles is 109.43°, which is not exactly $\cos^{-1}(-1/3)$; the determinant is now 0.005. For a more distorted fragment with, for example, five angles of 90° and one of 180° the average is 105° and the value of the determinant is 0.45.

One use of the symmetry coordinate classification is that it can tell us which types of distortion are expected to be coupled to other types. Each representative point can be associated with a value of the molecular potential energy, a structural invariant μ that must be independent of such matters as the choice of coordinate system or the way the labels of the various atoms or bonds have been chosen. Consider, for the spiro-ketal example, the general quadratic energy expression for the S_2 and S_8 coordinates, which both transform as B_2, symmetric with respect to the (yz) plane, antisymmetric with respect to the (xz) plane:

$$E = k_{22}(D_2)^2 + k_{88}(D_8)^2 + k_{28}D_2D_8 \ .$$

Interchanging the labels of the C−O bonds reverses the signs of both D_2 and D_8 and thus leaves the value of the energy cross-term unchanged. Since this cross-term will in general not be zero, the equipotential curves will be ellipses with their major and minor axes at some angle to the coordinate axes. In other words, a displacement along one symmetry coordinate is coupled with a displacement along the other.

With S_2 and S_7, on the other hand, the interchange of labels of the C−O bonds reverses the sign of the former but not of the latter. Hence it reverses the sign of the corresponding cross-term. But since the energy cannot depend on the way the bond labels are chosen, this implies that this cross-term must be zero; the major and minor axes of the equipotential ellipses must then run parallel to one symmetry coordinate and perpendicular to the other. Displacements along S_2 and S_7 are not expected to be coupled. This theme is discussed in more detail in Chapter 5.

2.4 Symmetry Aspects of Specific Types of Molecule

2.4.1 Tetrahedral MX_4, Molecules and Degenerate Irreducible Representations

We consider the case of the tetrahedral MX_4 molecule in more detail [8]. The different ways of labeling the four X ligands form a group of order 24, isomorphous to the point group T_d, which is the highest symmetry that a molecule of this type can attain. The character table for this group is given in Table 2.3. The E representation is doubly degenerate, i.e. the E deformation "space" is spanned by two equivalent basis vectors; the T representations are triply degenerate and the corresponding deformation spaces are three-dimensional.

The internal coordinates, which may be taken as $r_1, r_2, r_3, r_4, \theta_{12}, \theta_{13}, \theta_{14}, \theta_{23}$, θ_{24}, θ_{34} transform as $2A_1 + E + 2T_2$. Symmetry coordinates may be chosen in several ways, one of which is shown in Table 2.4.

Recall that the six angles must be related by the determinantal condition described in Section 1.3.1. For small angular displacements from tetrahedral symmetry the positive and negative deviations from 109.47° cancel, so that the totally symmetric angular coordinate is zero. For larger displacements its value is negative. In any case, it can be regarded as redundant since the six angles are fixed by the values of the other five angular symmetry coordinates plus the determinantal equation.

Table 2.3. Character table for the point group T_d

	I	$8C_3$	$3C_2$	$6S_4$	$6\sigma_d$		Kernel	Co-kernels
A_1	1	1	1	1	1		T_d	
A_2	1	1	1	-1	-1		T	
E	2	-1	2	0	0		D_2	D_{2d}
T_1	3	0	-1	1	-1	(R_x, R_y, R_z)	C_1	S_4, C_3, C_s
T_2	3	0	-1	-1	1	(T_x, T_y, T_z)	C_1	C_{2v}, C_{3v}, C_s

Table 2.4. One choice of symmetry coordinates for a tetrahedral MX_4 molecule, emphasizing the action of the twofold symmetry operations of the tetrahedron

$S_1(A_1) = (r_1 + r_2 + r_3 + r_4)/2$
$S_{3a}(T_2) = (r_1 + r_2 - r_3 - r_4)/2$
$S_{3b}(T_2) = (r_1 - r_2 + r_3 - r_4)/2$
$S_{3c}(T_2) = (r_1 - r_2 - r_3 + r_4)/2$
$S_{2a}(E) = (2\theta_{12} - \theta_{13} - \theta_{14} - \theta_{23} - \theta_{24} + 2\theta_{34})/\sqrt{12}$
$S_{2b}(E) = (\theta_{13} - \theta_{14} - \theta_{23} + \theta_{24})/2$
$S_{4a}(T_2) = (\theta_{12} - \theta_{34})/\sqrt{2}$
$S_{4b}(T_2) = (\theta_{13} - \theta_{24})/\sqrt{2}$
$S_{4c}(T_2) = (\theta_{14} - \theta_{23})/\sqrt{2}$
$S_5(A_1) = (\theta_{12} + \theta_{13} + \theta_{14} + \theta_{23} + \theta_{24} + \theta_{34})/\sqrt{6}$

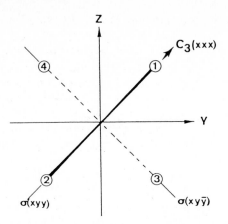

Fig. 2.2. Projection of MX_4 molecule with Cartesian coordinate system showing ligands 1 and 2 above the plane of the paper, ligands 3 and 4 below. The X axis runs perpendicular to the plane of the paper

The symmetry coordinates listed in Table 2.4 have been chosen so that both sets of T_2 coordinates correspond in a natural way to the view of the tetrahedron shown in Figure 2.2, along an S_4 axis. Both S_{3a} and S_{4a} are transformed into themselves by the same twofold axis $C_2(x)$ and mirror planes, $\sigma(xyy)$ and $\sigma(xy\bar{y})$; they both have C_{2v} co-kernel symmetry, and similarly for the other two matching pairs. Other ways of selecting the symmetry coordinates emphasize different aspects. The setting could be made to correspond to the view down one of the threefold axes, for example $C_3(xxx)$. This can be done by choosing suitable linear combinations of the previous coordinates, as follows:

$$S'_a = (S_a + S_b + S_c)/\sqrt{3}$$
$$S'_b = (2S_a - S_b - S_c)/\sqrt{6}$$
$$S'_c = (S_b - S_c)/\sqrt{2}$$

leading to the set of symmetry coordinates shown in Table 2.5.

Table 2.5. Another choice of symmetry coordinates for a tetrahedral MX_4 molecule, emphasizing the action of a threefold rotation operation of the tetrahedron (compare Table 2.4)

$S_1(A_1)$ $= (r_1 + r_2 + r_3 + r_4)/2$
$S'_{3a}(T_2) = (3r_1 - r_2 - r_3 - r_4)/\sqrt{12}$
$S'_{3b}(T_2) = (2r_2 - r_3 - r_4)/\sqrt{6}$
$S'_{3c}(T_2) = (r_3 - r_4)/\sqrt{2}$
$S_{2a}(E) = (2\theta_{12} - \theta_{13} - \theta_{14} - \theta_{23} - \theta_{24} + 2\theta_{34})/\sqrt{12}$
$S_{2b}(E) = (\theta_{13} - \theta_{14} - \theta_{23} + \theta_{24})/2$
$S'_{4a}(T_2) = (\theta_{12} + \theta_{13} + \theta_{14} - \theta_{23} - \theta_{24} - \theta_{34})/\sqrt{6}$
$S'_{4b}(T_2) = (2\theta_{12} - \theta_{13} - \theta_{14} + \theta_{23} + \theta_{24} - 2\theta_{34})/\sqrt{12}$
$S'_{4c}(T_2) = (\theta_{13} - \theta_{14} + \theta_{23} - \theta_{24})/2$

For this second set of symmetry coordinates, displacements along S'_{3a} and S'_{4a} are transformed into themselves by the same threefold axis; they have C_{3v} co-kernel symmetry. But we could equally well choose four equivalent (but now linearly dependent) symmetry coordinates, emphasizing the equivalence of the four threefold axes, by cyclic permutation of the four subscript numbers in $S'_{3a'}$ and $S'_{4a'}$. As shown in Figure 2.3, the resultant of two equal displacements along different C_{3v} co-kernel directions is a displacement along a C_{2v} co-kernel direction. In other words, a distortion of a tetrahedron conserving C_{2v} symmetry can be regarded as a combination of two distortions conserving C_{3v} symmetry, and similarly for other kinds of distortion.

Fig. 2.3. Scheme showing that the sum of two equal deformations along different C_{3v} co-kernel directions is a deformation with C_{2v} co-kernel symmetry

The choice of symmetry coordinates should be made according to convenience, but, if several molecules are to be compared, it is useful to label the atoms according to some specific convention, for example, in order of decreasing bond length, so that corresponding displacement vectors lie in the same asymmetric unit of the vector space defined by the symmetry coordinates. If this convention is adopted, then $D_{3a} \geq D_{3b} \geq D_{3c}$ in both the primed and unprimed systems, but no restrictions are imposed on the components of the angle deformation vectors. Alternatively, the labeling of the four M$-$X distances may be made in some random way (as is usually the case when the data are retrieved from crystallographic data files). There will then be no special relationship among the components D_{3a}, D_{3b}, D_{3c}, but there will always be a symmetry transformation that converts them into a sequence of decreasing magnitude.

It may be useful to consider a definite numerical example, a distorted PO_4 tetrahedron with internal coordinates:

$$r_1 = 1.645 \text{ Å} \quad \theta_{12} = 102.87° \quad \theta_{13} = 104.97° \quad \theta_{14} = 104.89°$$
$$r_2 = 1.558 \quad\quad\quad\quad\quad\quad\quad \theta_{23} = 113.12 \quad\quad \theta_{24} = 116.81$$
$$r_3 = 1.516 \quad\quad\quad\quad\quad\quad\quad\quad\quad\quad\quad\quad \theta_{34} = 112.5$$
$$r_4 = 1.449$$

The reference bond length in a phosphate tetrahedron can be taken as 1.534 Å [9]. The qualitative impression from this tabulation is that one bond length has been stretched, and another one shortened, both by large amounts; the three shorter bonds have been bent towards the stretched bond, thus diminishing the three bond angles involving this bond. The numerical analysis confirms this impression but makes it more precise. For the linear combinations corresponding to the symmetry coordinates of Table 2.4 we obtain:

$$D_1(A_1) = 0.016 \text{ Å} \quad D_{2a}(E) = -2.59° \quad D_5(A_1) = -0.66°$$

$$D_{3a}(T_2) = 0.119 \quad\quad D_{2b}(E) = \quad 1.89 \quad D_{4a}(T_2) = -6.84$$

$$D_{3b}(T_2) = 0.077 \quad\quad\quad\quad\quad\quad\quad\quad\quad D_{4b}(T_2) = -8.37$$

$$D_{3c}(T_2) = 0.010 \quad\quad\quad\quad\quad\quad\quad\quad\quad D_{4c}(T_2) = -5.81$$

The length of the two-dimensional $D_2(E)$ displacement vector is 3.20°. The length of the three-dimensional $D_3(T_2)$ vector is 0.142 Å, that of $D_4(T_2)$ is 12.28°. It turns out that $D_3(T_2)$ makes almost the same angle (33°) with the S_{3a} axis (C_{2v} co-kernel symmetry) as with the S'_{3a} axis (C_{3v} co-kernel symmetry). As far as the bond lengths are concerned, the distortion is thus intermediate between one conserving C_{2v} symmetry and one conserving C_{3v} symmetry. But the bond angle distortion is much closer to conserving C_{3v} symmetry, the angle between $D_4(T_2)$ and the S'_{4a} axis being only 9°. This makes it convenient to go over to the primed coordinate system of Table 2.5:

$$D'_{3a}(T_2) = 0.119 \quad D'_{4a}(T_2) = -12.14$$

$$D'_{3b}(T_2) = 0.062 \quad D'_{4b}(T_2) = \quad 0.21$$

$$D'_{3c}(T_2) = 0.047 \quad D'_{4c'}(T_2) = -1.81$$

Thus we may say that as far as bond angles are concerned "the fragment shows approximate C_{3v} symmetry".

We can see that equal and opposite deformations along S_{3a} or S_{4a} correspond to isometric structures and must therefore have the same energy. But equal and opposite deformations along S'_{3a} or S'_{4a} correspond to non-isometric structures. For small deviations from tetrahedral symmetry these have the same energy since the quadratic part of the potential energy in each of the triply degenerate T_2 sub-spaces must be spherically symmetric. For larger deviations, however, the energy need not be spherically symmetric (although it must still be totally symmetric with respect to all the T_d symmetry operations). For example, for a distortion along $+S'_{3a}$ (preserving C_{3v} symmetry) the unique bond is stretched and the other three shortened, whereas for one along $-S'_{3a}$ it is the other way round. There is no reason that the two deformations must have the same energy.

2.4.2 MX₅ Molecules

For any given MX₅ molecule we have 5 M−X distances, 10 X−X distances and $5! = 120$ ways of permuting the atomic labels. Because of the absence of any three-dimensional figure with the same symmetry properties as the permutation group $\mathbb{P}_5$, most structure correlation studies of molecules are based on those subgroups that do have an isomorphous counterpart among the point groups in three-dimensional space, e.g. D_{5h}, D_{3h} and C_{4v} [10].

First we take up the case of D_{3h} symmetry, i.e. of a trigonal bipyramidal arrangement of the five X's. This choice requires two of the X's to be classified as *axial* (*1,2*, say) and three (*3,4,5*) as *equatorial*. As long as the molecule is not too strongly distorted from D_{3h} symmetry, the distinction between axial and equatorial ligands is fairly obvious, but for strongly distorted structures such a distinction may be difficult or even meaningless. Nevertheless, it is always possible to choose one pair of X's as axial in such a way that the deviation from D_{3h} symmetry is minimal. For each of the 10 possible choices of an axial pair the displacements d_i along the angular D_{3h} symmetry coordinates are calculated. The total displacement is then the length of the resultant displacement vector obtained by summing over the various components for each choice. The choice leading to the smallest total displacement vector is the preferred one [10]. For this choice, there are still 12 possible permutations of the atomic labels and thus 12 permutational isomers that all show the same degree of distortion. We call a group of such isomers *distortionally equivalent.* Groups of permutational isomers with a different choice of an axial pair are called *distortionally inequivalent.* As we have seen, for a D_{3h} reference structure there are 10 groups of inequivalent isomers.

Similar considerations apply when the reference structure is taken to have C_{4v} or D_{5h} symmetry. In the former case there are 15 groups of distortionally inequivalent permutational isomers, corresponding to the 15 different ways of choosing the axial X and a pair of opposite equatorial X's. In the latter case there are 12 groups of distortionally inequivalent permutational isomers. This short discussion of distortionally equivalent and inequivalent structures in terms of permutations of atomic labels (or, alternatively, of unsigned geometric quantities, such as interatomic distances) ignores difficulties that arise from relations among enantiomorphic labelings (see Section 2.5).

2.4.3 MX$_6$ Molecules

For an MX$_6$ molecule or fragment there are 6 M$-$X distances, 15 X $\ldots$ X distances, and $6! = 720$ different ways of labeling the X atoms. Again, there is no three-dimensional point group that is isomorphous with the permutation symmetry group; to construct one, we need to go into five dimensions (six equivalent unit vectors making mutual angles of 101.54°). The most common reference structure is the regular octahedron (point group O_h). The characteristic feature of a structure that is classified as distorted octahedral is that the six X ligands can be identified as three "opposite" pairs, the *trans* pairs, the members of each pair not sharing any common face of the distorted octahedron. There are thus 15 groups of distortionally inequivalent permutational isomers, and one of them will show the smallest degree of distortion from the reference octahedron. Other possible reference structures are the regular hexagon (point group D_{6h}), and the trigonal prism (point group D_{3h}). To search for the minimum deformation from each of these reference structures it is necessary to test 60 permutationally inequivalent isomers for both the hexagon and trigonal prism.

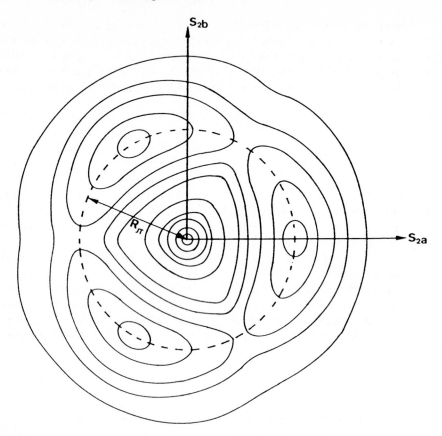

Fig. 2.4. Warped Mexican-hat type of potential surface for Jahn-Teller distortion of octahedral MX_6 molecule. The symmetry of the two-dimensional deformation space is $3m$, and displacements along the mirror lines correspond to distortions of the octahedron that preserve tetragonal symmetry (elongated or compressed octahedron)

Octahedral MX_6 molecules with M = Cu(*II*), Ag(*II*), low-spin Co(*II*) or high-spin Cr(*II*) have an electronically degenerate ^2E_g ground state and are thus expected to undergo Jahn-Teller distortion along displacement coordinates of E_g symmetry. Appropriate coordinates may be defined as:

$$S_a(E_g) = (2r_1 - r_2 - r_3 + 2r_4 - r_5 - r_6)/\sqrt{12}$$

$$S_b(E_g) = (r_2 - r_3 + r_5 - r_6)/2$$

where r_i and r_{i+3} are *trans* bond distances. The kernel symmetry of the E_g representation is D_{2h} and the symmetry of the two-dimensional E_g sub-space is $3m$ [7]. If the S_{2a} basis vector is chosen to lie in one of the mirror lines, then a displacement along this vector corresponds to tetragonal distortion of the reference octahedron

(D_{4h} co-kernel symmetry), a positive displacement corresponding to tetragonal elongation, a negative displacement to tetragonal compression. In view of the 3 m symmetry of the sub-space, the orthogonal S_{2b} vector does not lie in a mirror line and the corresponding displacement shows only the D_{2h} kernel symmetry of the representation. In terms of polar coordinates R and α, corresponding components are

$$R = (S_a^2 + S_b^2)^{1/2}$$

$$\cos \alpha = S_a/R \qquad \sin \alpha = S_b/R \ .$$

The energy surface for the E_g sub-space shows minima at $R = R_{JT} \neq 0$, $\alpha = 0$, 120° and 240° (tetragonal elongation) and saddle points at $R \approx R_{JT}$, $\alpha = 60°$, 180° and 300° (tetragonal compression). This type of potential is commonly described as a "warped Mexican hat" potential (Figure 2.4).

This way of describing Jahn-Teller distorted complexes is enlightening because it provides a coherent picture of the various geometries that have been observed and also of their interconversions [11]. These take place along a roughly circular pathway in the sub-space spanned by the E_g coordinates without the need to pass through the O_h symmetric reference structure. Similar situations will be encountered in the following discussion of the out-of-plane deformations of five- and six-membered rings.

2.4.4 Out-of-Plane Deformations of Five-Membered Rings

The choice of suitable coordinates to describe the puckering of a five-membered ring has been the subject of innumerable papers. The essential points were formulated in the classic paper introducing the concept of pseudorotation [12]:

"...the ring puckering motions are: first an ordinary vibration in which the amount of puckering oscillates about a most stable value and second, a pseudo one-dimensional rotation in which the phase of the puckering rotates around the ring. This is not a real rotation since the actual motion of the atoms is perpendicular to the direction of rotation and there is no angular momentum about the axis of rotation."

The out-of-plane distortion of a five-membered ring can always be described in terms of only two coordinates, since, of the nine ($3 \times 5 - 6$) independent coordinates, seven ($2 \times 5 - 3$) may be chosen in an arbitrary plane, the mean plane of the ring. These two coordinates can be chosen as a pair of symmetry displacement co-ordinates of a regular pentagon that transform together as the doubly degenerate E_2'' representation of D_{5h}.

$$S_a(E_2'') = \sqrt{(2/5)} \sum z_i \cos(4\pi i/5)$$
$$S_b(E_2'') = \sqrt{(2/5)} \sum z_i \sin(4\pi i/5) \ .$$

z_i being the displacement coordinate of the ith atom ($i = 1,2,3,4,5$) from the mean plane of the nonplanar pentagon. Alternatively, the displacement vector may be expressed in terms of a puckering amplitude q and a phase angle ϕ (polar coordinates) with components

$$q = (S_a^2 + S_b^2)^{1/2}$$
$$\cos\phi = S_a/q \qquad \sin\phi = S_b/q$$

Deformations along the S_a and S_b coordinates are shown in Figure 2.5. In this description, a displacement along S_a leads to a mirror-symmetric "envelope" form of the five-membered ring with four atoms in a common plane, while a displacement along the orthogonal coordinate S_b leads to a "twist" form with a dyad axis. The displacements of the individual atoms are, for these two symmetric forms

$$z_i = \sqrt{(2/5)}\, q \cos(4\pi i/5) \qquad \text{envelope form}$$
$$z_i = \sqrt{(2/5)}\, q \sin(4\pi i/5) \qquad \text{twist form}$$

For a general distortion the out-of-plane displacement z_i of the ith atom is

$$z_i = \sqrt{(2/5)}\, q \cos(4\pi i/5 + \phi) \ .$$

Rings described by ϕ angles of 0, 36°, 72°, ... are envelope forms, those described by ϕ angles of 18°, 54°, 90°, ... are twist forms. For an arbitrary distortion, rotation of the displacement vector through a multiple of $4\pi/5$ (144°, 288°, 72°, 216°, 360°) produces an equivalent out-of-plane distortion that differs from the original by rotation of the pentagon through $2\pi n/5$, while rotation of the displacement vector through 180° produces an equivalent out-of-plane distortion that differs from the original by reflection in the plane of the pentagon. Since the non-planar rings (except the envelope forms) are chiral, this operation produces the enantiomer of the

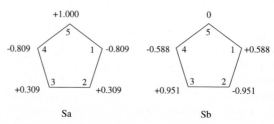

Fig. 2.5. Pattern of individual atomic displacements along the S_a and S_b symmetry coordinates describing the out-of-plane deformation of a regular pentagon. Alternatively, the symmetry coordinates can be constructed from torsion angles

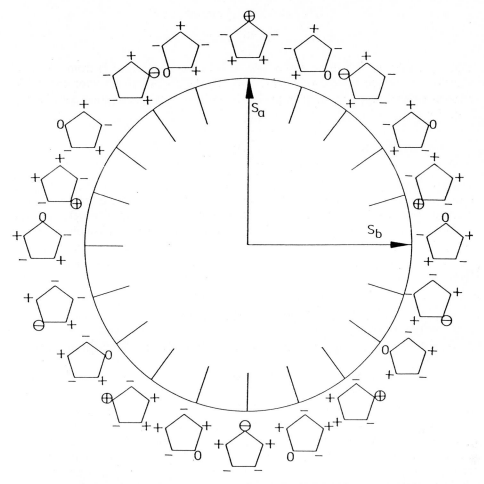

Fig. 2.6. Out-of-plane distortion of a pentagon as the phase angle coordinate ϕ goes through a complete cycle. Only forms with C_2 or C_s symmetry are shown; their sequence and relationships are explained in the text. For intermediate angles the corresponding forms have no symmetry

original. The complete itinerary along the cyclic ϕ coordinate is shown in Figure 2.6. It is curious that five-membered rings tend to be described as "envelopes" even when they are much closer to twist forms; from inspection of models or drawings, the approximate planarity of a four-atom grouping seems to spring to the eye more easily than the presence of an approximate or even an exact dyad axis.

This is essentially the group-theoretical background to the description of out-of-plane distortions of the cyclopentane ring by Pitzer and Donath [13]. Most other authors [14] prefer to use torsion angles ω_j about bonds instead of z_i's of atoms as the out-of-plane coordinates. The ω's and z's transform in the same way, except that the special forms corresponding to the S_a and S_b symmetry coordinates are interchanged. In the torsion angle description,

$$S_a(E_2'') = \sqrt{2/5} \sum \omega_j \cos (4\pi j/5)$$
$$S_b(E_2'') = \sqrt{2/5} \sum \omega_j \sin (4\pi j/5) .$$

Displacements along S_a now correspond to the twist form with the dyad axis, and those along S_b to the mirror-symmetric envelope form. This difference results from the difference in the way the two types of coordinate behave under these symmetry operations; in the envelope form z's related by the mirror-plane have the same sign while ω's have opposite sign, and vice versa for the behavior of the coordinates with respect to the dyad axis.

The three conditions that reduce the five z or ω coordinates to only two out-of-plane coordinates are of the same mathematical form:

$$S(A_1') = \sqrt{(1/5)} \sum p_j = 0$$
$$S_a(E_1'') = \sqrt{(2/5)} \sum p_j \cos (2\pi j/5) = 0$$
$$S_b(E_1'') = \sqrt{(2/5)} \sum p_j \sin (2\pi j/5) = 0 .$$

But the geometric meaning of the equations is different. For the z's (positional coordinates) they are the conditions of no net translation or rotation; for the ω's (internal coordinates) they are the ring-closure conditions for an equilateral pentagon. Although these equations are strictly valid only for infinitesimal puckering amplitudes of an equilateral pentagon, they hold well even for strongly puckered rings in which the bond lengths vary considerably from their average value.

For any given set of out-of-plane displacements z_i, the values of the puckering amplitude q and the phase angle ϕ can be calculated from

$$\tan \phi = \frac{-z_1 + z_2 - z_3 + z_4}{z_5 (\sin 36° + \sin 72°)} = \frac{-z_1 + z_2 - z_3 + z_4}{3.077 \, z_5}$$

$$q = \frac{\sqrt{(5/2)} \, z_5}{\cos \phi}$$

with a corresponding expression for the ω's. The value of ϕ depends on how the atoms are numbered. For a ring where the atoms are chemically non-equivalent (e.g. furanose) some definite numbering scheme should be adopted. If the atoms are chemically equivalent (e.g. as in cyclopentane), the value of ϕ can be expressed modulo 36°.

It is easily shown that for a five-membered ring, when ϕ is varied, the interdependence of pairs of torsion angles is given by the equation of an ellipse. Consider the sum and difference of torsion angles about two adjacent bonds:

$$X = (\omega_5 + \omega_1)/\sqrt{2} = \omega_0 [\cos \phi + \cos (144° + \phi)]/\sqrt{2} = \sqrt{2} \omega_0 \cos (\phi + 72°) \cos 72°$$
$$Y = (\omega_5 - \omega_1)/\sqrt{2} = \omega_0 [\cos \phi - \cos (144° + \phi)]/\sqrt{2} = -\sqrt{2} \omega_0 \sin (\phi + 72°) \sin 72°$$

or $X/[\sqrt{2} \cos 72°] = \omega_0 \cos (\phi + 72°)$ and $Y/[\sqrt{2} \sin 72°] = \omega_0 \sin (\phi + 72°)$

which can be combined to give

$$X^2/2\,(0.309)^2 + Y^2/2\,(1.809)^2 = (\omega_0)^2$$

the equation of an ellipse with major and minor axes in the ratio (sin 72°/cos 72°) or approximately 3 : 1. The actual distribution of torsion angles around adjacent C–O bonds in five-membered ketal rings is shown in Figure 2.7 [15]. The dependence between non-adjacent torsion angles is also described by an ellipse, this time with major and minor axes in the ratio (cos 36°/sin 36°) or approximately 1.4 : 1.

We mentioned earlier the possible pitfalls in taking average values of geometric parameters that are not independent variables. The cyclopentane ring is a good example. From an electron-diffraction analysis [16] the C–C bond lengths are equal at 1.546(1) Å, the average C–C–C bond angle is 104.46° and the average ring torsion angle is zero. However, these structural parameters cannot correspond to a D_{5h} symmetric pentagon, where all the angles must equal 108°. In fact, they do not correspond to any pentagon realizable in three-dimensional space.

As the cyclopentane molecule undergoes pseudorotation, the individual bond angles change (as well as the torsion angles) but the average value stays constant. For

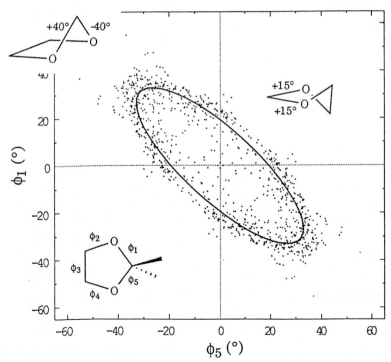

Fig. 2.7. Distribution of torsion angles around adjacent C–O bonds in five-membered ketal rings [15]

example, for the form with a pseudorotational phase angle of $10°$ the individual bond angles are: 103.191, 105.714, 106.373, 104.709, 102.309 °. It would be futile to try to explain these differences in bond angles in terms of a theory of chemical bonding; the reason is a purely geometric one. In a non-planar equilateral pentagon the angles *must* be different. The average angle is necessarily less than 108 ° and it is impossible to construct a non-planar pentagon with this average angle. In other words, the equilateral equiangular pentagon is planar [17]. Note that the pentagon is the only n-gon that has this property. For all other values of n (apart from the trivial case of the triangle), equilateral, equiangular n-gons can be constructed that are non-planar.

As in our earlier discussion of MX_5 fragments (Section 2.4.2), the problem of describing out-of-plane deformations of a five-membered ring involves one of the sub-groups of the permutation group $\mathbb{P}_5$, this time D_{5h}. There are 10 interatomic distances (analogous to the $10\,X \ldots X$ distances in MX_5), but now their identification in terms of nearest or next-nearest neighbours usually presents no problem because of the large difference between bonded and non-bonded distances. Although experimentally observed geometries of five-membered rings rarely show D_{5h} symmetry and often deviate quite strongly from it (e.g. furanose rings), the D_{5h} structure can usually be employed as a reference structure, and the analysis in terms of, and restricted to, the degenerate pair of E_2'' symmetry coordinates is usually informative because it provides a coherent picture of the continuum of observed (and possible) geometries and their interconversions. However, some reservations may be called for in the case of five-membered chelate rings in coordination compounds, where two ring distances are often very much larger than the other three; here the D_{5h} reference model may be stretched beyond its limits.

2.4.5 Out-of-Plane Deformations of Six-Membered Rings

In the same way, the following description of geometry of six-membered rings is related to the MX_6 problem described in Section 2.4.3. We usually take the reference structure to be a regular hexagon with D_{6h} symmetry, and the 15 interatomic distances can usually be divided into three groups corresponding to the 1,2-, 1,3-, and 1,4-distances in the hexagon. This makes it possible to restrict the analysis to the three out-of-plane coordinates transforming as the B_{2g} and E_{2u} irreducible representations. Nevertheless, the cyclohexane ring, with bond angles of about $110°$ instead of $120°$ (as in the planar hexagon), is quite strongly deformed from planarity, and the hexasulphur ring, with bond angles close to $90°$, is even more so. The six bond distances of these strongly non-planar rings could also be associated with six edges of a strongly deformed octahedron, but the hexagonal reference is usually found to be more appropriate.

The cyclohexane ring has a unique property that is familiar to everyone who has handled molecular models. There is one form, the "chair" form (point group D_{3d}), that appears to be rigid, and there is a family of flexible forms that includes the "boat" form (point group C_{2v}) and the "twist" form (point group D_2) as special cases. With the six bond lengths and six bond angles fixed in the model, the $3\times6-6 = 12$ degrees of freedom of the six-membered ring would appear to be exhausted, so the rigidity of the chair form may come as no surprise. Indeed, if the six atoms are regarded as vertices of an octahedron, as adumbrated above, the rigidity of this figure follows from a theorem due to Cauchy, provided that the 12 edges of the octahedron (the 6 bond distances and the 6 1,3-distances or bond angles) are regarded as fixed [18]. But the flexible forms seem to have at least one degree of torsional freedom. Where does it come from? Or is it illusory? Can one be sure that the passage from one flexible form to another does not involve small bond-angle deformations that might be difficult to detect from inspection of mechanical models?

A thorough analysis of this kind of problem is mathematically quite complicated [19, 20], but we can obtain some insight by considering the symmetry properties of the out-of-plane deformations of a regular hexagon (point group D_{6h}) [21]. The argument is similar to that already given for the five-membered ring. For the hexagon we have three out-of-plane coordinates, since, of the 12 degrees of freedom, nine $(2\times6-3)$ may be chosen in the mean plane. Of the three symmetry displacement coordinates, one is non-degenerate and the other two form a degenerate pair:

$$S(B_{2g}) = \sqrt{(1/6)} \sum z_i(-1)^i$$
$$S_a(E_{2u}) = \sqrt{(1/3)} \sum z_i \cos(4\pi i/6)$$
$$S_b(E_{2u}) = \sqrt{(1/3)} \sum z_i \sin(4\pi i/6)$$

z_i being the displacement coordinate of the ith atom ($i = 1,2,3,4,5,6$) from the mean plane of the puckered hexagon. The deformations corresponding to these coordinates are shown in Figure 2.8; they have the same form as the anti-bonding Hückel molecular orbitals of benzene. In this description, displacement along the non-degenerate B_{2g} coordinate gives the chair form, while displacements along the S_a and S_b coordinates give the boat and twist forms respectively. The only symmetry element these have in common is a twofold rotation axis normal to the mean plane; this is the kernel symmetry of the E_{2u} irreducible representation. We refer to this axis as a non-intersecting axis since it does not pass through any atoms or any bonds.

Just as for the pentagon, any orthogonal pair of linear combination of the degenerate pair of E_{2u} coordinates is also a symmetry-adapted pair of deformation

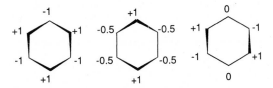

Fig. 2.8. Patterns of individual atomic displacements for the three symmetry coordinates describing the out-of-plane deformation of a regular hexagon. Alternatively, the symmetry coordinates may be constructed from torsion angles

coordinates transforming as E_{2u}. The out-of-plane displacements for such an arbitrary combination may be expressed as:

$$z_i = \sqrt{(1/3)}\, q \cos (4\pi i/6 + \phi)$$

where q is the puckering amplitude. The extra degree of freedom can now be identified with the phase angle ϕ. Rings described by ϕ values of 0, 60°, 120°, ... are boat forms with C_{2v} symmetry, those described by ϕ angles of 30°, 90°, 150°, ... are twist form with D_2 symmetry. Intermediate values of ϕ give forms that have only C_2 symmetry. The variable phase angle confers an extra degree of flexibility — the ring undergoes pseudo-rotation. As with the pentagon, the torsion angles transform in the same way as the out-of-plane displacements; however if z is interpreted as a torsion angle the identification of special ϕ values with the boat and twist forms is reversed.

The non-degenerate out-of-plane displacement transforming as B_{2g} is not associated with any phase angle; once its magnitude is fixed the ring is rigid. Although the symmetry coordinate description applies strictly only for infinitesimal out-of-plane deformations of a regular hexagon, it holds reasonably well for quite large out-of-plane deformations of rings in which the atoms are not chemically equivalent.

Indeed, for any given set of out-of-plane displacements or torsion angles of a six-membered ring, the displacements along the three out-of-plane symmetry coordinates can be computed. Clearly, a chair form will have zero or very small components along the E_{2u} coordinates, while twist and boat forms will have zero or very small components along the B_{2g} coordinate. For forms with lower symmetry, both types of symmetry coordinate will be involved, and the symmetry coordinate description can thus lead to a quantitative dissection of the total out-of-plane deformation into its components.

The application of the symmetry coordinate approach to larger rings is possible. One can always define $N-3$ out-of-plane coordinates with reference to the mean plane of the N-membered ring. For a regular N-gon, these occur in degenerate pairs of symmetry coordinates (with one non-degenerate coordinate left over for the even-membered rings), but the usefulness of this approach obviously diminishes as the ring geometry deviates more and more from that of the corresponding regular planar polygon.

2.5 Configuration Spaces for Molecules with Several Symmetrical Reference Structures

So far, we have been assuming that the choice of a suitable reference structure is obvious. There is no problem as long as the distortions from some symmetrical structure are small, at any rate not large enough as to suggest comparison with an alternative symmetrical structure that could equally well serve as the reference structure. However, a frequent problem in structure correlation concerns the transformations between equivalent structures of one symmetry through one or more intermediate structures of another symmetry.

We return to the simple example of a linear triiodide fragment and consider the possible interchange of the central atom with a terminal one. There are various ways to achieve such an interchange (Figure 2.9). One involves a shift of the atoms in a line; this is perfectly feasible geometrically although it is not a very likely possibility from a physical point of view since it leads to configurations with two atoms in the same place. An alternative is to bend the triiodide fragment through a series of scalene triangles into an equilateral triangle and then to straighten it out again into a different linear arrangement. Imagine that such a process takes place in two-dimensional space, e.g. on a surface; the linear fragment may then be bent in either of two directions to produce two families of scalene triangles which are isometric but non-

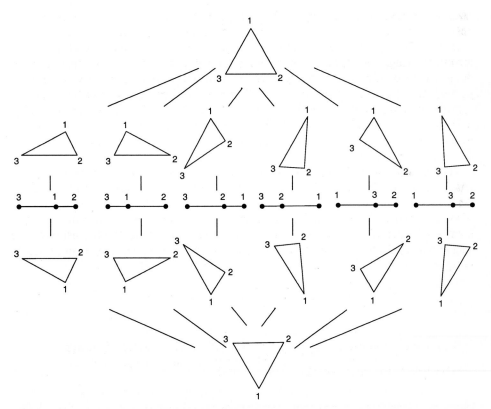

Fig. 2.9. Linear triatomic molecules: interchange of the central atom with a terminal one by bending through a series of scalene triangles into an equilateral triangle and then straightening out again. The two alternative modes of bending lead to two families of scalene triangles that are isometric in pairs but not superimposable in two dimensions

superimposable, i.e. enantiomeric, in two dimensions. For each such scalene triangle there are 6 permutations of the three interatomc distances a, b, c, as described in Section 2.1. Thus, while the relevant permutation group for a given scalene triangle is of order 6, the order of the group becomes doubled if interconversion between the two enantiomeric families of triangles is taken into account. If the 6 rearrangements of the permutation group alone are associated with the symmetry operations of the point group D_3, then the 12 elements of the doubled group can be associated with those of D_{3h}. For each symmetry operation of the first kind contained in D_3, there is an exactly analogous one of the second kind in the doubled group; for each labeled scalene triangle with sides a, b, c in one family, there is an enantiomeric triangle in the other.

It is not obvious how to define a configuration space with the required symmetry properties. One possibility is to start from symmetry adapted linear combinations of the three distance coordinates:

$$S_1 = (a+b+c)/\sqrt{3}$$
$$S_{2a} = (2a-b-c)/\sqrt{6}$$
$$S_{2b} = (b-c)/\sqrt{2}$$

With distance coordinates we can define the size and shape of a triangle but not its sense of chirality. To achieve this, vectors are needed. We define vectors d_{12}, d_{23}, d_{31}, from atoms 1 to 2, 2 to 3, and 3 to 1, respectively, whose lengths correspond to any permutation of the distances a, b, c. Next we replace S_1 by a coordinate S_1' which includes chirality by evaluation of the vector product $d_{12} \times d_{23}$ (or $d_{23} \times d_{31}$ or $d_{31} \times d_{12}$, since all three vector products are equal by virtue of $d_{12} + d_{23} + d_{31} = 0$). These products are each in magnitude twice the area F of the triangle, but their sign s is different for a clockwise and an anti-clockwise arrangement of the three vectors. Thus, we can choose

$$S_1' = sF .$$

There are 12 symmetry equivalent general points in this space, defining uniquely the 12 isometric triangles. All degenerate triangles with three atoms in a line are represented by the plane $F = 0$, which dissects configuration space into two halves. All equilateral triangles have zero components along S_{2a} and S_{2b}. The point at the origin of the configuration space corresponds to three superimposed atoms, the only arrangement whose coordinates in configuration space are unchanged by any of the six permutations of $a, b,$ and c, and in which there is no difference between a clockwise and an anti-clockwise arrangement of the three atoms. Not all points in this configuration space are of chemical interest, but all configurations of chemical interest are contained in it [22].

The two-dimensional problem with the triangles is analogous to the chemically more interesting three-dimensional problem concerning inversion of configuration of a tetrahedron. For a distorted tetrahedral MX_4 molecule there are 24 ways of labeling the vertices, i.e. 24 isometric distortions from T_d symmetry; and there are also 24 ways of labeling the vertices of the enantiomeric tetrahedron. Thus, the relevant group for describing distortions large enough to include enantiomerization processes is of order 48. One such distortion involves the D_{2d} coordinate (see Section 2.4.1) which interconnects two tetrahedral arrangements via a square planar one. One may arbitrarily choose one of the two enantiomeric tetrahedral arrangements or the square planar one as the reference configuration; alternatively, one may try to define a configuration space with the required symmetry of order 48. As in the previous example, enantiomeric tetrahedra cannot be distinguished by distance coordinates alone; vectors are required. For example, if the four vertices are labeled $1, 2, 3, 4$ then the volume of the tetrahedron, $d_{12} \cdot (d_{13} \times d_{14})/6$, can be taken as positive or negative, depending on whether the vectors d_{12}, d_{13}, d_{14} form a right- or a left-handed coordinate system. If the vectors lie in a common plane then the volume is zero.

Analogous considerations apply to interconversions of MX_5 molecules between trigonal bipyramidal structures via, for example, square pyramidal ones, or to interconversions of MX_6 molecules between octahedral structures via trigonal prismatic ones (see Chapter 8), and, quite generally, to all interconversion processes between isometric structures. A more rigorous treatment of these problems would have to take into account the concept of the permutation-inversion group [23].

In these examples, isometric structures are interconnected by large distortions that trace possible interconversion pathways between alternative reference structures. Instead of defining configuration space relative to a single symmetric reference structure it is advantageous in these cases to have a symmetrical description of the relevant portion of configuration space and thereby of the entirety of geometric transformations between isometric structures.

2.6 Internal Rotation in Non-Rigid Molecules

Similar problems arise when we deal with possible reference structures that can be interconverted by internal rotations. Consider, for example, the diphenylmethane molecule (Scheme 2.2), where conformations with the planes of both phenyl groups either parallel or perpendicular to the $C-CH_2-C$ plane have C_{2v} symmetry. Rotation of a phenyl group through π produces a conformation that is isometric with the initial one, but such a rotation is not a symmetry operation of the C_{2v} point group. However, it is clear that such operations have to be included in the group of all operations that leave the reference structure invariant. We shall not attempt to review here the work that has been done on the classification of such "symmetry operations" on non-rigid molecules [23], nor the controversies that have ensued from it. Instead we present four representative examples of such molecules.

2.6.1 Ethane, One Internal Rotational Degree of Freedom

We consider the ethane molecule as a simple example. In Figure 2.10 the Newman projection **E** shows an arbitrary conformation characterized by some value of the torsion angle about the $C-C$ bond. The point group of the molecule is D_3, order 6, so there are 6 symmetry equivalent arrangements of the same figure. Projections **R** and $\mathbf{R}^2$ show the result of rotating the distal methyl group relative to the proximal one by 120° and 240°. Rotation by 360° leads back to **E**. Conformations **R** and $\mathbf{R}^2$ are isometric with **E** and hence indistinguishable from it; each of them also gives rise to 6 symmetry equivalent arrangements, making 18 in all. However, since the initial arrangement **E** is chiral, there is a matching set of 18 isometric structures that are enantiomorphic to **E**. Thus the order of the isometric symmetry group is 36. Note that rotation of one methyl group with respect to the other is *not* a point symmetry operation of the ethane molecule regarded as a rigid figure. It is, however, a symmetry operation of the isometric group and hence of the configuration space of the molecule.

If we plot some property of the ethane molecule, say, its potential energy, as a function of the torsion angle, we obtain a curve with threefold periodicity. In crystallographic parlance we have a repeating one-dimensional pattern with the line group *pm* and periodicity $t = 120°$. The special positions (fixed points) at $\alpha = 0°$ and 60° (modulo 120°) correspond to structures with special symmetry: at 0° the eclipsed conformations with D_{3h} symmetry, at 60° the staggered conformations with D_{3d} symmetry. A general position corresponds to a chiral conformation with D_3 symmetry only. Note that the order of D_3 is half that of D_{3h} or D_{3d}, but the number of isometric D_3 conformations is double the number of isometric D_{3h} or D_{3d} conformations.

The example is trivial, but it suggests that the conformational dependence of properties of more complicated molecules with cyclic degrees of freedom (in particular, torsional degrees of freedom) can be described using the ideas of space-group symmetry. With this kind of description, there is a one-to-one correspondence between familiar crystallographic concepts and molecular symmetry. For example, a general position will always correspond to the lowest molecular symmetry admitted in a conformational analysis, while a special position will always correspond to a higher molecular symmetry. The exact nature of this correspondence needs to be made more precise, and we shall address some aspects of this problem later in this chapter.

2.6.2 Simplified Symmetry Analysis of Conformationally Flexible Molecules

Given the constitution (connectedness) of a molecule, the internal coordinates are separated into those which are regarded as being fixed (at least for the purpose of the analysis) and those which, for one reason or another, may vary over a large range

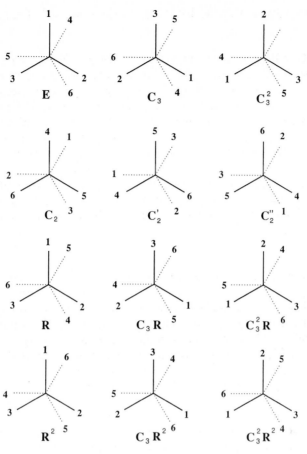

Fig. 2.10. Isometric structures of the ethane molecule, obtained by symmetry operations of the point group D_3 and by rotation of one methyl group with respect to the other

of values. In this way the molecule is divided conceptually into a number of rigid fragments whose relative positions or orientations may be varied within certain limits. Given the flexible degrees of freedom and the ranges over which they are allowed to vary, we ask for all possible sets of values of the relevant internal parameters that produce isometric structures. The relationships among such sets of parameter values can be expressed in terms of a set of transformations (or the corresponding transformation matrices) that form a group. It turns out that this group can often be decomposed into two sub-groups, one concerned with the flexible degrees of freedom, the other with the symmetry properties of a definite fragment of the molecule – we call it the molecular frame. The first sub-group is related to crystallographic translation groups (in n dimensions, with n the number of flexible degrees of freedrom), the second to point group symmetries.

2.6.3 Two Internal Rotational Degrees of Freedom

Molecules that could be described in this way include diphenylmethane (Scheme 2.2, **A**), 2,2′-dichlorodiphenylmethane (**B**), 4-chlorodiphenylmethane (**C**), propane (**D**), ethylbenzene (**E**), dicyclopentadienyltitaniumdichloride (**F**). These molecules can be regarded as having a rigid frame on which the phenyl, methyl, or cyclopentadienyl groups are free to rotate. We discuss the diphenylmethane example in detail [7], but some aspects of the discussion will be pertinent to all molecules with two degrees of internal rotation.

An arbitrary conformation of the diphenylmethane molecule can be characterized by the torsional coordinates ω_A and ω_B with specific components ϕ_1 and ϕ_2 (Figure 2.11). We define the zero value of ϕ_1 when C(2A) eclipses the *ipso*-carbon C(1B), and similarly for ϕ_2; increasing torsion angle corresponds to clockwise rotation of the phenyl group in question, viewed along the bond from the central atom to the phenyl group.

We now enumerate the ways in which this arbitrary conformation can be converted into an isometric structure. First, the enantiomorphic conformation can be obtained by reversing the signs of the torsion angles: $\phi_1 \rightarrow -\phi_1, \phi_2 \rightarrow -\phi_2$. Second, because the phenyl groups are equivalent, the values of ϕ_1 and ϕ_2 may be interchanged: $\phi_1 \rightarrow \phi_2, \phi_2 \rightarrow \phi_1$. Combination of these operations gives: $\phi_1 \rightarrow -\phi_2, \phi_2 \rightarrow -\phi_1$. These transformations (Figure 2.11) form a group that is isomorphous to C_{2v}. Since the elements of this group (Table 2.1) can be identified with the symmetry elements of the central $C-CH_2-C$ frame, we refer to this group as the frame group *F*.

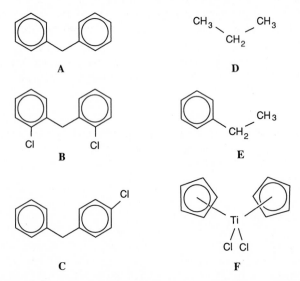

Scheme 2.2. Structural formulas of diphenylmethane (**A**), 2,2′-dichlorodiphenylmethane (**B**), 4-chlorodiphenylmethane (**C**), propane (**D**), ethylbenzene (**E**), dicyclopentadienyltitaniumdichloride (**F**)

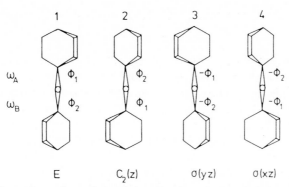

Fig. 2.11. Isometric conformations of the diphenylmethane molecule. The conformations **1, 2, 3** and **4** can be transformed into one another by point-group symmetry operations that leave the central frame unaltered or by appropriate rotations of the phenyl groups

A second type of operation that transforms an arbitrary conformation of the diphenylmethane molecule into an isometric one is physical rotation of one or both phenyl groups through the angle π (ring flips). Like the internal rotation in ethane, these flip operations are not point symmetry operations; they have been called "isodynamic" operations [24]. Nevertheless, they are analogous to symmetry operations and can be assigned analogous symbols (E, R_A, R_B, R_{AB}) that form a group called the rotor group R (provided that the torsion angles are interpreted modulo 2π). In our example, the two groups F and R happen to be isomorphous, but this will not generally be the case.

A complete list of conformations isometric to the original one is obtained by combining each of the rotor group operations with each of the frame operations. The combination of R and F leads to a super-group G of order 16 (Table 2.6). Note that this group is non-Abelian, i.e., the order of combination is important; for example, $C_2 R_A = R_B C_2 \neq R_A C_2$. Since it contains four elements of order 4 ($R_A C_2, R_B C_2, R_A \sigma(xz), R_B \sigma(xz)$) and eleven of order 2, the group described in Table 2.6 is isomorphous with D_{4h}.

Table 2.6. Group of 16 operations obtained by combining elements of the frame group $F(E, C_2, \sigma(xz), \sigma(yz))$ and rotor group $R(E, R_A, R_B, R_{AB})$ for the diphenylmethane example

Operation	ω_A	ω_B	Operation	ω_A	ω_B
E	ϕ_1	ϕ_2	$\sigma(xz)$	$-\phi_1$	$-\phi_2$
R_A	$\phi_1 + \pi$	ϕ_2	$R_A \sigma(xz)$	$-\phi_1 + \pi$	$-\phi_2$
R_B	ϕ_1	$\phi_2 + \pi$	$R_B \sigma(xz)$	$-\phi_1$	$-\phi_2 + \pi$
$R_{AB} \sigma(xz)$	$\phi_1 + \pi$	$\phi_2 + \pi$	$R_{AB} \sigma(xz)$	$-\phi_1 + \pi$	$-\phi_2 + \pi$
C_2	ϕ_2	ϕ_1	$\sigma(yz)$	$-\phi_2$	$-\phi_1$
$R_A C_2$	$\phi_2 + \pi$	ϕ_1	$R_A \sigma(yz)$	$-\phi_2 + \pi$	$-\phi_1$
$R_B C_2$	ϕ_2	$\phi_1 + \pi$	$R_B \sigma(yz)$	$-\phi_2$	$-\phi_1 + \pi$
$R_{AB} C_2$	$\phi_2 + \pi$	$\phi_1 + \pi$	$R_{AB} \sigma(yz)$	$-\phi_2 + \pi$	$-\phi_1 + \pi$

The partitioning of G into R and F is a consequence of the way we have dissected the molecule into a rigid frame and two groups that are allowed to rotate with respect to this frame. There is a striking analogy between it and the partitioning of a space group S into a semi-direct product [25] of an infinite translation group T and a unit cell group U [26]. We represent each isometric structure by a point with components ϕ_1 and ϕ_2 along the two torsion angle coordinates. The transformation corresponding to R_A can now be thought of as a *translation* of a representative point $P(\phi_1, \phi_2)$ by an amount π along the ω_A coordinate. Repetition of this operation produces a translation of 2π, a complete revolution of the torsion angle, equivalent to the identity operation. Similarly, the operation R_B corresponds to translation of π along the other coordinate ω_B, and R_{AB} corresponds to translations of π along ω_A and ω_B. If we remove the restriction that the torsion angles are to be interpreted modulo 2π, the group R is no longer of order 4 but becomes a translation group, of infinite order. To complete the analogy, the genuine symmetry operations of F correspond to transformation of $P(\phi_1, \phi_2)$ into a set of general positions of a unit cell.

A pattern of this kind, a conformational map [27], containing a set of 16 points corresponding to the 16 isometric conformations of diphenylmethane is shown in Figure 2.12. A primitive unit cell containing four equivalent points may be defined by $0 \le \omega_A < \pi$ and $0 \le \omega_B < \pi$. Its special positions (fixed points) are the diagonal

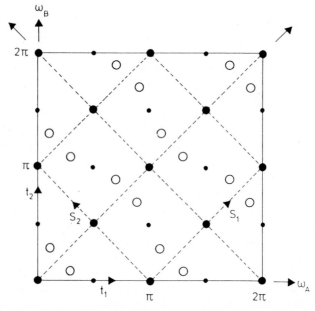

Fig. 2.12. Conformational map of diphenylmethane. The 16 equivalent positions (open circles) are images of the 16 isometric conformations with different values of the two torsion angles. The unit cell shown is non-primitive, the primitive lattice having translation distances of π along ω_A and along ω_B. The plane group is cmm, with translation vectors $S_1 = \omega_A + \omega_B$, $S_2 = \omega_A - \omega_B$. The general positions of this plane group are images of arbitrary conformations, the special positions images of conformations with point-group symmetry

mirror lines (m), the twofold rotation points at $(0, \pi/2)$ and $(\pi/2, 0)$, and the four-fold mm positions at $(0,0)$ and $(\pi/2, \pi/2)$. These special positions correspond to diphenylmethane conformations with non-trivial point-group symmetry. An alternative non-primitive unit cell can be based on the symmetric and antisymmetric combinations of the torsion angle coordinates: $S_1 = \omega_A + \omega_B$ and $S_2 = \omega_A - \omega_B$. This is the standard unit cell for the plane group cmm as given in International Tables for X-Ray Crystallography [28].

For the point $(0,0)$, all C atoms lie in a plane and the molecular symmetry is C_{2v}; we can say that this point is the image of a conformation with C_{2v} symmetry. The point $(\pi/2, \pi/2)$ is the image of another, non-equivalent C_{2v} conformation. The points $(0, \pi/2)$ and $(\pi/2, 0)$, lying on the twofold rotation points of the plane group, are images of conformations with C_s symmetry. The two mirror lines, $\omega_A + \omega_B$ and $\omega_A - \omega_B$ are not equivalent. Points lying on the first line are images of conformations with C_2 symmetry, while those lying on the second line are images of another class of conformations with C_s symmetry. The map thus provides a one-to-one correspondence between points in the plane (ϕ_1, ϕ_2) and the molecular conformations described by torsion angle pairs (ϕ_1, ϕ_2) and hence gives a vivid picture of the symmetry properties of the two-dimensional torsional potential energy surface. Note, however, that the site symmetry of a particular point, line or plane in the conformational map is not necessarily the same as the point group symmetry of the corresponding molecular conformation — the two are different representations of the same abstract group.

The map not only portrays the conformational space of a single isolated molecule but can also be used to plot observed torsion angles in particular structures. A collection of observed conformations will produce a scatter plot, and, according to the underlying structural correlation principle (see Chapter 5), the points will tend to congregate in regions of the conformational space associated with low energies.

As an example, we can refer to the results of an analysis [29] along these lines of the observed torsion angles in 38 crystallographically independent benzophenones (frame group C_{2v}, the same as diphenylmethane) retrieved from the Cambridge Structural Database (Figure 2.13). The points corresponding to the observed structures are spread, fairly evenly, along a curve close to the $(90°, 0°)$; $(0°, 90°)$ diagonal, i.e. both torsion angles have the same sign, indicating that the two phenyl rings are twisted in the same sense. There is a pronounced clustering of the points around $(30°, 30°)$, and one can discern the symmetry-related paths leading from this region to the points $(90°, 0°)$ and $(0°, 90°)$, where one ring lies in the same plane as the C−CO−C frame, the other perpendicular to it. The entire path can be interpreted as the reaction path for the transformation of an idealized symmetric (point group C_2) chiral molecule with torsion angles $(30°, 30°)$ into its enantiomer with torsion angles $(150°, -30°) = (-30°, -30°)$. Starting at the symmetric structure, one ring is twisted towards its perpendicular orientation while the other ring is twisted by half the amount in the opposite sense, i.e. towards the planar orientation. The interconversion process thus corresponds to what has been termed a one-ring flip mechanism [30]. These conclusions are also in accord with results of molecular mechanics calculations on an isolated benzophenone molecule. The more detailed analysis [29] shows that there are systematic differences between molecules that show

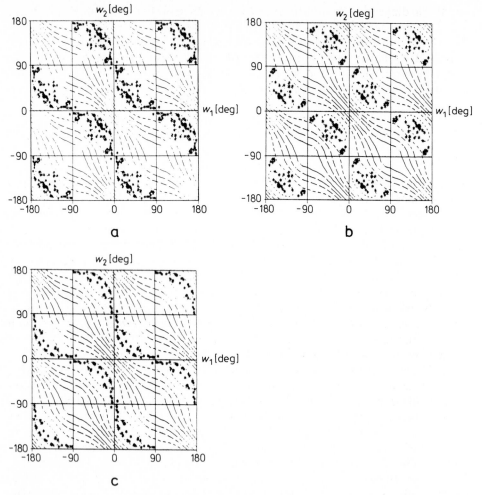

Fig. 2.13. Conformational map for benzophenones. The points correspond to observed structures, the contours to calculated equipotentials for benzophenone, spaced by 1.35 kcal mol^{-1}. a: All observed molecules. b: Non-hydrogen bonded molecules. c: Molecules with intramolecular hydrogen bonding

intramolecular hydrogen bonding (e.g. those with *ortho*-hydroxy substituents) and those that do not.

At this point, we may return to the other molecules mentioned in Scheme 2.2. What they have in common with diphenylmethane is that they can all be regarded as consisting of two rotors attached to a frame. In **B, D** and **F** this frame has the same C_{2v} symmetry as in diphenylmethane, and the corresponding groups G therefore contain the same frame group F, differing only in the periodicity of the rotor groups R. This is $2\pi \times 2\pi$ for **B**, $2\pi/3 \times 2\pi/3$ for **D**, and $2\pi/5 \times 2\pi/5$ for **F**. Maps showing isometric conformations of these molecules will thus have the same primitive unit cell as for diphenylmethane but different periodicities. For molecule **C** and **E** the rotors are non-equivalent and the frame has a lower symmetry, only C_s instead of C_{2v}. There are now only two points in the primitive unit cell, (ϕ_1, ϕ_2) and $(-\phi_1, -\phi_2)$, which thus has the plane group $p2$. The periodicities along the ω_A and ω_B directions are π and π for **C**, and π and $2\pi/3$ for **E**.

2.6.4 Three Internal Rotational Degrees of Freedom

The previous analysis led to a description in terms of two-dimensional crystallographic plane groups; we are now dealing with three internal rotations, i.e. with a three-dimensional conformational space involving three-dimensional space groups. As example, we consider the stereoisomerization path for triphenylphosphine oxide and similar molecules, Ph$_3$PX. If we assume that bond lengths and angles stay relatively invariant then an arbitrary conformation of such a molecule is specified by three torsion angle coordinates $\omega_A, \omega_B, \omega_C$ with components ϕ_1, ϕ_2, ϕ_3, where the labels A, B, C each refer to a given phenyl group, the sequence being taken as clockwise when viewed down the X$-$P direction, as in Figure 2.14. We still need to define the position of zero torsion angle; this is taken to be when the phenyl group in question eclipses the P$-$X bond. Increasing ϕ corresponds to clockwise rotation of the phenyl group, viewed along the P$-$C bond from the central P atom.

The frame symmetry is C_{3v}; this is the highest symmetry that an idealized Ph$_3$PX molecule can have (when all three torsion angles are 0 or 90°). The symmetry operations of this frame are shown in Table 2.7 together with the operations of the rotor group R, produced by 180° rotations of the phenyl groups, individually, in pairs, or all three at a time. The operations shown in Table 2.7 convert a given conformation into a set of equivalent, isometric conformations, or, what is the same, they convert a given point (ϕ_1, ϕ_2, ϕ_3) into a set of equivalent points. Crystallographers will note that the set of points produced by the frame group *F* corresponds exactly to the general positions (modulo π) of the three-dimensional space group $R32$. The com-

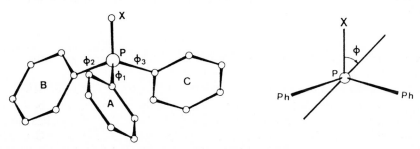

Fig. 2.14. Left: An arbitrary conformation of a Ph$_3$PX molecule is specified by three torsion angles ϕ_1, ϕ_2, ϕ_3, for the phenyl groups labeled A, B, C in a clockwise sequence when viewed down the X$-$P direction. Right: Definition of torsion angle ϕ

Table 2.7. The transformations associated with the frame group F and the rotor group R on the torsion angles of a Ph$_3$X molecule. The complete set of isometric conformations is obtained by combining the eight rotor group transformations with the six frame group transformations to obtain a group of order 48

F	E	C_3	C_3^{-1}	σ_A	σ_B	σ_C
ω_A	ϕ_1	ϕ_3	ϕ_2	$-\phi_1$	$-\phi_3$	$-\phi_2$
ω_B	ϕ_2	ϕ_1	ϕ_3	$-\phi_3$	$-\phi_2$	$-\phi_1$
ω_C	ϕ_3	ϕ_2	ϕ_1	$-\phi_2$	$-\phi_1$	$-\phi_3$

R	E	R_A	R_B	R_C	R_{AB}	R_{BC}	R_{CA}	R_{ABC}
ω_A	ϕ_1	$\phi_1+\pi$	ϕ_1	ϕ_1	$\phi_1+\pi$	ϕ_1	$\phi_1+\pi$	$\phi_1+\pi$
ω_B	ϕ_2	ϕ_2	$\phi_2+\pi$	ϕ_2	$\phi_2+\pi$	$\phi_2+\pi$	ϕ_2	$\phi_2+\pi$
ω_C	ϕ_3	ϕ_3	ϕ_3	$\phi_3+\pi$	ϕ_3	$\phi_3+\pi$	$\phi_3+\pi$	$\phi_3+\pi$

complete symmetry group G is obtained by combining each of the six operations of F with each of the eight operations of R to produce a group of order 48 [31]. Observe that a noncyclic permutation of the torsion angles, e.g. $\phi_1, \phi_2, \phi_3 \rightarrow \phi_1, \phi_3, \phi_2$, does not correspond to a symmetry operation of the frame group, and neither does a reversal of the signs of the three torsion angles $\phi_1, \phi_2, \phi_3 \rightarrow -\phi_1, -\phi_2, -\phi_3$; in other words, molecules described by ϕ_1, ϕ_2, ϕ_3 and by ϕ_1, ϕ_3, ϕ_2 (or $-\phi_1, -\phi_2, -\phi_3$) are, in general, not isometric.

For visualization, it is more convenient to use hexagonal rather than rhombohedral axes (Figure 2.15). The transformation from rhombohedral to hexagonal axes is:

$$a_H = \omega_A - \omega_B$$
$$b_H = \omega_B - \omega_C$$
$$c_H = \omega_A + \omega_B + \omega_C$$

the corresponding transformation of coordinates (vector components) is:

$$x_H = (2\phi_1 - \phi_2 - \phi_3)/3$$
$$y_H = (\phi_1 + \phi_2 - 2\phi_3)/3$$
$$z_H = (\phi_1 + \phi_2 + \phi_3)/3$$

the inverse transformation is:

$$\phi_1 = (x_H + z_H)$$
$$\phi_2 = (-x_H + y_H + z_H)$$
$$\phi_3 = (-y_H + z_H)$$

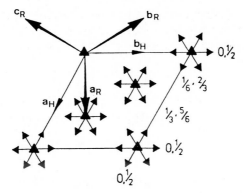

Fig. 2.15. Primitive rhombohedral cell (space group $R\,32$) in hexagonal coordinate system. The hexagonal cell has lattice points at $0,0,0$; $2/3,1/3,1/3$; $1/3,2/3,2/3$. The threefold axes are indicated by triangles. The twofold axes, whose directions are indicated by broad arrows, occur in pairs separated by $c_H/2$; the numbers give the fractional z_H coordinates

A molecule with equal torsion angles ϕ,ϕ,ϕ (point symmetry C_3) corresponds to a point with hexagonal coordinates $0,0,\phi$ situated on the threefold axis passing through the origin of the unit cell.

The distribution of such points for 62 Ph_3PO fragments (all with unsubstituted phenyl groups), retrieved from the Cambridge Structural Database [32], is shown in stereoscopic view in Figures 2.16 and 2.17. There is an obvious clustering of points around the threefold axes at $0,0,z_H \approx 40°$ and symmetry-equivalent regions. These are images of molecules with approximate C_3 symmetry. Moreover, the paths that interconnect these clusters are clearly discernible. These are then the paths for the stereoisomerization of a chiral Ph_3PO molecule into its enantiomer. Figure 2.16 shows that the path from $(0,0,40°)_H = (40°,40°,40°)_R$ to $(120°,60°,20°)_H = (140°,-40°,-40°)_R$ runs nearly along the line between the two points but curves

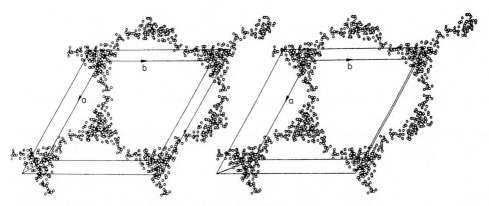

Fig. 2.16. Stereoscopic view of the distribution of Ph_3PO sample points in a section of the hexagonal cell with $0<z_H<60°$. The unit translations of the hexagonal cell are $180°$

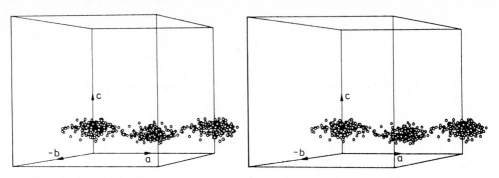

Fig. 2.17. As in Figure 2.16, but showing clusters of points centred at $0, 0, 40°$; $120°, 60°, 20°$; $180°, 0, 40°$, viewed along a direction approximately perpendicular to the a_H, c_H-plane

slightly to avoid the midpoint at $(60°, 30°, 30°)_H = (90°, 0, 0)_R$. Similarly, from Figure 2.17 it is evident that the path runs at nearly constant z_H. This means that as one torsion angle, say ω_A, increases from its initial 40° value, the other two decrease so as to keep the sum of the torsion angles roughly constant. As ω_A approaches 90°, the retrorotation of the other two rings is no longer synchronous. The midpoint of the actual path is close to $(60°, 40°, 30°)_H = (90°, 10°, -10°)_R$. The passage of one ring through torsion angle 90° and of the other two rings through zero corresponds to what has been termed a "two-ring flip" mechanism for the stereoisomerization of the cognate molecule trimesitylmethane [33].

2.6.5 Four Internal Degrees of Freedom: Tetraphenylmethane and Cognate Molecules

We turn now to the more difficult problems connected with stereoisomerization paths of molecules of the type Ph_4X. The difficulties are of two kinds. To begin with, the symmetry properties of the tetraphenyl systems are much more complicated than those of the triphenyl ones. The molecular frame consists of a skeleton that has T_d symmetry, but the highest point symmetry attainable by the molecule is D_{2d}, with all four phenyl rings lying in the mirror planes or perpendicular to them (Figure 2.18). If all four rings are rotated in the same sense and by equal amounts from one of these orientations, the molecular symmetry is lowered to D_2. However, when the common rotation angle reaches 60°, the D_{2d} symmetry is recovered and the S_4 axis appears in a new direction at right angles to the original direction [34]. The symmetry aspects of Ph_4X systems have been discussed in some detail from a permutational approach [35–37], but we prefer to discuss them in terms of the symmetry properties of the conformational space defined by the rotation angles of the four phenyl groups, i.e. in terms of a four-dimensional space group. This permits us, at least in principles, to trace the paths followed in the course of the stereoisomerization processes.

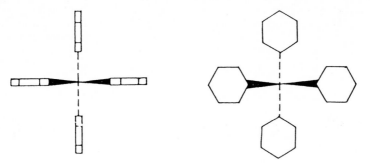

Fig. 2.18. Conformations of Ph_4X molecule showing conformations of highest attainable symmetry D_{2d}

The second difficulty is perhaps more serious. It is the one of recognizing patterns in four-dimensional space. Whereas the three-dimensional distributions Figures 2.16 and 2.17 can be *seen*, this is not possible in four dimensions. One possibility is to give up the attempt to visualize the distributions and concentrate instead on purely mathematical techniques for pattern recognition, i.e. cluster analysis and similar techniques (see Chapter 4). The trouble here is that available programs for cluster analysis make little provision for handling symmetry and are especially unsuited for dealing with the translational symmetries that occur in periodic distributions [38]. Here we try to utilize the visual faculty as far as possible [39] and adopt the following procedure [40]: for three of the dimensions we construct a stereoscopic view [41], while distance from the viewer along the fourth dimension is indicated by the size of a circle, decreasing with increasing distance [42]. An example is given in Figure 2.19, which shows an approximately linear distribution of points. The cluster of points at the lower right hand part of the diagram is farthest from the viewer along the third dimension vertical to the page but it is closest along the fourth dimension. For visualization of particular distributions it is often convenient to transform first to principal inertial axes of the distribution (see Chapter 4) and then take the fourth dimension along the direction of the eigenvector corresponding to the smallest eigenvalue, i.e. along the direction where the distribution is "thinnest".

We proceed as before. A Ph_4X molecule is regarded as a set of four rigid rotors on a frame of T_d symmetry. An arbitrary conformation is then specified by four torsion angles, $\phi_1, \phi_2, \phi_3, \phi_4$ along coordinates $\omega_A, \omega_B, \omega_C, \omega_D$, where the labels A, B, C, D each refer to a given phenyl group. Each observed conformation then corresponds to a point $(\phi_1, \phi_2, \phi_3, \phi_4)$ in four-dimensional space, and the distribution of such points, with allowance for the symmetry aspects of the problem, can be interpreted, as before, in energy terms. There are two questions that have to be settled before we start: How do we choose the zero positions of the torsion angles? And how do we choose the sequence of labels?

To discuss the angle problem it is convenient to consider first the slightly simpler tetrahydroxymethane molecule (Figure 2.20, top), where the rotor group has been reduced to a single hydrogen atom. In the arrangement shown in Figure 2.20 (bottom) the torsion angle ϕ_1 about bond a could be taken as $-90°$, $+30°$, or $+150°$,

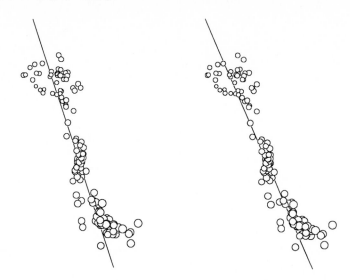

Fig. 2.19. Hyperstereoscopic view of a four-dimensional distribution of points. Distance along the fourth axis is indicated by the size of the circles, larger circles being closer to the viewer

depending on whether it is defined with respect to bonds c, b, or d. For definiteness, we choose to measure the a torsion angle ω_A against c, the c angle ω_C against a, the b angle ω_B against d, and the d angle ω_D against b. With the axes labeled as shown and in an obvious notation, the following relationships then hold:

$$\omega_A = \omega_A(c) = \omega_A(b) - 120° = \omega_A(d) + 120°$$
$$\omega_B = \omega_B(d) = \omega_B(c) - 120° = \omega_B(a) + 120°$$
$$\omega_C = \omega_C(a) = \omega_C(d) - 120° = \omega_C(b) + 120°$$
$$\omega_D = \omega_D(b) = \omega_D(a) - 120° = \omega_D(c) + 120°$$

With this choice, $\phi_1 = \phi_2 = \phi_3 = \phi_4 = 0$ corresponds to the situation shown in Figure 2.20 (top), where H_a is syn-planar to bond c, H_b to bond d, H_c to bond a, and H_d to bond b. In this situation the molecule has its highest possible symmetry, D_{2d} (order 8). For the Ph$_4$X molecules we adopt the same convention, noting that we may here add or subtract a multiple of $180°$ to any of the torsion angles to obtain equivalent conformations. For the Ph$_4$X molecules there are two conformations of D_{2d} symmetry, one with all torsion angles zero, the other with all torsion angles equal to $90°$ (Figure 2.18). The one with zero angles shows extreme steric overcrowding between pairs of phenyl groups, leaving the one with $90°$ angles as a possible low-energy conformation. For tetraphenylmethane it was dubbed the "open" D_{2d} (or $°D_{2d}$) conformation [43].

In order to label a given arbitrary conformation in a standard way we first use the above relationships to compute all possible values of the four torsion angles. Ring A is chosen as the one with torsion angle closest to $+90°$, this defines ring C;

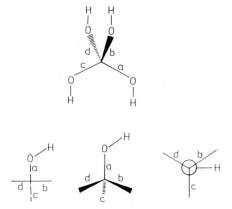

Fig. 2.20. Tetrahydroxymethane $C(OH)_4$ (top); three projections of a specific conformation about bond *a* (bottom)

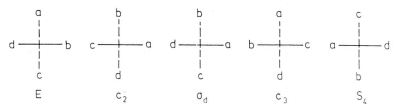

Fig. 2.21. Some symmetry operations of the T_d frame expressed as changes of labels. For the operations illustrated, we take C_2 and S_4 along the bisector of *ab*, C_3 along *a*, and σ_d in the *cd* plane, with respect to the initial axial frame

ring B is chosen so that the bonds from the central atom to rings A, B, and C form a right-handed coordinate system.

The symmetry operations of the tetrahedral frame can be expressed as interchanges of the labels *a, b, c, d* in Figure 2.20. The effect of some of these changes is shown in Figure 2.21. A list of the torsion-angle transformations corresponding to all 24 possible interchanges of labels is provided in Table 2.8. These transformations define the frame group $F(T_d)$ and the corresponding unit-cell group U for $C(OH)_4$.

For the Ph_4X molecules an additional set of 16 pure translation operations has to be included to allow for the additional equivalent conformations obtained by applying 180° rotations of the phenyl groups. These translations are:

$$
\begin{array}{llll}
0, 0, 0, 0 & 0, 0, 0, \pi & 0, 0, \pi, 0 & 0, \pi, 0, 0 \\
\pi, 0, 0, 0 & 0, 0, \pi, \pi & 0, \pi, 0, \pi & 0, \pi, \pi, 0 \\
\pi, 0, 0, \pi & \pi, 0, \pi, 0 & \pi, \pi, 0, 0 & 0, \pi, \pi, \pi \\
\pi, 0, \pi, \pi & \pi, \pi, 0, \pi & \pi, \pi, \pi, 0 & \pi, \pi, \pi, \pi \\
\end{array}
$$

The group corresponding to the complete set of equivalent, isometric conformations is thus of order $24 \times 16 = 384$. However, since the complete conformational space has a translational period of π along each of the four directions, we can always consider a cell with this smaller periodicity. In terms of this cell the general positions of Table 2.8 still apply but have to be taken modulo π instead of 2π.

The highest symmetry attainable by the $C(OH)_4$ and Ph_4X molecules is D_{2d} of order 8, whereas the T_d frame group, and the corresponding unit cell group, is of order 24. This means that the special position of highest symmetry must occur three times in the primitive four-dimensional cell with unit translations of 2π and π, respectively, in all four directions. Of the 24 general positions in four-dimensional space, only 8 are related by pure point-symmetry operations, the remainder being related to the general point $(\phi_1, \phi_2, \phi_3, \phi_4)$ by operations that include translations. These obviously correspond to the frame operations that add or subtract $2\pi/3$ from the torsion angles.

For a molecule like neopentane, with four equivalent threefold rotors, special conformations with point group symmetry higher than D_{2d} are possible. For example, conformations with arbitrary but equal torsion angles, imaged by points (ϕ, ϕ, ϕ, ϕ), have point group symmetry T, while those with torsion angles all zero or multiples of $60°$ have the full frame symmetry T_d. With the new set of lattice translations at intervals of $2\pi/3$ along the coordinate axes, the general positions of Table 2.8 can be expressed modulo $t = 2\pi/3$ instead of modulo π or 2π, so that the translation components asssociated with some of the frame operations disappear. For such molecules, the complete set of isometric conformations comprises an even greater number of operations, the complete group being of order $24 \times 81 = 1944$.

As in the earlier three-dimensional example, where it was convenient to use hexagonal axes involving linear combinations of the rhombohedral basis vectors, it may also be useful here to use an alternative coordinate system to bring out certain symmetry properties. For molecules with a T_d frame, one choice is to take:

$$S_{1a}(T_2) = \omega_A + \omega_B - \omega_C - \omega_D$$
$$S_{1b}(T_2) = \omega_A - \omega_B + \omega_C - \omega_D$$
$$S_{1c}(T_2) = \omega_A - \omega_B - \omega_C + \omega_D$$
$$S_2(A_2) = \omega_A + \omega_B + \omega_C + \omega_D$$

where $\omega_A, \omega_B, \omega_C, \omega_D$ define the primitive cell with unit translations $2\pi, \pi$, or $2\pi/3$, depending on the nature of the rotors. The determinant of the transformation matrix equals 16, so the primed cell is 16 times as large as the primitive one and contains 16 lattice points. The corresponding transformation of the coordinates is:

$$x_{1a} = (\phi_1 + \phi_2 - \phi_3 - \phi_4)/4$$
$$x_{1b} = (\phi_1 - \phi_2 + \phi_3 - \phi_4)/4$$
$$x_{1c} = (\phi_1 - \phi_2 - \phi_3 + \phi_4)/4$$
$$x_2 = (\phi_1 + \phi_2 + \phi_3 + \phi_4)/4$$

These combinations are symmetry coordinates that are transformed by the symmetry operations of T_d in the same way as the irreducible representations (see Table 2.3).

Table 2.8. Torsion angle transformations associated with the 24 frame symmetry operations of a CR_4 molecule with a tetrahedral frame group F. For tetraphenylmethane, the complete set of isometric conformations is obtained by combining these 24 frame group transformations with a rotor group of order 16 to obtain a group of order 384

Frame symmetry element	Axis arrangement	Cycle	ω_A	ω_B	ω_C	ω_D
E	abcd		ϕ_1	ϕ_2	ϕ_3	ϕ_4
C_2	badc	$(ab)(cd)$	ϕ_2	ϕ_1	ϕ_4	ϕ_3
	cdab	$(ac)(bd)$	ϕ_3	ϕ_4	ϕ_1	ϕ_2
	dcba	$(ad)(bc)$	ϕ_4	ϕ_3	ϕ_2	ϕ_1
σ_d	cbad	(ac)	$-\phi_3$	$-\phi_2$	$-\phi_1$	$-\phi_4$
	adcb	(bd)	$-\phi_1$	$-\phi_4$	$-\phi_3$	$-\phi_2$
	bacd	(ab)	$-\phi_2+t$	$-\phi_1+t$	$-\phi_3+t$	$-\phi_4+t$
	dbca	(ad)	$-\phi_4-t$	$-\phi_2-t$	$-\phi_3-t$	$-\phi_1-t$
	acbd	(bc)	$-\phi_1-t$	$-\phi_3-t$	$-\phi_2-t$	$-\phi_4-t$
	abdc	(cd)	$-\phi_1+t$	$-\phi_2+t$	$-\phi_4+t$	$-\phi_3+t$
C_3	acdb	(bcd)	ϕ_1+t	ϕ_4+t	ϕ_2+t	ϕ_3+t
	adbc	(bdc)	ϕ_1-t	ϕ_3-t	ϕ_4-t	ϕ_2-t
	cbda	(acd)	ϕ_4-t	ϕ_2-t	ϕ_1-t	ϕ_3-t
	dbac	(adc)	ϕ_3+t	ϕ_2+t	ϕ_4+t	ϕ_1+t
	bdca	(abd)	ϕ_4+t	ϕ_1+t	ϕ_3+t	ϕ_2+t
	dacb	(adb)	ϕ_2-t	ϕ_4-t	ϕ_3-t	ϕ_1-t
	bcad	(abc)	ϕ_3-t	ϕ_1-t	ϕ_2-t	ϕ_4-t
	cabd	(acb)	ϕ_2+t	ϕ_3+t	ϕ_1+t	ϕ_4+t
S_4	bcda	$(abcd)$	$-\phi_4$	$-\phi_1$	$-\phi_2$	$-\phi_3$
	dabc	$(adcb)$	$-\phi_2$	$-\phi_3$	$-\phi_4$	$-\phi_1$
	bdac	$(abdc)$	$-\phi_3-t$	$-\phi_1-t$	$-\phi_4-t$	$-\phi_2-t$
	cadb	$(acdb)$	$-\phi_2-t$	$-\phi_4-t$	$-\phi_1-t$	$-\phi_3-t$
	cdba	$(acbd)$	$-\phi_4+t$	$-\phi_3+t$	$-\phi_1+t$	$-\phi_2+t$
	dcab	$(adbc)$	$-\phi_3+t$	$-\phi_4+t$	$-\phi_2+t$	$-\phi_1+t$

[a]The translation component $t = 2\pi/3$.

Thus, x_2 transforms as the A_2 representation, and the other three transform as the triply degenerate T_2 representation. The relationship between internal and symmetry coordinates for Ph_4X molecules is best illustrated by a few examples. One of the closed and open D_{2d} conformations is represented by the special positions $(0,0,0,0)$ and $(90°,90°,90°,90°)$ respectively in the ω coordinate system and by $[0,0,0,0]$ and $[0,0,0,90°]$ in the symmetry coordinate system [44] with symmetry equivalent points displaced by $120°$ along the S_2 axis. The points (ϕ,ϕ,ϕ,ϕ), which image conformations with D_2 symmetry, lie on a line $[0,0,0,\phi]$ parallel to the S_2 axis, while points $(\phi,\phi,-\phi,-\phi)$, $(\phi,-\phi,\phi,-\phi)$, $(\phi,-\phi,-\phi,\phi)$, which image conformations with S_4 symmetry lie on a set of lines $[\phi,0,0,0]$, $[0,\phi,0,0]$, $[0,0,\phi,0]$ parallel to S_{1a}, S_{1b}, and S_{1c} but perpendicular to S_2 and the lines $[0,0,0,\phi]$ imaging D_2 conformations.

Tetraphenylmethane adopts a conformation with S_4 symmetry in the crystal [45], but this information is hardly sufficient in itself to make a structure correlation analysis. However, the Cambridge Structural Database contains many structures containing tetraphenylborate anions, as well as tetraphenylphosphonium and tetraphenylarsonium cations. Distributions of corresponding points in the hypercubic unit cell are shown in Figure 2.22 for these three systems and are evidently broadly similar [46]. The boron distribution contains two types of clusters. One type is centered at the special positions that image $°D_{2d}$ conformations $(30°, 30°, 30°, 30°)$, $(90°, 90°, 90°, 90°)$, $(150°, 150°, 150°, 150°)$ or $[0, 0, 0, 30°]$, etc. There are eight symmetry equivalent general positions around each of these special positions, and passing through each of the special positions is one of the special lines that images conformations with S_4 symmetry. The $°D_{2d}$ clusters are elongated along these lines, which lead to a second set of six clusters centered around S_4 positions with components approximately $(90°, -30°, -30°, 90°)$, $(150°, 30°, 150°, 30°)$, $(210°, 210°, 90°, 90°)$ or $[0, 0, 60°, 30°]$, $[0, 60°, 0, 90°]$, $[60°, 0, 0, 150°]$, etc. Empty regions occur between the clusters and around the borders of the unit cell shown, especially at the cell corners, which are images of closed D_{2d} conformations with torsion angles $(0, 0, 0, 0)$ (Figure 2.18), and which evidently correspond to high-energy regions of the conformational map. As the central atom becomes larger, the $°D_{2d}$ clusters disappear and the S_4 clusters become more diffuse, an indication of the greater torsional flexibility that develops as the phenyl groups move apart and non-bonded interactions between them become less severe. For the tetraphenylborates, the paths that interconvert the equivalent $°D_{2d}$ conformations are considerably more intricate than the analogous interconversion paths for triphenylphosphine oxide and its congeners (Section 2.6.4, Figures 2.16 and 2.17). As Figure 2.22 suggests, it is easy to lose one's way in the four-dimensional forest. However, the puzzled reader may derive comfort from the observation that the change from three to four dimensions is not only one of visualization and greater conceptual difficulty — it also involves an intrinsically more complex conformational interconversion (see Chapter 9).

Fig. 2.22. Hyperstereoscopic views of four-dimensional distributions of sample points for PhX_4 molecules, all along the same directions (1111) or [0001] of the hypercubic unit cell of conformational space. In this view the projected cell is seen to have the shape of a rhombic dodecahedron. Distance along the viewing direction (1000) or [1110] is indicated by normal stereoscopy, while distance along the (1111) or [0001] direction (S_2) is indicated by the size of the circles, larger circles being closer to the viewer, as in Figure 2.19. The special position that images the open $°D_{2d}$ conformation (Figure 2.18) with torsion angles $(90°, 90°, 90°, 90°)$ is situated at the center of the cell $[0, 0, 0, 90°]$ in all the views.

The top view shows the primitive cell with translations of π (180°) along all four axes. Corners (vertices, V) of the hypercube are labeled and the 24 general positions of a representative point with coordinates $(70°, 60°, 50°, 130°)$ are shown.

The following three views show distributions of sample points for Ph_4B, Ph_4P, and Ph_4As within the central section of the cell, between x_2 values of 60° and 120°. Although x_2 ranges over only one third of its repreat distance, this section contains the sample points in nearly three quarters (20/27) of the cell

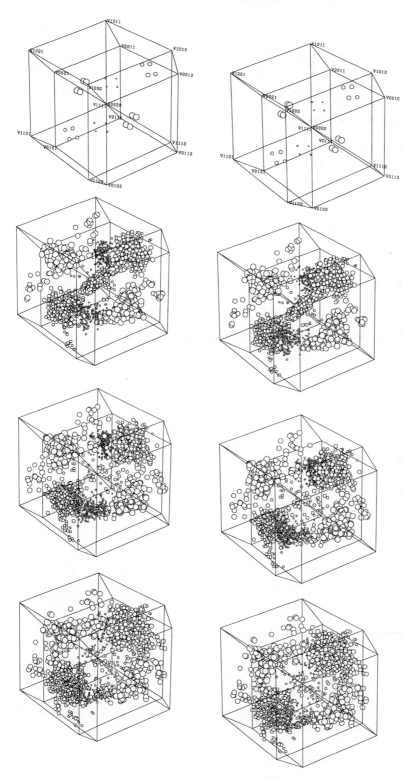

2.7 Summary

In this chapter we have discussed symmetry aspects of molecular structure, with emphasis on the symmetry of *small distortions* from a symmetric reference structure and on the *interconversions* between isometric structures, which mostly involve large distortions. In the language of configuration space, such interconversions take the form of pathways between equivalent points. In Sections 2.4.3, 2.6.1 and 2.6.3 – 5 we have described specific examples of such pathways and related them to the molecular potential energy surface. A more detailed discussion of the use of symmetry arguments in deriving properties of the energy surface is included in Chapter 5.

One of the topics we have touched on is the concept of "approximate symmetry" [7]. Generalizing the discussion in Section 2.4.1, we can say that for any point in configuration space we can calculate the shortest distance to any given special position (point, line, plane or hyperplane) in that space, i.e. to a configuration with some or all of the symmetry of the reference structure. This distance is a measure of the deviation of the structure in question from this particular symmetry. Whether a distance calculated in this way is significant or not must be decided on the background of the problem under investigation and in relation to the available experimental data. This kind of approach is being used in connection with the problem of quantifying the degree of chirality of a geometric figure [47].

References

[1] For a more precise definition see Bauder, A., Meyer, R., Günthard, H. H., *Molec. Phys.* **1974**, *28*, 1305 – 1343

[2] See, for example, Wilson, E. B. Jr., Decius, J. C., Cross, P. C., *Molecular Vibrations*, McGraw-Hill, New York, **1955**, pp. 115 ff

[3] So many introductory texts on group theory are available that any selection we might make would be invidious

[4] Character tables for the common point groups are to be found in any book on group theory

[5] In more complicated cases, these combinations may not be so obvious but they can always be found with the help of projection operators, as described in many books on group theory

[6] Chadwick, D. J., Dunitz, J. D., Schweizer, W. B., *Acta Cryst.* **1977**, *B33*, 1643 – 1645

[7] Murray-Rust, P., Bürgi, H. B., Dunitz, J. D., *Acta Cryst.* **1979**, *A35*, 703 – 713

[8] Murray-Rust, P., Bürgi, H. B., Dunitz, J. D., *Acta Cryst.* **1978**, *B34*, 1787 – 1793

[9] Murray-Rust, P., Bürgi, H. B., Dunitz, J. D., *J. Am. Chem. Soc.* **1975**, *97*, 921 – 922

[10] Auf der Heyde, T. P. E., Bürgi, H. B., *Inorg. Chem.* **1989**, *28*, 3960 – 3969

[11] Ammeter, J., Bürgi, H.-B., Gamp, E., Meyer-Sandrin, V., Jensen, W. P., *Inorg. Chem.* **1979**, *18*, 733 – 750

[12] Kilpatrick, J. E., Pitzer, K. S., Spitzer, R., *J. Am. Chem. Soc.* **1947**, *69*, 2483 – 2488

[13] Pitzer, K. S., Donath, W. E., *J. Am. Chem. Soc.* **1959**, *81*, 3213 – 3218

[14] For example, Altona, C., Geise, H. J., Romers, C., *Tetrahedron* **1968**, *13*, 13 – 22

[15] Irwin, J. J., *Doctoral Dissertation No. 9397*, ETH Zürich, **1991**

[16] Adams, W. J., Geise, H. J., Bartell, L. S., *J. Am. Chem. Soc.* **1970**, *92*, 5013–5019

[17] Waser, J., Schomaker, V., *J. Am. Chem. Soc.* **1945**, *67*, 2014–2020. See Dunitz, J. D., Waser, J., *El. Math.* **1972**, *27*, 25–32 for a collection of proofs

[18] For a statement of Cauchy's theorem see Lyusternik, L. A., *Convex Figures and Polyhedra*, translated from the Russian by T. Jefferson Smith, Dover Publications, New York, **1963**, p. 60 ff

[19] See, for example, Crippen, G. M., Havel, T. F., *Distance Geometry and Molecular Conformation*, Research Studies Press, Taunton, England, **1988**, especially Chapter 4

[20] For the analysis of six- and eight-membered rings in particular see Dunitz, J. D., Waser, J., *J. Am. Chem. Soc.* **1972**, *94*, 5645–5650

[21] Pickett, H. M., Strauss, H. L., *J. Am. Chem. Soc.* **1970**, *92*, 7281–7288

[22] There remains the problem of calculating a,b,c from the components F,x,y along the coordinates S'_1, S_{2a}, S_{2b}. We make use of the relation $F^2 = s(s-a)(s-b)(s-c)$ where $s = (a+b+c)/2$. Expressing a,b,c in terms of s,x,y we obain $F^2 = s^4/27 - (x^2+y^2)s^2/6 - (x^3/3+xy^2)s/\sqrt{6}$. With the real positive solution for s, $[a\ b\ c]^T = [1/\sqrt{3}\ \ 2/\sqrt{6}\ \ 0\ \ |1/\sqrt{3}\ \ -1/\sqrt{6}\ \ 1/\sqrt{2}\ \ |1/\sqrt{3}\ \ -1/\sqrt{6}\ \ -1/\sqrt{2}]\ [2s/\sqrt{3}\ x\ y]^T$

[23] Beginning with the classic paper of Longuet-Higgins, H. C., *Molec. Phys.* **1963**, *6*, 445–460

[24] Altmann, S. L., *Molec. Phys.* **1971**, *21*, 587–607

[25] The semi-direct product G of R and F (written $G = R \wedge F$) has the properties: $F_x R_y = R_z F_x$, $F_x \in F$, $R_y, R_z \in R$; $R = (F_x)^{-1} R F_x$ is an invariant subgroup of G; $R_x R_y = R_y R_x$ (R is Abelian; $F \neq (R)^{-1} F R$ is not an invariant subgroup of G; F is isomorphous with the factor group G/R. See Lomont, J. S., *Applications of Finite Groups*, Academic Press, New York London, **1959**, p. 29 ff for further definitions, proofs and discussion

[26] The unit cell group U is the group of all elements of the space group S reduced modulo the unit cell translations. In group theoretical language, U is the factor group of S with respect to the translational subgroup T of S. The unit cell need not be primitive. Its order is equal to the number of elements it contains, whereas the order of S (and of T) is infinite

[27] The term "conformational map" has been applied to two-dimensional energy diagrams or scatter plots describing the expected or actual distribution of torsion angles $\phi(C'NC^\alpha C')$ and $\psi(NC^\alpha C'N)$ in polypeptide chains. See Ramachandran, G. N., Sasikharan, V., *Adv. Protein Chem.* **1968**, *23*, 284–437

[28] *International Tables for X-ray Crystallography*, Vol. 1, *Symmetry Groups*. Kynoch Press, Birmingham, 1st ed., **1952**; 2nd ed., **1965**; 3rd ed., **1969**; Vol. A, *Space Group Symmetry*. Reidel, Dordrecht, Boston, **1983**

[29] Rappoport, Z., Biali, S. E., Kaftory, M., *J. Am. Chem. Soc.* **1990**, *112*, 7742–7748

[30] A term introduced by Kurland, R. J., Schuster, I. I., Colter, A. K., *J. Am. Chem. Soc.* **1965**, *87*, 2279–2284. The ring that "flips" passes through the plane perpendicular to the planar frame

[31] This group contains 8 operations $(6 R_i C_3, 2 R_{ijk} C_3)$ of order 6; 12 operations $(6 R_{ij}\sigma_i, 6 R_i \sigma_j)$ of order 4; 8 operations $(2 C_3, 6 R_{ij} C_3)$ of order 3; 19 operations $(3 R_i, 3 R_{ij}, R_{ijk}, 3 R_i \sigma_i, 3 R_{ijk}\text{-}\sigma_i, 3 R_{ij}\sigma_k, 3\sigma_i)$ of order 2. It is isomorphous with the octahedral group O_h

[32] Bye, E., Schweizer, W. B., Dunitz, J. D., *J. Am. Chem. Soc.* **1982**, *104*, 5893–5898

[33] Andose, J. D., Mislow, K., *J. Am. Chem. Soc.* **1974**, *96*, 2168–2176

[34] This property of the Ph_4X system, along with others, is discussed in [36]. At first sight it appears somewhat mysterious (rather like a conjuring trick), but a few minutes examination of a molecular model should be sufficient to convince anyone about its authenticity. As discussed in [37], it is due the fact that D_{2d} is not a normal subgroup of T_d

[35] Strohbusch, F., *Tetrahedron* **1974**, *30*, 1261–1282

[36] Hutchings, M. G., Nourse, J. G., Mislow, K., *Tetrahedron* **1974**, *30*, 1535–1549

[37] Nourse, J. G., Mislow, K., *J. Am. Chem. Soc.* **1975**, *97*, 4571–4578

[38] Nørskov-Lauritsen, L., Bürgi, H. B., *J. Comput. Chem.* **1985**, *6*, 216–228. The solution adopted here for a problem involving 8 angular coordinates was to bend each parameter axis into a circle with a circumference corresponding to one period. Each angle was then specified by the values of its sine and cosine. The data points now lie on the "8-D surface" of a torus in 16-D space

[39] For a brief discussion of intuitive reasoning in four-dimensional space see Davis, P.J., Hersh, P., *The Mathematical Experience*, Birkhauser, Boston, Mass., **1981**, pp. 400–405

[40] Developed for this and similar problems by Schweizer, W.B., Dunitz, J.D., unpublished work

[41] Using program PLUTO 78, obtainable from the Cambridge Crystallographic Data Centre, University Chemical Laboratories, Cambridge, England

[42] A related approach has been used in a different context by Whittaker, E.J.W., *Acta Cryst.* **1973**, *A 29*, 678–684; *idem, ibid*, **1983**, *A 39*, 123–130; *idem, ibid*, **1984**, *A 40*, 58–66; *idem, An Atlas of Hyperstereograms of the Four-Dimensional Crystal Classes*, Clarendon Press, Oxford, **1985**

[43] Hutchings, M.G., Andose, J.D., Mislow, K., *J. Am. Chem. Soc.* **1975**, *97*, 4553–4561

[44] We use curved brackets ( ) for components along the ω axes, the individual torsion angles, and straight brackets [ ] for components along the symmetry axes, the linear combinations of torsion angles

[45] Sumsion, H.T., MacLauchlan, D., *Acta Cryst.* **1950**, *3*, 217–219

[46] Bye, E., Schweizer, W.B., Bürgi, H.-B., Dunitz, J.D., unpublished results

[47] This problem is discussed by Buda, A.B., Auf der Heyde, T.P.E., Mislow, K., *Angew. Chem.* **1992**, *104*, 1012–1030; *Angew. Chem. Int. Ed. Engl.* **1992**, *31*, 989–1007; *idem, J. Math. Chem.*, **1991**, *6*, 243–253; Auf der Heyde, T.P.E., Buda, A.B., Mislow, K., *ibid*, 255–265

3 Crystallographic Databases: Search and Retrieval of Information from the Cambridge Structural Database

Frank H. Allen, Olga Kennard and David G. Watson

3.1 Introduction

It is now eighty years since the first crystal structures were determined by X-ray diffraction techniques. Since that time, X-ray crystallography has developed into the most powerful method available for the study of three-dimensional (3D) chemical structure at atomic resolution. It is worth remembering, though, that even the earliest results made an immediate and significant impact, due to the extraordinary richness of the information provided by each new analysis. For molecular crystals, the primary crystallographic results (3D atomic coordinates, unit-cell parameters and space-group symmetry) define the atomic connectivity, provide a depiction of the configurational, conformational and stereochemical features, and yield a precise geometrical description of the 3D atomic arrangement. Further, the crystallographic method is unique in providing direct experimental observations of the hydrogen-bonded and non-bonded interactions that govern molecular aggregation in the extended crystal structure.

Given the significance and wide applicability of their results, it is not surprising that crystallographers have a long and distinguished history of self documentation. *Strukturbericht* was first published in 1929, as a printed compendium of primary structural data and text descriptions. This series developed into *Structure Reports*, which formed a part of the publication program of the International Union of Crystallography founded in 1948. Many other printed compendia, with both general and specific coverage, have appeared over the past sixty years and their availability and information content have been summarized by Watson [1].

The fundamental value of crystallographic data, and of the printed compendia themselves, is nowhere more eloquently illustrated than in the classic text, *The Nature of the Chemical Bond* by Pauling [2]. Here for the first time we see the systematic correlation of geometrical results from the relatively small number of structural studies available at that time. Despite this paucity of data, at least by modern day standards, the impact of this volume on chemical thinking has been enormous. The results presented there for the covalent and van der Waals radii of common elements remain in use today. It was not until the late 1950's that a further major compilation of geometrical data was accomplished by Sutton et al. [3]. These

Tables of Distances and Configuration in Molecules and Ions appeared as Special Publications of the Chemical Society of London, and drew data from some 1000 crystal structures available at that time.

Over the past thirty years, X-ray crystallography has seen rapid advances in the experimental methods available for recording the diffraction intensities, and in the theoretical basis of structure solution. Underpinning both of these developments has been the dramatic impact of the digital computer on the subject as a whole. Thus, the range of structures that can now be studied has been extended from small inorganics and organics, through medium-size molecules, to proteins and viruses containing many thousands of atoms. Most importantly, the speed with which structure analyses can now be performed has been reduced from years and months to a matter of days for routine small-molecule studies: the X-ray technique is now the method of choice rather than the method of last resort.

Taken together, the theoretical, experimental and computational developments have generated an exponential growth in published crystallographic structural results: of the 130000 crystal structures now in the public domain, more than 60% have been published since 1980. Despite the success of the crystallographic method or, rather, because of it, systematic studies of large numbers of related structures in the tradition of Pauling [2] or Sutton [3] were rarely undertaken. The updating of the existing printed compendia became an increasingly labour-intensive task, whilst their value in terms of rapid entry to the available literature was diminished. Most importantly, the personal labour involved in retrieving and organizing the printed data into a form which was suitable for automated processing acted as a major barrier to systematic studies. For these reasons, and in view of their computational background, crystallographers were amongst the first scientists to turn to the digital computer to solve their information needs. The printed sources, so carefully compiled and maintained, formed ideal starting points for the creation of computerized databases of crystallographic structural results.

Applications of crystallographic databases developed slowly at first, but with a gathering impetus over recent years with improved worldwide accessibility and gradually improving software systems. Indeed, with the databases themselves now well established, the development of sophisticated and user-friendly software is currently a major preoccupation. This software must accomplish not only the bibliographic and 2D structural searches normally available in chemical database systems, but also provide facilities for 3D geometric searches, the generation of specific geometrical parameters, and an ability to display and manipulate 3D structural images. In addition, statistical and numerical techniques must be applied to the large volumes of numerical data that may result from such searches. Only in this way can we form classifications or recognize patterns of structural behaviour that may be interpreted in chemical terms.

The main purpose of this chapter, then, is to describe the mechanics of database use: the processes of search, retrieval and data organization that are essential precursors to any experiment in structure correlation. In particular, we show that a clear understanding of database organization and information content is vital to a successful experiment of this kind. We illustrate these points by indicating a variety of pitfalls that may beset the unwary and show how they can be circumvented by careful use of the available database software. In Chapter 4, we go on to describe the various statistical and numerical methods that have already proved useful in structure correlation work. In view of the contents of other chapters in this volume, we restrict our discussion to the Cambridge Structural Database, which covers organics, organometallics and metal complexes, and to its associated software system [4]. We begin, however, with a brief general survey of all four crystallographic structural databases.

3.2 Crystallographic Databases

3.2.1 Overview

Four major databases now record the primary 3D numerical results of all single crystal structure analyses obtained using X-ray or neutron diffraction. Taken together, they cover the complete chemical spectrum comprising (a) metals and alloys, (b) inorganics and minerals, (c) organics, organometallics and metal complexes, and (d) biological macromolecules. Each database grew originally from the interests of academic-subject specialists and each was started early enough so that all databases are fully retrospective and are updated on a current basis. Although created as archives of 3D coordinate data, the four crystallographic databases all contain bibliographic and chemical information; this is essential to the search process and as a direct link back to the original publication.

In this section, we provide brief details concerning the origin, information content, overall statistics and availability of each of the four databases. More complete details have been published in a recent monograph [5]. However, in view of the rapid development of these databases, and of their associated software systems and methods of dissemination, we would recommend that potential users contact the database originators in the first instance to obtain the latest available information.

3.2.2 The Metals Crystallographic Data File (MCDF)

The MCDF covers metals and intermetallic phases for which unit-cell parameters have been determined by diffraction methods. The database also includes hydrides and some binary oxides, but excludes elements of Groups VIIA and VIIIA. The database contains over 11 000 entries, of which some 6 000 have atomic coordinates, displacement parameters and site occupation data recorded. The remaining 5 000 entries have been assigned to known structure types. Software for search and retrieval is available for distribution. The database is compiled and distributed by the *Canadian Institute of Scientific and Technical Information* (CISTI), National Research Council of Canada, Ottawa, Ontario, Canada K1A OS2.

3.2.3 The Inorganic Crystal Structure Database (ICSD)

The ICSD [6] covers inorganics and minerals whose atomic coordinates have been fully determined. The database currently contains 32 000 entries with some 1 200 new structures added each year. Software for search, retrieval and structure display is available. The database is compiled by the *Institut für Anorganische Chemie der Universität Bonn*, Gerhard-Domagk-Strasse 1, D-53121 Bonn, Germany, in conjunction with the *Fachinformationszentrum Energie Physik Mathematik*, D-76344 Eggenstein-Leopoldshafen 2, Germany, and the Gmelin Institut, Varrentrappstrasse, D-60486 Frankfurt/Main, Germany.

3.2.4 The Cambridge Structural Database (CSD)

The CSD [4] covers organics, organometallics and metal complexes for which atomic coordinates have been fully determined. The database currently contains 95 617 entries, and software for search, retrieval, display, and analysis of retrieved data is available. The CSD is distributed to academics via 31 National Affiliated Centres and to companies through individual licensing arrangements. The CSD is compiled by the *Cambridge Crystallographic Data Centre* (CCDC), 12 Union Road, Cambridge CB2 1EZ, England.

3.2.5 The Protein Data Bank (PDB)

The PDB [7] covers biological macromolecules (proteins and nucleic acids) for which crystallographic results have been directly deposited with the compilers. The database currently contains 655 entries. Applications software is not distributed with the PDB. However, the data bank has been recast in a variety of forms for search and retrieval purposes [8, 9]. Many molecular modelling systems (both commercial and academic in origin) can also read the PDB format. The database is compiled at the *Department of Chemistry, Brookhaven National Laboratory*, Upton, New York 11973, USA.

3.2.6 Areas of Structural Overlap

The definitions of information content given above often give rise to "edge-effects", which are irritating to some users. In particular, the CSD has obvious areas of overlap with both the ICSD and with the PDB. This problem is now being solved via the inclusion of certain structures in both of the databases concerned. Thus, the CSD now contains a number of structure types (e.g. pure metal carbonyls, boranes, etc.) which also occur in the ICSD. Likewise, a number of oligonucleotides may be found in both the PDB and in the CSD. The precise definition of overlap areas is currently under active discussion.

3.2.7 Data Acquisition and Data Integrity

The MCDF, ICSD and CSD currently acquire the vast bulk of their input data by visual scanning of the literature, followed by manual keyboarding of the relevant information. Only the PDB is compiled in its entirety from data submitted in machine readable form. Whilst all four databases perform extensive checks of data integrity, both computerized and visual, to ensure the accuracy of their final products, this function is particularly important for data acquired by manual methods. Statistics from the CSD show that some 15% of published structural papers contain at least one numerical error, and this is, of course, compounded by the inevitable additional risk of error due to the manual keyboarding.

In order to reduce the problems of error detection, whether they occur in the journal article itself or in the keyboarding process, a collaborative initiative involving both journals and the databases is now beginning. In particular, the International Union of Crystallography (IUCr) now accepts structural results in machine-readable form in an effort both to simplify submission procedures and minimize the typographical errors occurring in *Acta Crystallographica*. These machine-readable files are then made available to the crystallographic databases. For such a scheme to be successful, a single common format for primary structural data must replace the myriad formats currently in use. The Crystallographic Information File (CIF) [10] fills that role, and has been developed by the IUCr as the standard format for the interchange of structural data in electronic form. A number of crystallographic program packages have agreed to provide primary structural results in CIF format, and it is hoped that the related problems of data acquisition and data integrity will gradually improve over the next decade. As an added bonus, certain numerical data, such as atomic displacement parameters for CSD entries, for which re-keyboarding was simply too labour intensive, can now be added to the database directly from the machine-readable CIF submissions.

3.3 Overview of the Cambridge Structural Database (CSD)

3.3.1 Coverage

The CSD stores the primary results of full three-dimensional X-ray and neutron-diffraction studies of organics, organometallics, and of metal complexes having organic ligands. The database is fully retrospective (earliest reference 1930) and is updated regularly by some 700 new structures (entries) per month. The primary literature is abstracted together with any associated supplementary (deposited) data. The CSD itself acts as a computerized depository for large-volume numerical results for some 30 journals. A total of 584 primary sources are now referenced in the CSD, of which 74 are regularly scanned in-house to provide ca. 80% of current input. Remaining references are located via a scan of secondary sources, particularly *Chemical Abstracts*.

3.3.2 The Reference Code System

Each entry in the CSD relates to a specific crystal structure determination of a specific chemical compound. Each entry is identified by a CSD reference code (refcode). This consists of up to eight characters: the first six are alphabetic and identify the chemical compound, the last two characters are digits which trace the publication history and define: (a) whether the paper is a re-publication by the same authors, perhaps reporting an improved coordinate set, denoted by incrementation of the first digit, or (b) whether the paper is an independent determination by a different set of authors, denoted by incrementation of the second digit. This topic is discussed further in Section 3.7.1.

3.3.3 Information Content

The information recorded for each entry may conveniently by categorized according to its "dimensionality".

▶ *1D Information*

This consists of bibliographical and chemical text strings, together with certain individual numerical items:

- Compound name(s)
- Molecular formula
- Authors' names
- Journal name, journal code number, volume, page and year
- Qualifying phrase(s), e.g. neutron study, absolute configuration, etc.
- Text comment on e.g. the experimental method (if unusual), errors located in the publication, details of disorder, etc.
- Chemical class assignment(s) in the range $1-86$, e.g. class 15 is benzene nitro compounds, 51 is steroids, 74 is organometallic π-complexes (arene ligands), etc.
- Indicators of experimental precision
- Flags which summarize the check and evaluation processes

▶ *2D Information*

This is a representation of the 2D chemical structural formula, encoded as a connectivity table (Chapter 1, Section 1.1), i.e. as a graph, in which the nodes are atoms and the edges are bonds; hence the structure may be described in terms of atom and bond properties, as illustrated in Figure 3.1 a.

- Atom properties are:
  Atom sequence number (n)
  Element symbol (el)
  Number of connected non-H atoms (nca)
  Number of attached terminal hydrogens (nh)
  Net atomic charge (ch)
  Display coordinates (x, y) for graphical output of structural diagram
- Bond properties are:
  Pair of sequence numbers (i, j) for the atoms connected by the bond.
  Bond type (bt) for the connection $(i-j)$. The available bond types are: 1 = single, 2 = double, 3 = triple, 4 = quadruple, 5 = aromatic, 6 = polymeric, 7 = delocalized double and 9 = π-bond.
  Bond-type cyclicity indicator $(bt$ is negative for cyclic bonds).

The wide range of bond types is necessary to cover the broad spectrum of compounds included in the CSD. It is planned that other types, describing hydrogen bonds and short non-bonded interactions, will be added in the future. Since 1981 the chemical connectivity tables have been derived from digitized diagrams entered graphically by CCDC staff. This provides for direct input of the x, y-coordinates noted above, as well as the atom- and bond-property information. Coordinates for connectivity tables entered before 1981 (some 35,000 entries) were generated either algorithmically, followed by a graphical edit in many cases, or by re-digitization of the diagram. The bond-type cyclicity flags are added by program, using an algorithm described by Wipke and Dyott [11].

The chemical connectivity tables are, of course, a formalized representation of 2D chemistry, and, in some situations, rules must be applied in creating the table. These rules must be known to the user when a chemical substructure search query is being formulated and are further discussed in Section 3.7.4.

► *3D Information*

This consists of the numerical crystallographic results:

- Unit-cell parameters
- Space group and symmetry operators
- Atomic coordinates, as fractional x, y, z referred to crystal axes
- Atomic coordinate e.s.d.'s (since 1985)
- Covalent radii, and crystallographic connectivity established using these radii

The atomic coordinates recorded for any entry refer to the bonded crystal chemical unit (Figure 3.1 b), which may consist of one or more molecules or ions, each of which is termed a residue in the CSD. Any symmetry-generated atoms that are bonded to the crystallographic asymmetric unit are included in the crystal chemical unit represented by the CSD coordinate set (Figure 3.1 b). By this means, the two connectivity representations, i.e. the chemical representation entered by CCDC staff and the crystallographic representation derived by use of covalent radii criteria, should be equivalent. The direct matching of these two representations (Figure 3.1 a with Figure 3.1 b) is an important recent development in the CSD and will be further discussed in Section 3.6.6. This connectivity matching is not only essential for search purposes, but also provides a valuable additional check during database building.

3.3.4 Checking and Evaluation

These procedures are designed to ensure the accuracy of transcription of each CSD entry. They include visual scans of text fields, together with programmed internal consistency checks. Thus, the related chemical formulation, chemical connectivity table, unit cell dimensions and cited density should all be consistent. The unit cell dimensions, atomic coordinates and symmetry operators should together generate bond lengths which are in agreement with those listed in the publication, and which are input by the CCDC staff for this evaluation process. The matching of chemical and crystallographic connectivity representations, noted above, forms another powerful evaluation tool. The checks are performed not only to ensure the accuracy of CCDC-keyboarded input but also the correctness of the data presented in the relevant publication. Some 15% of all publications contain one or more numerical errors. The raw input file is continually edited and rechecked, and evaluation flags and text comment are added during the process.

3.3.5 Registration and Archiving

Each new CSD entry is registered against the current master archive file. The crystallographic unit cell parameters, the chemical formula and a key derived from the chemical connection table are used to find identities with existing entries. These

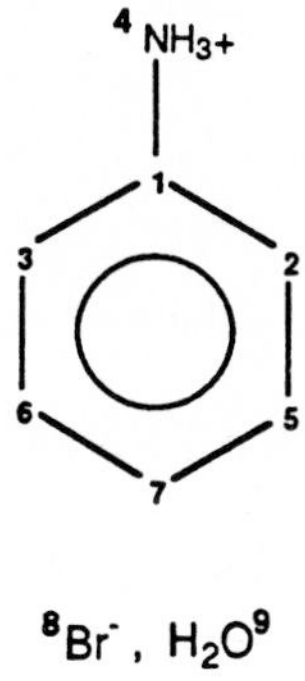

(a)   Chemical connectivity table

Atom properties

residue 1

n	el	nca	nh	ch
1	C	3	0	0
2	C	2	1	0
3	C	2	1	0
4	N	1	3	1
5	C	2	1	0
6	C	2	1	0
7	C	2	1	0

residue 2

8	Br	0	0	-1

residue 3

9	O	0	2	0

Bond Properties

i	j	bt
1	2	-5
1	3	-5
1	4	1
2	5	-5
3	6	-5
5	7	-5
6	7	-5

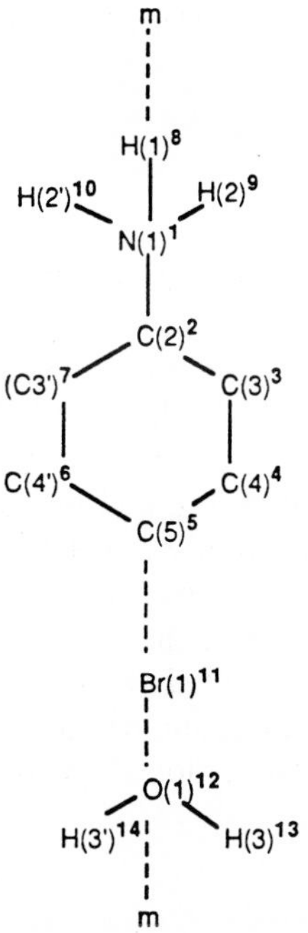

(b)   Crystallographic connectivity table

Atom properties

residue 1

n	el	nca
1	N	4
2	C	3
3	C	2
4	C	2
5	C	2
6	C	2
7	C	2
8	H	1
9	H	1
10	H	1

residue 2

11	Br	0

residue 3

12	O	2
13	H	1
14	H	1

Bond Properties

i	j
1	2
1	8
1	9
1	10
2	3
2	7
3	4
4	5
5	6
6	7

12	13
12	14

Fig. 3.1. a: Chemical connectivity table for the three-residue structure $C_6H_8N^+$, Br^-, H_2O. Here, *n* is the canonical atom number assigned by the Morgan algorithm [19];
b: Crystallographic connectivity for structure (a), assumed to adopt m-symmetry in the crystal structure. Atoms H(2'), C(3'), C(4'), H(3') are symmetry-related to the asymmetric unit and are included in the "crystal chemical unit" of the CSD. Here *n* is the sequence number of each atom in the CSD atomic coordinate listings

identities are examined and the initially assigned CSD refcode is altered appropriately before the entry is archived. The check, evaluation and registration steps result in a considerably upgraded version of the initial raw free-format input file. The data are then converted to binary format and certain data items are added at this stage (e.g. the bond cyclicity flags, the set of numbers which link the chemical and crystallographic connection tables, etc.). The binary entries are then merged with the master archive file, ANEW, used for internal storage at the CCDC.

3.3.6 Database Statistics

On 1st October 1991 the CSD contained 95,617 structural entries relating to 84,962 unique chemical compounds. Of these 87% had 3D atomic coordinates present, the remainder being conference reports and short communications, yielding a total well

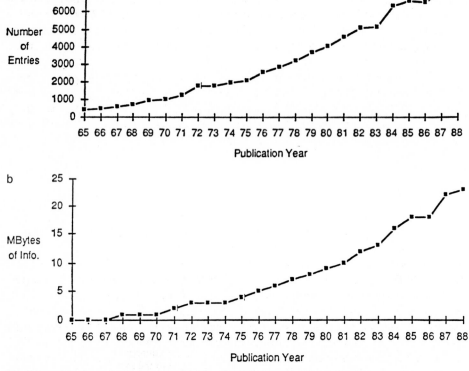

Fig. 3.2. a: Number of new entries added to the CSD vs year of publication
b: Data content of the CSD as megabytes of information vs year of publication

in excess of 4 million atoms. Only 2.3% of entries with coordinates contained unresolved numerical errors after the evaluation process. The rate of increase in size of the CSD is illustrated in two ways in Figure 3.2. The standard plot of new entries against year of publication is augmented by a plot of the number of megabytes of information entering the database for each publication year. Whilst the annual publication rate (Figure 3.2a) continues to increase, the upward slope of Figure 3.2b is even greater, indicating that the mean size of each entry is also increasing every year. The amount of information processed (in megabytes) during 1987 was approximately equal to that processed in the whole decade 1965 to 1974, and almost double that processed in 1981.

3.4 The CSD Software Systems

The CSD is distributed as a complete system, incorporating the database itself together with software for search, retrieval, analysis and display of information. At the time of writing (September 1991) two systems are available, Version 3 and Version 4, whilst a further enhancement to a Version 5 system is planned for distribution during 1992 (Release: October 1992). Complete details of these software developments have been published [4] so that only brief and necessary background material is given here.

A flowchart which covers the operations of Versions 3 and 4 is given in Figure 3.3. Both of these systems operate on a database file known as ASER, which is derived from the CCDC internal archive file and is specially structured for rapid searching. The structure of the basic entry-sequential and machine-independent ASER file is further modified to take advantage of specific systems architectures, e.g. VAX/VMS and Silicon Graphics 4D workstations.

▶ *Version 3*

This system represents an integration and extension of the functions of two earlier programs (BIBSER and CONNSER) that searched the bibliographic information and the chemical connectivity information separately. These programs formed the core of the earlier, and now obsolete, Versions 1 and 2 of the CSD system. The Version 3 program QUEST (Figure 3.3) now performs integrated searches of the 1D (text and individual numerical items) and 2D (chemical connectivity) information. Version 3 is driven by an alphanumeric query language, and hits may be examined interactively. The program QUEST will generate a number of output subfiles relating to the search hits obtained. The most important of these is the file denoted as FDAT in Figure 3.3. This (card-image) subfile contains retrieved 3D information for use by the programs GSTAT and PLUTO, and also by external software. GSTAT permits 3D searching by use of geometric constraints, and will generate listings of geometry for

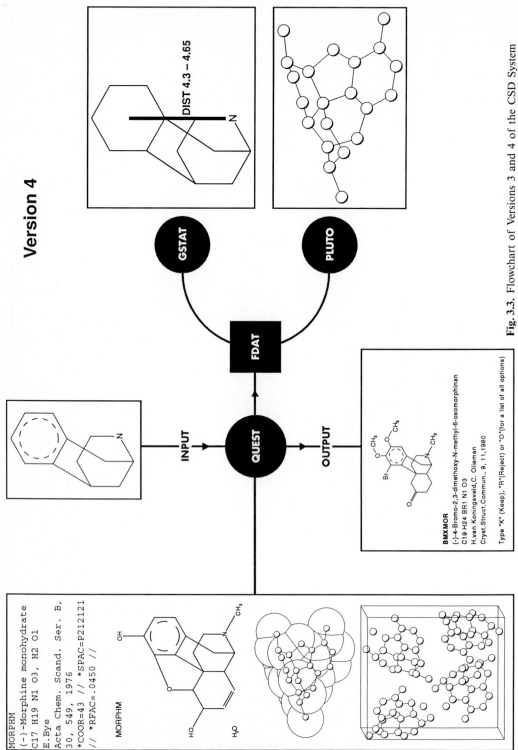

Fig. 3.3. Flowchart of Versions 3 and 4 of the CSD System

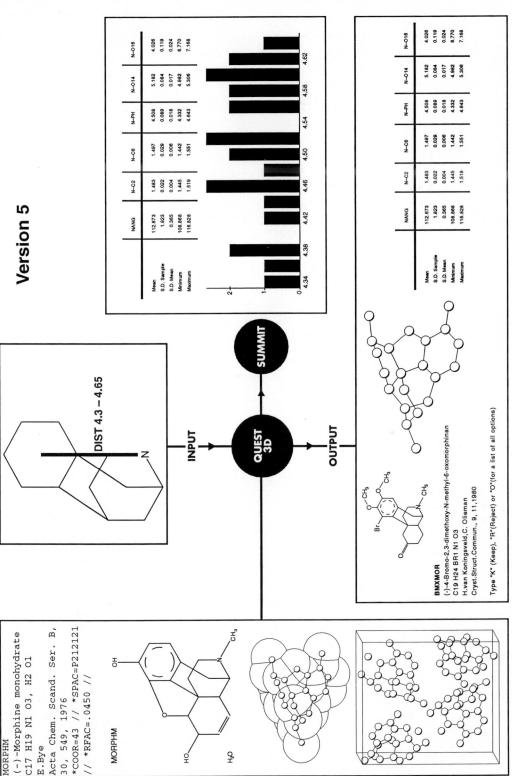

Fig. 3.4. Flowchart of Version 5 of the CSD System

both complete molecules and for chemical fragments. It also contains facilities for many of the statistical analyses discussed in detail in Chapter 4 of this volume. GSTAT is an upwards development of GEOM 78, included in Version 1 of the CSD, and GEOSTAT (see e.g. [12, 13]) included in Version 2. PLUTO is the well-known public-domain crystallographic plotting package originally developed at the CCDC by Motherwell [14].

► *Version 4*

This system incorporates an interactive menu-driven graphics interface for query construction in QUEST (Figure 3.3). The underlying Version 4 database also contains, for the first time, complete sets of x, y-coordinates which permit the display of 2D chemical diagrams for each hit as it is located by QUEST. The PLUTO package is now being blended into QUEST so that, in later releases of Version 4, it will also be possible to examine interactively a 3D plot of each hit in addition to the 2D representation of the chemical structure. Other facilities of Version 4 (subfiles, interface to GSTAT, etc.) are as for Version 3.

► *Version 5*

The system flowchart now changes to that in Figure 3.4, in which the 3D (geometric) search capabilities of GSTAT are transferred to QUEST to permit integrated searching of all information items. The statistical functionality of GSTAT is contained in a separate program, SUMMIT, driven by a new interface file, TABLES. This software development is underpinned by the matching of the chemical and crystallographic connectivity representations, and is fully described in Section 3.6.6.

3.5 Bibliographic, Numerical, and Chemical Searching

Complete operational details of all CSD System programs are given in the relevant documentation [15]. In this section we give a brief overview of the search and retrieval facilities for the 1D and 2D information items described in Section 3.3.3. These details apply to all Versions of the CSD software but, where examples are necessary, we use the alphanumeric query coding of Version 3 in the text below.

3.5.1 The Query Language

A complete query is defined by a number of test statements Tn, each defining a search of a specific information item. The search itself is initiated by use of a single

QUEStion command which defines how the individual tests are to be combined using the logical operators .AND., .OR., .NOT. Thus:

```
T1 *CLASS 51
T2 *YEAR 1986
T3 *AUTHOR Smith
T4 *AUTHOR Jones
QUES (T1 .AND. T2) .AND. (T3 .OR. T4)
```

would locate entries in CSD chemical class 51 (steroids) that were published in 1986 by either Smith or Jones. The paragraphs below summarize the types of Tn statements that can be constructed with the QUEST program for various categories of information.

3.5.2 Numerical Information Items

There are 38 possible individual numerical search fields, e.g. *R*-factor, chemical class, journal code number, space-group number, etc. Numerical tests are specified using the general construct:

Tn *FIELD .LO. nval

where .LO. is one of the logic operators: .EQ. (default), .NE., .LT., .LE., .GT. and .GE. The specification of a numerical range is also permitted as:

Tn *FIELD nval 1−nval 2

Crystallographic unit-cell searches use a variant of this procedure. Here, the crystal system must be specified, together with six unit-cell parameters and a tolerance value to be used in cell-edge comparison. The searches operate on the Niggli reduced cell stored in the CSD and the search cell is similarly reduced prior to carrying out the test.

3.5.3 Text Information Items

There are 19 text fields available for searching, e.g. compound name, synonym name, authors names, journal volume and page "numbers", etc. The general test construct is:

Tn *FIELD string

A special feature is provided for searches of compound names and their synonyms. These fields may be searched separately (as *COMP or *SYNO) or together under the composite field name *XNAM. Here, the two names are first stripped of their non-alphabetic characters (blanks, numbers, punctuation symbols) and then concatenated. This feature removes some, but not all, of the possible vagaries of compound name searching and is discussed further in Section 3.7.3.

3.5.4 Molecular Formula Information Items

The standard molecular formula ($C_x H_y$ followed by other element/count combinations in alphabetical order) may be searched for specified elements or for element/count combinations in two basic ways: (a) within the discrete bonded residues or ions of the crystal chemical unit, or (b) within a formula that has been summed over all such residues and ions. The test:

$$Tn \ *RESF \ Os\,8{-}10$$

is typical, and would locate all bonded residues that contain between 8 and 10 osmium atoms. These searches can be generalized by use of a pre-defined set of 30 element-group symbols, where a non-standard element symbol is used to represent a vertical or horizontal sub-division of the periodic table, e.g. TR = any transition element, 7A = any halogen, etc.

3.5.5 2D Chemical Connectivity Information

The location of chemical substructures (fragments) within the 2D connection tables is the single most important search of the CSD, particularly for structure correlation purposes. In Version 3, the fragment is specified by a packet of instructions that commences with a Tn *CONN record and terminates with an END record. As with other tests, multiple use of the Tn *CONN . . . END construct is permitted, e.g. to locate fragment A but .NOT. in the presence of fragment B. The instructions that can be included within a Tn *CONN packet are summarized briefly in Table 3.1. They provide wide flexibility, both in the definition of the fragment itself, and in the definition of its environment. A complete example is shown in Figure 3.5.

Even with the advent of interactive menu-driven query construction (see Section 3.5.8 below), it is still advisable to draw the required fragment carefully on paper and to consider the query definition under the following four subheadings.

▶ *Definition of basic fragment topology*

At its simplest level, this consists of the definition of the atomic nodes for the fragment, in terms of element type only, and of the simple connections that must exist

Table 3.1. Alphanumeric instruction set for 2D substructure search definition within CSD System Versions 3 and 4

---

a: Main Keywords and Summary Definitions

Tn *CONN	Start of instruction "packet"
C [text]	Optional user comment on query
ELDEF	User definition of "element-group" symbol and group composition (see text)
ATp	Definition of atom properties [see (b) below]
BO	Definition of bond properties [see (c) below]
NFRAG [nval]	Number of occurrences of fragments required
SAMERES	Locate fragments in same bonded residue
DEFB [nval]	Reset default bond type from single to [nval]
NOCR [nval(s)]	The defined fragment may not have substituents which form part of a cyclic route (see text), [nval(s)] restricts NOCR to operate at the atom(s) specified, otherwise it is global
NOLN [nval(s)]	No direct links shall exist between atoms in the fragment, except those defined by the BOnd records (see text), [nval(s)] restricts NOLN to operate at the atoms specified, otherwise it is global
ALLBOND [A or C]	All bonds in fragment are to be acyclic (A) or cyclic (C)
END	End of instructions for this test

b: ATom property definitions

el	Element symbol: individual symbol, pre-set group symbol, user-defined group symbol
mca	Minimum number of connected atoms excluding terminal H atoms. Tests on mca can be avoided by using a value of **99**. The "minimum" criterion can be altered by use of:
E	The mca value must be matched exactly
nh	Number of terminal H atoms to be attached to this atom, can be an OR list as n1,n2,n3, etc.
nch	Integral charge assigned to atom, can be a single +ve or −ve value, an OR list, or special values: **49** (any −ve), **50** (any +ve or −ve), **51** (any +ve)
Tn	Total coordination number (see text)
C or A	Atom must be part of a cyclic or acyclic unit

c: BOnd properties:

n1 n2	Defines a connection between AToms n1 and n2, n2 may be an OR list to allow for variable point of attachment of one subfragment to another (see Figure 3.5)
bt	Bond type: may be a single +ve number, an OR list, or set to **99** to allow any bond type. Assumed to be single if not set. Allowed values are: **1** (single), **2** (double), **3** (triple), **4** (quadruple), **5** (aromatic), **6** (polymeric, as in catena-structures), **7** (delocalized double), **9** (π-bond)
C or A	Bond must be part of a cyclic system

Structure (a)

```
T1 *CONN
AT1 N 1 2
AT2 C 2
AT3 C 3
AT4 O 1
AT5 O 1
AT6 Cu 1
BO 1 2 1
BO 2 3 1
BO 3 4 1
BO 3 5 2
BO 6 1,4 1
END
```

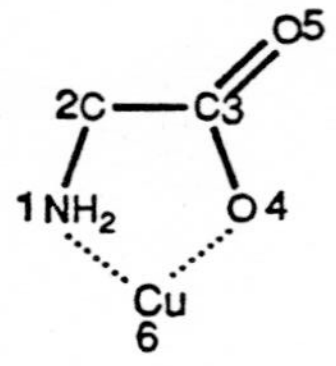

(a)

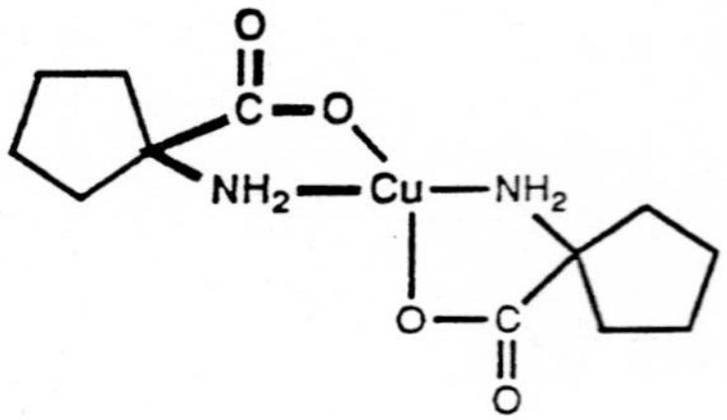

(i)

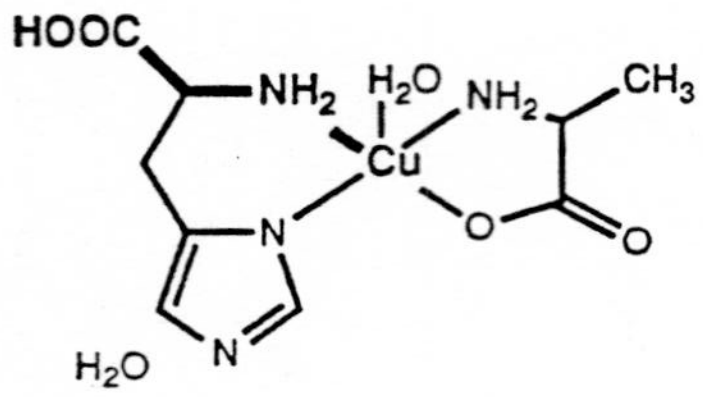

(ii)

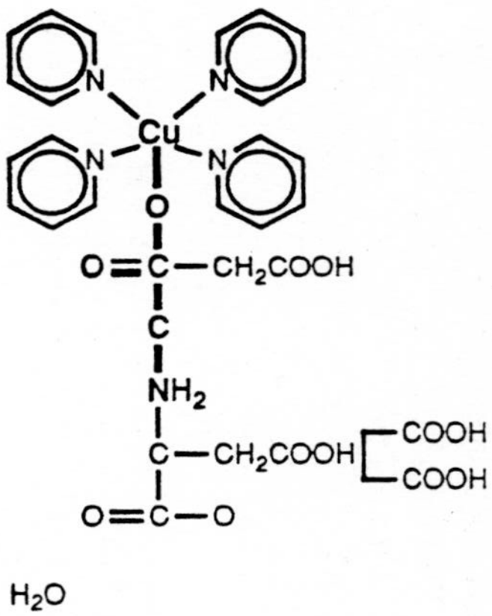

(iii)

Fig. 3.5. Example of a chemical substructure search in the CSD System (see text)

between these atoms. Normally, one might begin with the atoms being defined by individual standard element symbols. However, to generalize the search two other possibilities exist: (a) use of one of the 30 pre-defined element-group symbols within the CSD software, or (b) user-definition of an element-group symbol, e.g. Pk = C, N, S, CL, via the ELDEF command, in those cases where the CSD pre-definitions are inappropriate.

Normally, each bond in this basic topological description has fixed points of attachment, i.e. it is defined by two atom numbers *i* and *j*. It is possible to generalize the fragment by defining *j* as a set of atom numbers in the BOnd record, e.g. BO 1 2, 3, 4 would permit atom 1 to be bonded to atoms 2 or 3 or 4 in the fragments located in the database. This is useful in the location of metal complexes, where there may be uncertainty as to which ligand atom(s) are directly connected to the metal, as illustrated in Figure 3.5. This facility can be used to retrieve all nitroanilines, for example: here the nitrogen of the nitro group (atom 1, say) would be permitted to bond to either the *ortho* (atom 2), *meta* (atom 3) or *para* (atom 4) position of the aniline moiety.

▶ *Definition of atom-property constraints*

The chemical nature of each atomic node can be further refined to specify requirements concerning (a) degree of substitution by non-H atoms, (b) number(s) of terminal H atoms, (c) the net charge value(s) that are permitted for this atom, and (d) whether the atom must be located in a cyclic or acyclic environment. Items (b, c, d) are self-explanatory from the descriptions given in Table 3.1. Frequently, however, care must be taken in specifying the required degree of atomic substitution, for which a number of options exist. Firstly, the specifications of bonds within the fragment obviously implies a minimum number of connected non-H atoms; thus the mca value for each atom can be left blank or set equal to this minimum, and any further substitution of the atom is allowed. It is perfectly permissable to set mca to some value higher than the minimum value implied by the bonding specifications, i.e. to set mca to 3 or higher, even though only two bonds to that atom are explicitly entered. Whilst mca is treated as a minimum value by default, this definition can be changed, for any atom, to an exact value by use of the E qualifier (Table 3.1). Remember, however, that mca and mca E refer to connected non-hydrogen atoms. In many cases, we wish to specify the total coordination number of a given atom, where the total must include terminal H atoms. For the common "organic" elements, e.g. C, N, O, this requirement corresponds to a definition of the hybridization state of the atom; for organometallics and metal complexes, definition of the coordination number for a central metal is a common requirement. Both of these cases are coded by use of the Tn qualifier of Table 3.1 b.

▶ *Definition of bond-property constraints*

The chemical nature of each of the connections in the basic diagram can be further refined to specify requirements concerning (a) the chemical bond type(s) and (b) whether the bond must be located in a cyclic or acyclic environment. The application of these constraints is self-explanatory from the description given in Table 3.1.

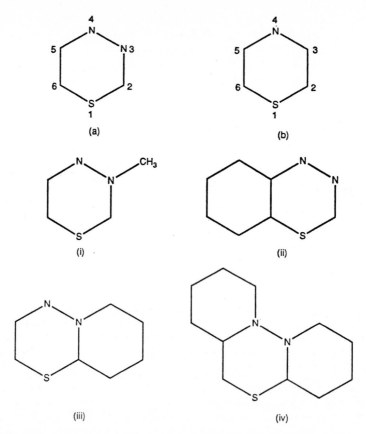

Fig. 3.6. Diagrams to show use of No Cyclic Routes (NOCR) option. NOCR requires that no atoms of the search fragment can be connected to atoms outside the fragment by cyclic routes (bonds). NOCR ijk ... requires that atoms i,j,k ... of the search fragment cannot be connected to atoms outside the fragment by cyclic routes (bonds). Consider the search fragment (a) above, with target structures (i)–(iv). NOCR: This will register hits only for (i). NOCR 3: This will register hits for (i) and (ii). NOCR 6: This will register hits for (i) (iii), and (iv). NOCR 4 5: This will register hits for (i) and (iii). Suppose the search fragment was changed to (b). This has topological symmetry. Therefore, if you wish to apply NOCR to atom 2, it must also be applied to atom 6, i.e. NOCR 2 6

▶ *Definition of fragment environment*

The local chemical environment of the chosen fragment can be defined through specification of (a) the degree of substitution permitted by the atom property constraints, or (b) by extending the overall query construction to specify permitted substituent groups. Additional facilities are provided by the keywords NOCR and NOLN, as illustrated in Figures 3.6 and 3.7. NOCR is used to exclude fusion or bridging of the defined fragment. This command can be applied globally to the whole fragment, or restricted to certain subsections of the fragment defined by atom numbers. NOLN demands that the bonded links specified in the fragment definition

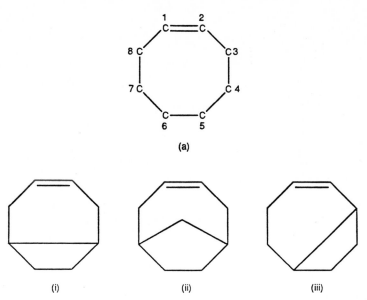

Fig. 3.7. Diagrams to show use of No Direct Links (NOLN) option. It can take two forms: NOLN requires that no atoms of the search fragment can be directly linked to any other atoms of the search fragment except by the bonds specified in the bond property records. NOLN ijk . . . requires that atoms i,j,k . . . cannot be directly linked. Consider the search fragment (a) above, with target structures (i)–(iii). NOLN: This will register hits only for (ii). Atoms 4 and 7 are connected by an indirect link through a bridging C atom. NOLN 4 7: This will register hits for (ii) and (iii). NOLN 3 8: This will register hits for (i), (ii), and (iii). NOLN 3 6 8 5: This will register hits for (i) and (ii). Note that the search fragment has topological symmetry. Thus a link 3–6 would be equivalent to a link 8–5. Therefore, to exclude (iii) we must code both possibilities, i.e. NOLN 3 6 8 5

are the only direct links that are permitted in the retrieved fragments. Again, this can be applied in a global sense, or by atom specification.

The flexibility permitted in fragment definition is necessary in order to make full use of the encoded 2 D chemistry in the CSD. However, this flexibility, together with certain specific conventions used in the database, can lead to apparent misconceptions of the search process. Some of these problem areas are discussed in detail in Section 3.7.4 below.

3.5.6 Bit-encoded Information Items

In many database systems it is common practice to record the presence/absence of certain specified features in terms of the on/off (1/0) setting of bits in a bit-map. Items in the CSD that are suitable for such space-saving treatment are e.g. yes/no flags indicating a neutron diffraction study, absolute configuration determination, residual errors in entry, etc. A total of 124 bits record information in this way and can be accessed via the **SCREEN** search command. In practice this is an extremely powerful and rapid search technique, either in its own right, or in combination with the test constructs Tn described above.

Bit settings are also used in all current Versions of the CSD (as in most modern chemical databases) as heuristic functions, or screens, to enhance search speeds. Thus, bit screens encode the presence or absence of a large set of pre-defined chemical structural characteristics (428 in all) in complete entries in the CSD. These cover such items as, e.g. presence of $C=O$ bond, six-membered ring, charged nitrogen, etc. Identical bit settings are generated from the input chemical connectivity query, and these query bits are compared with those stored for each database entry. All bit-encoded features of the query must be found in a CSD entry before it can be regarded as a potential hit, e.g. before we need embark on the time-consuming atom-by-atom, bond-by-bond 2D pattern matching process. Similar treatments are also applied to element searching, and to some of the text searches. Leading references to this important topic, and full details of its application to the CSD, can be found in Allen et al. [4].

3.5.7 Chemical Similarity Searching

The chemical connectivity searches described in Section 3.5.5 match the explicit chemical requirements of the query with the explicit chemical description of 2D structure held in CSD. Normally, they are used to locate substructural fragments but these methods can, of course, be used to locate a specific complete molecule. In this case, the molecule will either be present exactly as coded, or it will not. In the past few years, methods have been developed [16, 17] for locating molecules that show a high degree of 2D structural similarity with an input query molecule. The methods are based upon the calculation of similarity indices derived from a comparison of the bit screens for the query molecule and the bit screens stored for each database entry. By this means, it is possible to rank the resulting output in terms of the decreasing similarity of each database entry to the input query. An example of the output from a similarity search of the CSD is illustrated in Figure 3.8. Since these searches are at the full-molecule level, it is much easier to encode the query molecule within the graphical menus of the CSD Version 4, rather than via the alphanumeric system of CSD Version 3. Current research efforts [18] are directed towards the development of methods for 3D similarity searching.

3.5.8 Interactive Menu-driven Graphics

All of the search functionality described above can be accessed in CSD Version 4 via an integrated set of interactive, graphical menus. These cover the numerical, text, unit-cell and bit-screen searches but were chiefly designed to simplify the encoding

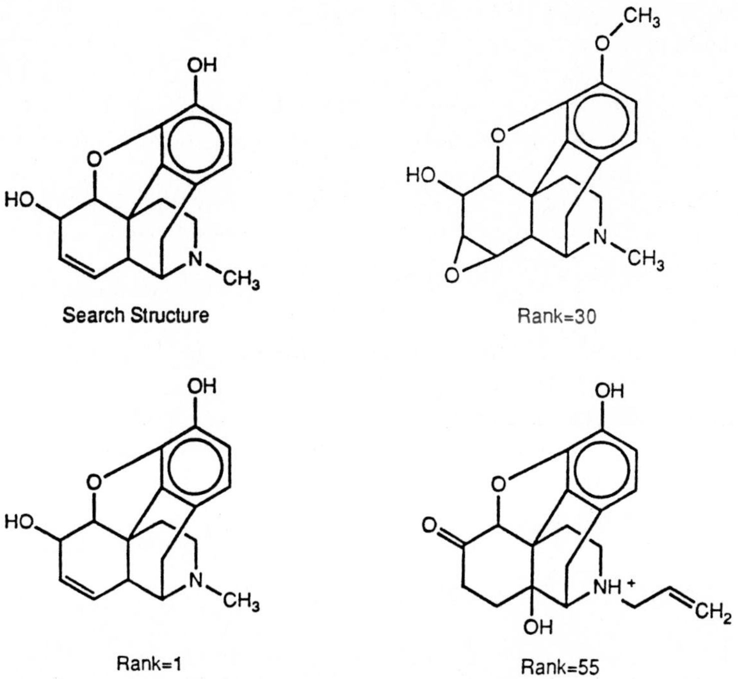

Fig. 3.8. Example of a 2D chemical similarity search for morphine

of 2D fragment searches. Thus, the BUILD, CONSTRAIN and PERIODIC TABLE menus cover all of the options listed in Table 3.1, whilst two further menus permit the selection of common functional GROUPS and of a wide variety of basic 2D structural TEMPLATES. Full details are in the CSD User Manuals [15].

3.6 3D Searching and Geometry Tabulations

The flowcharts of Figures 3.3 and 3.4 show that 3D information is handled in different ways in Versions 3/4 and in Version 5. Hence, we begin this section with an overview of the functions of the GSTAT program within Versions 3/4 and develop this theme to show how the functionality of this program can be divided, to the benefit of the whole system, in Version 5.

3.6.1 Overview of the GSTAT Program

The program GSTAT has two primary modes of operation. Firstly, there is an *entry-by-entry mode* in which intramolecular geometry (bond lengths, valence angles and torsion angles), metal coordination sphere geometry (distances and angles), or inter-molecular distances between specified elements, are calculated systematically for every entry in a retrieved FDAT subfile (Figure 3.3). These listings are presented in terms of the crystallographic atom-labelling schemes, and can only be interpreted in conjunction with a set of labelled 3D plots, generated using PLUTO, or by reference to the original paper. The variety of labelling schemes employed in the literature, even for closely similar molecules (or fragments), militates against the use of entry-by-entry geometry listings for systematic studies. Their use is, therefore, restricted to the examination of one or very few specific entries.

It is the *fragment mode* of GSTAT that provides the facilities for 3D searching, i.e. the addition of further geometrical constraints to the original chemical fragment definition, and for the systematic statistical and numerical analyses that underpin so many of the research projects based upon the CSD. In fragment mode, the GSTAT program permits:

(a) The definition and location of a substructural fragment within the crystallo-graphic connectivity stored in FDAT, a procedure that automatically imposes the user's own atomic numbering onto each hit.
(b) The calculation of a wide variety of user-defined geometrical parameters (internal coordinates) for each occurrence of the fragment.
(c) The selection of fragments on the basis of geometrical criteria.
(d) The generation of a final systematic tabulation of the N_p defined geometrical parameters for each of the N_f fragments that survive any selection procedures at (c).
(e) The statistical and numerical analysis of the geometry table, expressed in the form of a raw data matrix $G(N_f, N_p)$.

Statistical and numerical analysis (e) forms the subject matter of Chapter 4. It is the purpose of this current section to summarize the aspects of GSTAT that are itemized at (a–d) above. These four functions of the GSTAT program of Versions 3 and 4 are transferred to QUEST in Version 5 and are illustrated in Figure 3.9 for a simple 3-coordinate metal fragment.

3.6.2 Location of Fragments in GSTAT

The object of fragment location in GSTAT is threefold: (a) to impose a user-specified atomic enumeration onto every instance of that fragment in the CSD, (b) to associate the 3D crystallographic coordinates stored in the CSD with each atom of the fragment according to the prescribed atomic enumeration, and (c) to accept or reject

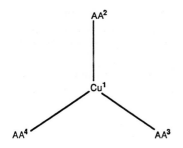

```
C Specify Fragment
FRAG
AT1  CU  3 E
AT2  AA  1
AT3  AA  1
AT4  AA  1
BO  1 2
BO  1 3
BO  1 4
END
C SETup plane through
C ligand atoms
SETUP P1  2  3  4
C DEFine valence angles
C at the metal centre
DEF A1  2  1  3
DEF A2  2  1  4
DEF A3  3  1  4
C DEFine distance of
C Cu from ligand plane
DEF DEV1  1  P1
C Use TRAnsform
C Parameter_names starting
C with # are temporary names
C which do not appear in
C final tabulation
TRA  #DA1  =  120.0 - A1
TRA  #DA2  =  120.0 - A2
TRA  #DA3  =  120.0 - A3
TRA  SA1 = #DA1 - #DA2
TRA  #S4 = 2.0 * #DA3
TRA  SA2  =  SA1 + #S4
TRA  #S5  =  #DA1 + #DA2
TRA  SA3  =  #S5 + #DA3
```

Nfrag	Refcod	A1	A2	A3	DEV1	SA1	SA2	SA3
1	BAFFOD	112.890	112.890	134.220	0.000	0.000	-28.440	0.000
2	BETDAF	121.097	121.097	117.766	0.026	0.000	4.468	0.041
3	BETDAF10	121.097	121.097	117.766	0.026	0.000	4.468	0.041
4	BETDEJ	119.761	122.533	117.707	0.000	2.772	7.359	0.000
5	BETDEJ10	119.761	122.533	117.707	0.000	2.772	7.359	0.000
6	BILLOX	112.149	124.400	114.368	0.399	12.251	23.515	9.082
7	BILLOX	112.297	130.179	115.401	-0.191	17.882	27.080	2.123
8	BILLOX	117.020	126.646	116.332	-0.007	9.625	16.962	0.003
9	BILLOX	121.676	114.957	113.345	-0.423	-6.719	6.592	10.022
10	BILLOX	120.651	119.041	116.331	-0.261	-1.610	5.728	3.978
11	BILLOX	111.426	133.217	113.844	-0.159	21.790	34.103	1.514
12	BILLOX	118.445	113.188	126.482	0.180	-5.257	-18.221	1.885
13	BILLOX	134.872	99.694	110.393	-0.500	-35.178	-15.964	15.042
14	BINVAV	155.953	99.130	100.060	0.207	-56.822	-16.943	4.857
15	BOBVET	102.412	104.256	151.221	0.147	1.843	-60.598	2.111
16	BOBVET	99.728	150.753	104.494	0.226	51.025	82.037	5.026
17	BOBVET10	102.402	104.243	151.224	0.148	1.841	-60.606	2.131
18	BOBVET10	99.599	104.476	150.853	0.226	4.877	-56.828	5.073
19	BOMCEL	130.970	108.957	119.999	-0.035	-22.013	-22.011	0.074

Fig. 3.9. Example of GSTAT operating in fragment mode
Top left: The Cu(AA)$_3$ fragment. Top right: Annotated instruction set. Bottom: Resultant geometry table

fragments on the basis of limiting values of geometrical parameters that can be calculated from the 3D fragment coordinates. Fragment topology is defined in terms of atom and bond properties, as in the initial step of a 2D connectivity search. However, the chemical constraints available in 3D searches are severely restricted by comparison with the 2D case. These restrictions are entirely due to the underlying connectivity representation that is being searched.

The full 2D connectivity description is exemplified in Figure 3.1a. It has no direct links to the crystallographic coordinate set: the 2D structure, which is seldom explicit in crystallographic publications, is entered by CCDC staff and the atoms are enumerated by use of an algorithmic canonical scheme due to Morgan [19]. The enumeration of atoms in the crystal structure is taken directly from the publication. The 3D coordinate set is then used to establish a crystallographic connectivity referred to this published atomic enumeration. Here, links are established between atoms A and B if the distance $d(AB)$ lies in the range $r(A) + r(B) \pm t$, where $r(A), r(B)$ are covalent radii, and t is a tolerance value. Suitable covalent radii have been established over the years as part of the CSD check procedures; t is normally 0.4 Å. It is this crystallographic connectivity that is associated with the 3D coordinates, and which must be used by GSTAT in fragment location.

The crystallographic connectivity for the chemical molecule in Figure 3.1a is illustrated in Figure 3.1b. By comparison with (a), the crystallographic representation lacks bond-type information and atomic charges. Most importantly, there is a significant difference in the treatment of H atoms. These are normally implicit in the 2D representation, through the use of the nh atom property. In the 3D case, hydrogens are included as explicit atoms in the connectivity table, if they are present in the published coordinate lists. If coordinates are not available, then the crystallographic connectivity has no knowledge of H counts. The sum result of these connectivity differences is that fragment definitions in GSTAT may only contain (a) atom properties in the form of element types and mca values, and (b) bond properties that simply indicate that two atoms are connected: bond type information must be imposed through the use of appropriate geometrical constraints, particularly those based on bond lengths.

In practice, these connectivity differences are not as serious as might be thought. Many fragments are chemically distinctive even at the simple topological level, i.e. the pattern of chemical elements and their connectedness. This is particularly true if elements other than C, N and O are present. However, even here, it is often a simple matter to distinguish identical topological patterns by use of one, or a very few, geometrical tests. Recent compilations of standard bond lengths ([20, 21] and Appendix A to this volume) in a wide variety of chemical environments are valuable in making these decisions. For example, a C=O bond might be approximated by a distance range of 1.15 − 1.24 Å etc.

One benefit of the distance-based connectivity approach is that it can be extended to include selected intermolecular interactions by use of the space-group symmetry. For studies of H-bonding, for example C−O−H . . . O=C, we may designate all O . . . O distances of less than, say, 3.2 Å as a "connection" and locate H-bonded fragments through the normal atom-by-atom, bond-by-bond pattern matching process, using bond length constraints to distinguish C−O from C=O in the located fragments.

Finally, we note that fragment location is exhaustive in 3D searches. We must locate all instances of the fragment in a given CSD entry, since each is (normally) regarded as an independent example in structure correlation work. Obviously, it is possible for two fragments to overlap to some extent, and the retention/elimination of overlapping fragments is a user decision. This need for exhaustivity in 3D searching contrasts with the 2D case, where the location of a single instance is normally sufficient to classify the entry as a hit.

3.6.3 Calculation of Fragment Geometry

The program is able to calculate a wide variety of internal coordinates for each fragment located in the search procedure. The subset of parameters required for a particular study is entirely under the control of the user. Each individual parameter is specified in terms of the (user-defined) atomic enumeration of the fragment. Three major commands, DEFine, SETup and TRAnsform are used to specify geometrical parameters.

The form of the DEFine command is:

$$\text{DEF } \langle\text{parameter\_name}\rangle \ \langle\text{obj 1}\rangle \ \langle\text{obj 2}\rangle \dots \langle\text{objn}\rangle$$

where ⟨parameter_name⟩ is a parameter identifier supplied by the user, and ⟨obj 1⟩, etc. define geometrical objects that must be operated on by GSTAT to generate the required parameter. The simplest object is an atom, thus:

```
DEF  C=O  1  2
DEF  OCO  3  1  2
DEF  COCO  4  3  1  2
```

where the commands request the calculation of a bond length, a valence angle and a torsion angle, respectively. In some cases, it is necessary to use more complex objects to generate the required parameter. Within GSTAT, additional objects are: a centroid (denoted X), a vector (denoted V) and a plane (denoted P). These objects must be established in program memory by use of the SETup command:

```
SET  X1  1  2  3  4 ...
SET  V1  1  2
SET  P1  1  2  3  4  5 ...
```

where the integers are atom numbers. These objects may then be used within the DEFine command in a variety of self-explanatory ways:

```
DEF  DIST1  1  X1  (a distance)
DEF  ANG1  V1  V2  (an angle)
DEF  DIST2  2  P1  (a distance)
DEF  ANG2  P1  P2  (an angle)
DEF  TOR1  1  2  3  X1  (a torsion)
```

The TRAnsform command is used to perform unitary and binary operations on parameter(s) established using SETup/DEFine. A set of FORTRAN-like operators, e.g. SIN, COS, ABS, +, −, *, etc., are provided which, when used in successive TRAnsform statements, enable the user to construct linear combinations of the initial parameters. For example, we may use TRAnsform to obtain the mean valence angle at nitrogen, from the individual valence angles obtained using DEFine, for use, perhaps, as a single-parameter measure of pyramidality.

The use of TRAnsform to calculate linear combinations of angular deformation coordinates is illustrated in Figure 3.9. The basic valence angles A1, A2, A3 are first TRAnsformed to give their differences from 120° as DA1, DA2, DA3. Some linear combinations (not normalized) might be:

$$SA1 = DA1 - DA2$$
$$SA2 = DA1 - DA2 + 2 DA3$$
$$SA3 = DA1 + DA2 + DA3$$

which are included in the final tabulation.

In normal use, DEFine will generate a single (named) parameter for inclusion in the final geometry tabulation. However, there are a number of cases where special functions are provided, so that two or more parameters will result from a single DEFine statement. So far, these special functions apply to the generation of (a) the two parameters that describe pseudorotation in a five-membered ring in the definition of Altona and Sundaralingam [22], (b) the $n-3$ parameters that describe the pucker in an n-membered ring in the definition of Cremer and Pople [23], and (c) the spherical polar angles [24, 25] that define the direction of approach of a D(onor)-H vector to the putative lone pairs on an A(cceptor) atom; constructs exist for generating these angles with respect to either sp^2 or sp^3 lone-pair geometry.

3.6.4 Fragment Selection: the 3D Search Process

Any geometrical parameter that can be obtained using the DEFine or TRAnsform commands can be used in fragment SELection via:

SEL ⟨parameter_name⟩ ⟨value 1⟩ ⟨value 2⟩

where ⟨value 1⟩ and ⟨value 2⟩ define an (ascending) numerical acceptance range for the named parameter. This procedure is entered after the fragment has been located and all geometrical parameters have been calculated. In order to save computing time, the program will also perform numerical tests of the basic geometrical parameters: bond lengths, valence angles and torsion angles, during the fragment search process itself.

3.6.5 Tabulation of Fragment Geometry

The final output of GSTAT operating in fragment mode is in the form of a table (Figure 3.9) of the N_p geometrical parameters established using DEFine and TRAnsform for each of the N_f occurrences of a fragment that survive any SELection processes. The user-supplied parameter names are used as column headings, whilst the fragment number and a CSD refcode identify the lines of the table. This table can be output as a subfile for input to external software, or retained internally in GSTAT, as a raw data matrix to be operated on by the various statistical routines that are available in that program.

3.6.6 The CSD Version 5 Upgrade

The retrieval and organization of numerical data for structure correlation work is a two-stage process in Versions 3 and 4 of the CSD system. This is due to the two discrete representations of connectivity in the file (Figure 3.1 a, b), one containing the formal chemical description of the residues and the other linked to the coordinate-based description of 3D structure. If we can associate the 3D coordinates with the chemical description, then the second fragment location process (Section 3.6.2) is eliminated. This can be effected by mapping one of the connectivity representations onto the other, i.e. by expressing both using a common atom numbering scheme.

This mapping process has now been completed for the majority (some 90%) of CSD entries, using the canonical enumeration algorithm [19] as a basis. However, the process has not been straightforward due to the nature of the crystallographic results. The chief problems were: (a) the presence of more than one residue in the asymmetric unit of the crystal structure; the residues may be chemically different, in which case a 1:1 mapping is preserved, or they may be chemically identical, whence a 1:many mapping is required, (b) some or all of the hydrogen atoms may be missing in the crystal structure, (c) there may be local or complete topological symmetry in the simple crystallographic connectivity, which is not present in the 2D representation, where symmetry is broken by the bond typing, (d) all or part of the crystal structure may be disordered, giving a 1:many mapping for the affected atoms, and (d) the crystal structure may be polymeric using the crystallographic distance criteria, in which case reduction to the monomer prior to mapping can present difficulties.

The matching of the chemical and crystallographic connectivities has resulted in the addition of a new field to the CSD, which links each atom in the crystallographic coordinate list to one of the atoms in the chemical connection table. Thus, once a chemical fragment has been located in the 2D connection table, we may directly assign one (or more) sets of coordinates to the atoms of the fragment. The 3D coordinates can be regarded as additional properties of the chemical atoms. It is this development in the database itself that underpins the Version 5 software upgrade. Thus, the effects of the GSTAT fragment location (Section 3.6.2) are now implicit in the results of the QUEST search, and the functionality described in Sections 3.6.3 – 3.6.5 can be transferred to QUEST itself. Any chemical imprecisions introduced by the geometry-based fragment location in GSTAT are eliminated, and the GSTAT commands can be added to the interactive menus of QUEST. This gross simplification of CSD system operations also means that the cpu-intensive statistical functionality of GSTAT can be developed and released within a separate menu-driven program, linked with QUEST through a simple interface file.

3.7 Special Considerations in Using the CSD System

The CSD is now approaching 100,000 entries. Hence, it covers a very broad range of both chemical and crystallographic situations. All of these must be encoded in ways which conform to the basic structure of the database, and certain conventions must, perforce, be employed. In this section, we highlight certain aspects of the database content that have not been discussed elsewhere, and which may affect the outcome of certain searches and, hence, the interpretation of numerical results derived from such searches. In particular, we hope to alert readers to areas in which a careful reading of the CSD User Manuals [15] may be a necessity rather than a chore.

3.7.1 The Reference Code System

The 6-letter part of the refcode "defines" the chemical compound and two numerals are appended to identify multiple studies of the same compound. Thus, three completely independent studies of compound X, e.g. at different temperatures, or crystallizing in different space groups, might be represented by entries having refcodes ABCDEF, ABCDEF01 and ABCDEF02. If the authors of ABCDEF01 publish a later paper on this work it will have refcode ABCDEF11. Thus ABCDEF11 is said to supersede ABCDEF01 and ABCDEF10 supersedes ABCDEF, etc. A set of entries having the same 6-letter component constitutes a refcode family.

Until 10–15 years ago it was common practice to publish a short note followed later by the full paper which superseded the earlier note. Nowadays, multiple publication of a study is still widespread, but very often the two (or more) papers complement each other, rather than one superseding the other. In the early days of CSD construction, superseded entries were deleted from the database. Now all entries are retained. This means that a refcode family may contain two (or more) identical data sets if the author has published or deposited the same set of atomic coordinates in more than one article. Such duplicates should be eliminated in any numerical analysis project which involves the calculation of statistical measures.

Since the search file, ASER, is ordered alphanumerically by refcode, all members of a refcode family are clustered together. However there are two situations which could result in the loss of relevant hits when using QUEST. Firstly, *stereoisomers* are assigned different 6-letter identifiers. Since mid-1991, this situation has been recognized and an appropriate refcode cross-reference is included as a qualifying phrase in the database. However, as with all database upgrades, this leaves an immediate and large backlog of non-upgraded entries! A project is now under way, using a graph-theoretical comparison of stereoisomers, to extend the cross-referencing system to all existing database entries. The second potential problem area is the presence of *deuterated species* in the database. These are also given 6-letter identifiers which differ from those of their parent compounds, and also differ for varying degrees of deuteration. A pre-defined element symbol, corresponding to H or D, is therefore provided within the QUEST program.

3.7.2 Searches Using the SCREEN Command

The existence of a large number of yes/no bit-screen settings in the CSD was described in Section 3.5.6. Of these, only 124 bits may be accessed directly by the user. These bits represent a considerable amount of summary information concerning (a) the error status of an entry, (b) the experimental conditions, results and their precision, (c) summary details of the cell, lattice and symmetry, and (d) a summary of the non-mandatory information actually present for the entry. Often an individual screen represents the result of a quite complex search in its own right, e.g. that the space group is a member of that small subset in which an absolute configuration determination is possible, etc. A careful, and often combinatorial, use of these screen settings will impose quite complex entry selection criteria, either as a complete search in its own right, or in placing global restrictions on entries to be examined by other search tests. This is particularly important in selecting entries for a later analysis on the basis of general criteria, such as presence of coordinates, lack of residual errors, acceptable indicators of structural precision, etc.

In principle, any of the 682 screen settings in the CSD bitmap may be accessed by users. However, those bits connected with the text and chemical connectivity searches should be used with care, and results examined in detail. They are set algorithmically within the program in response to the overall search query formulation, for the heuristic purposes indentified earlier. Whilst their definition in the available documentation is as accurate as possible, space limitations mean that these descriptions are necessarily brief.

3.7.3 Compound Name Searching

In searching for a specific chemical compound, there is a natural tendency to think in terms of its compound or synonym name as a search query. There are a number of pitfalls in this approach. Firstly, the CSD carries one full compound name and, where appropriate, a single synonym. For some compounds, particularly drug molecules, a number of synonyms may be in common use, but may not correspond to that recorded in the database. Secondly, the syntactic structure of the compound names themselves must be considered carefully in framing search queries. The compound names in the CSD are either taken from the paper itself or constructed using IUPAC rules by CCDC staff. The database therefore contains names from a variety of sources and syntactic consistency cannot be guaranteed, particularly in the positioning of numerical locant identifiers, hyphens, and other punctuation marks. It was for these reasons that the concatenated **XNAM** search field was created in QUEST. Thus, the possible name-syntax alternatives 1,2-butadiene and buta-1,2-diene will both match the search string "butadiene" in **XNAM**, but the latter would be missed in a **COMP**ound name search. Despite this improvement, the splitting of longer alphabetic search strings into short syllables, followed by the use of a number of tests involving logical .**AND**. operations, remains a sensible option.

3.7.4 Chemical Connectivity Searching

The location of substructural fragments within the 2D connection tables is the most powerful chemical search facility within the CSD system. Searches of this type form the basic starting point in most structure correlation studies, where the ultimate aim is to correlate the experimental crystallographic observations of the fragment with the underlying formal chemistry. In general, these searches are not difficult to formulate, particularly with the advent of interactive menu-driven query coding and the introduction into the QUEST program of an extensive query analysis procedure designed to warn of chemically unlikely requirements in a given query. However, as in all chemical databases, the rigidly structured connectivity representation must accommodate all possible flexibilities used in the description of chemical structure. In some specific situations, it is necessary to impose rules in order to formalize a given structure type within the basic framework of the database. It is important that these situations, and their associated rules, are taken into consideration in the search process, else misconceptions will result and potential hits will be missed.

We preface a description of these special situations with some comments on query construction that are generally relevant, but particularly so for queries posed in the special areas. Earlier, we identified four stages in query construction involving definition of (a) basic topology, (b) atom-property constraints, (c) bond-property constraints, and (d) the fragment environment. Because the constraints at (b), (c), (d) are so easy to impose through the graphics system, there is a tendency to over-define a search query. Further, the intimate chemical relationship between the atom and bond constraints is not always realized. It is well worth examining the results of query coding after stage (a), the definition of basic connectedness, and asking the question "Is this basic chemical pattern so discriminatory that further constraints are unnecessary?" If elements other than C, N, O are present, then the answer to this question is often "yes". In cases where detailed bond-type constraints must be imposed, the notes below must be taken into account, since bond typing is a major problem in some areas. Thus, the use of a variable bond type (for example $bt = 2, 5, 7$) is recommended to avoid possible vagaries in CSD content. It is easier to pose a more general query, examine the results, and refine the query in the light of the observed bonding conventions, than to discover missing hits by accident after some detailed, restrictive analysis has already been performed.

The degree to which a query fragment needs to be "refined", i.e. the degree of chemical definition of the fragment environment, depends on the type of study being undertaken. In studying intermolecular interactions, hydrogen bonding, reaction pathways, etc., it is often advantageous to sample a variable fragment environment. However, in studies of chemical bonding, structure-reactivity correlations, and in some conformational studies, a more precise definition of the fragment is required, and often develops from the analysis itself. We return to this topic, and cite some case histories, in Chapter 4.

3.7.4.1 Bond-Type Assignments

Considerable effort has been made, over many years of CSD development, to standardize bond type assignments in most major structural areas. However, it is inevitable that problems arise in some cases due to (a) the chemical interpretations of the crystallographic results by the original authors, or by CCDC staff, and (b) the occurrence of totally new chemical situations in incoming papers. Because crystallographic analysis is usually employed to characterize novel compounds, the uncommon tends to be commonplace within the CSD. Special care should be taken in any area where chemical and/or structural ambiguity may exist, particularly over the localization or delocalization of unsaturated bonds. For example, an ionized carboxylate group, or its thio-analogue may be coded with one single and one double C...O or C...S bond, or with two delocalized ($bt = 7$) bonds. The use of the aromatic bond ($bt = 5$) may also be unclear in fused 5- and 6-membered ring systems, particularly those involving hetero-atoms. Here, alternate single/double bonding, or use of the delocalized bond may well occur, and specification of a generalized unsaturated bond (bt = 2,5,7) is to be recommended in search query construction.

A particular difficulty arises in determining connectedness in metal coordination spheres: "when is a bond not a bond?" In assigning connectivity in σ or π systems, the assessment of bonding (or not) depends on distance criteria, far more than it does in organic moieties. It has been general CCDC policy to formulate compounds so as to reflect the conclusions of the authors of the paper. Hence, because of current imprecisions in basic chemistry, it is possible that the CSD contains different bonding patterns for structures that some users might regard as identical or substantially similar. Consequently, the E(xact) or T(otal coordination) flags (Table 3.1) should be used with care in defining the atom properties of metals. These strictures can make it difficult to retrieve all instances of, for example, 5-coordinate nickel complexes according to the user's conception of 5-coordination, as opposed to that encoded in the CSD. The problem is alleviated by allowing the user to define additional connectivity in distance terms using GSTAT. This function has been transferred to QUEST during Version 5 development.

A special type of bond-type assignment problem concerns *tautomerism*. Some chemical databases provide specific tautomeric bond types, or analyse a query for either the keto or enol constructions and add the missing one (as an .OR. option) automatically. In the CSD, only the single tautomeric form identified by the crystal structure determination is encoded in the connectivity tables. The software has not yet been upgraded to accommodate searches for both forms, hence the onus is on the user to encode both representations if that is the desired goal of a query.

Finally in this section, we must mention the problem of *polymeric bonding*, primarily in metal-organic systems, through which chains, sheets and nets are formed in the extended crystal structure. In the CSD, the principal links emanating from the monomeric unit that fall within the covalent distance criteria, are encoded as polymeric bonds (bt = 6).

Connectivity representations reflect the close interplay between chemical and structural concepts, particularly distance concepts. Now that the mapping of chemical and crystallographic representations has been effected within the CSD, we are in a much better position to commence a thorough reappraisal of known problem areas where bond-typing is less than consistent. Furthermore, the integrated 2D and 3D search facilities of Version 5, coupled with increasing published knowledge of geometrical systematics, permit greater control over fragment location than hitherto.

3.7.4.2 Treatment of Hydrogen Atoms

The hydrogen atoms in the vast majority of structures in the CSD are terminal, single-connected atoms and are represented via the atom property (nh) assigned to the relevant non-H atom. In those few cases where hydrogen is connected to more than one other atom, it is treated explicitly in the connectivity tables, i.e. it is regarded as a non-H atom. Deuterium is always treated in this way. The principal source of multi-connected H atoms arises in bridging hydride complexes of transition metals. This situation also occurs, but rarely, where the crystal structure indicates a symmetrical hydrogen-bonded dimer: $-O\ldots H\ldots O-$ where the $O\ldots H$ distances are equal and it is impossible to localize the H atom. The only case of explicit treatment of terminal hydrogens involves the methanide ion CH_3^-, which must be located via a similar explicit treatment in the query. However, all examples of this rare unit exist in chemical class 12, which may also be specified to speed up the search. It is for these reasons that the QUEST program was designed to handle both explicit and implicit hydrogen atom definitions in the query coding.

3.7.5 The Crystallographic Data

3.7.5.1 Accuracy and Precision

The extensive checks that are performed on raw input data have been detailed earlier. The results of this evaluation process are encoded within the database, both in the form of bit settings and as text statements. These statements are printed by GSTAT so that the extent of the errors located can be assessed during any analysis of molecular geometry. Entries containing residual errors may be eliminated, either using the bit-screen settings in QUEST or a similar feature in GSTAT.

The precision of any entry may be judged through two primary indicators, the crystallographic R-factor and a flag (AS) which indicates ranges of estimated standard deviations (esd's) for $C-C$, $C-N$, $C-O$ bonds in the structure (AS1 = $0.001-0.005$ Å, AS2 = $0.006-0.010$ Å, AS3 = $0.011-0.030$ Å, AS4 = 0.031 Å and up, AS0 indicates that the value was not available in the paper). These ranges were established very early in CSD development, and serve as broad guidelines only. Today, coordinate esd's for post-1985 entries are available, and these data will be used to improve this aspect of entry selection within the CSD software. Other features of the entry must also be taken into account in assessing precision, e.g. the atomic number of the heaviest element in the structure, etc., along lines suggested in a preliminary study [26]. The general levels of precision of entries in the CSD can be judged from the summary statistics of R-factor versus AS flag settings presented in Table 3.2.

The selection of entries for inclusion in a particular study on the basis of structural precision depends on the purpose of the study. In cases where structural changes are likely to be several times larger than the experimental uncertainties, then precision criteria should not be restrictive. If, however, we are looking for very small geometrical variations, as in structure-reactivity correlations, then only the most precise structures are satisfactory. Indeed, the degree of experimental precision and the degree of precision with which the chemical fragment should be defined (see Section 3.7.4) are interrelated criteria in structure correlation work.

3.7.5.2 Disorder

Given the improving precision of X-ray diffraction measurements and the wide availability of sophisticated structure refinement software, an increasing number of

Table 3.2. Distribution of the AS flag (see text) for ranges of the crystallographic R-factor. AS encodes the mean e.s.d. of a C−C bond length in each structure in the following ranges: AS0 = not available, AS1 = ⩽0.005 Å, AS2 = 0.006−0.010 Å, AS3 = 0.011−0.030 Å, AS4 = >0.030 Å

R-factor (%)	AS0	AS1	AS2	AS3	AS4	Total
a: Organic Compounds						
<3	201	771	305	98	8	1 383
3−4	591	3 207	1 324	312	37	5 471
4−5	949	4 032	2 602	746	69	8 398
5−6	854	2 108	2 639	1 031	87	6 719
6−7	661	858	1 879	1 216	102	4 716
7−8	530	324	1 114	1 049	94	3 111
8−9	428	128	574	859	115	2 104
9−10	337	52	341	637	133	1 500
>10	1 094	55	318	1 051	463	2 981
Total	5 645	11 535	11 096	6 999	1 108	36 383
%	15.5	31.7	30.5	19.2	3.0	
b: Main Group Compounds, Organometallics and Metal Complexes						
<3	462	1 425	1 773	661	33	4 354
3−4	999	1 739	3 464	2 652	170	9 024
4−5	1 270	1 075	3 104	3 665	468	9 582
5−6	1 174	410	1 993	3 484	675	7 736
6−7	931	156	1 001	2 594	719	5 401
7−8	689	66	436	1 626	699	3 516
8−9	492	20	203	957	524	2 196
9−10	294	12	73	545	393	1 317
>10	764	18	56	659	788	2 285
Totals	7 075	4 821	12 103	16 843	4 469	45 411
%	15.6	10.8	26.7	37.1	9.8	
c: All entries a + b						
<3	663	2 196	2 078	759	41	5 737
3−4	1 590	4 946	4 788	2 964	207	14 495
4−5	2 219	5 107	5 706	4 411	537	17 980
5−6	2 028	2 518	4 632	4 515	762	14 455
6−7	1 592	1 014	2 880	3 810	821	10 117
7−8	1 219	390	1 550	2 675	793	6 627
8−9	920	148	777	1 816	639	4 300
9−10	631	64	414	1 182	526	2 817
>10	1 858	73	374	1 710	1 251	5 266
Totals	12 720	16 456	23 199	23 842	5 577	81 794
%	15.6	20.1	28.4	29.1	6.8	

papers now report detailed results for disordered structures. Within the CSD, this problem has been handled in two different ways. In the early development, two situations were recognized: (a) total disorder, and (b) partial disorder of a ring, side chain, etc. In the former case, all coordinates were omitted and a remark inserted in the entry. In the case of partial disorder, only the coordinates for the ordered part of the structure were retained. As structure determinations improved, it became policy to retain the ordered atoms, together with coordinates corresponding to the major disordered atomic sites, i.e. those sites with the largest site occupation factors. Minor site atoms are also retained, but are "suppressed" in the generation of crystallographic connectivity. It remains for the software to catch up with these decisions, and give indications of geometrical variation over the various disordered units. Entries which contain disordered atoms can be avoided completely by use of bit-screen settings in QUEST, or via a text command in GSTAT.

3.7.5.3 Hydrogen Atoms

No distinction is made between H atoms and non-H atoms in generating the crystallographic connectivity (cf. Section 3.7.4.3 above). If hydrogens are to be used in a GSTAT fragment specification, they must be explicitly coded. Further, and perhaps regrettably, the CSD does not contain an indication whether H-atom positions have been observed directly or placed geometrically. Indications of the latter can be gained from an analysis of mean, minimum and maximum $C-H$, $O-H$ and $N-H$ bond lengths in each entry, and such an analysis is presented in the GSTAT output for each entry.

If hydrogens are present, then their geometry may be extremely variable. In particular, an $X-H$ vector may well have the correct direction, but the $X-H$ distance is often shortened in the X-ray results. The GSTAT program is able to "normalize" H-atom positions, by moving the H along the $X-H$ vector so that its bond distance corresponds to the mean $C-H$, $N-H$, $O-H$ value derived from neutron studies. Suitable values for $X-H$ distances ($X \neq C, N, O$) may be input by the user.

3.7.6 Geometric Searching

Our knowledge of the systematics of 3D geometrical structure is vastly less than our knowledge of conventional 2D chemistry. This is only to be expected: 2D chemical conventions are a human invention, whilst the geometrical descriptions are invented by the molecules themselves. Indeed, the purpose of many of the statistical and numerical studies discussed in Chapter 4 is to discover what the molecules *are* telling us about 3D structure, and how this information may be classified in relation to chemical concepts. In setting geometrical selection criteria, we must be sure that we can express the structural concept we wish to locate in geometrical terms, i.e. that we are in possession of the necessary geometrical knowledge. Even then, we may get some surprises. A suitable way of approaching this problem is to perform a preliminary study, in which histograms are generated for the parameters of interest. This visual representation of, for example, a distance distribution, is invaluable in assessing precise selection criteria and, indeed, may well add to (or even alter) our *a priori* assumptions concerning the fragment under study.

3.7.6.1 Bond Length Constraints

The use of bond length constraints within the GSTAT program of Versions 3 and 4 of the CSD system is a common necessity, in order to approximate bond type in fragment selection. This problem is reduced, but not entirely removed, by the rationalization of connectivity representations in Version 5. The considerable body of published knowledge on bond lengths in many chemical environments ([20, 21] and Appendix A to this volume) assists in these decisions. Nevertheless, it is possible that details concerning the precise environment of interest are not specifically stated there, and it may be inappropriate to use a mean value from some "similar" situation. We repeat, the histogram is a simple and invaluable aid in setting constraints of this type.

3.7.6.2 Torsion Angle Constraints: Stereochemical Searching

Some databases of 2D chemical structures contain explicit stereochemical descriptors, involving wedge and broken bonds, and also contain R, S chirality assignments as additional atom properties, incorporated through human effort during database building. Within the CSD, the relationship of the 3D coordinates to the conventional chemical structural description provides all of this information, albeit in numerical terms. Thus, chemical descriptors such as *cis-, trans-, axial, equatorial, gauche, anti-periplanar*, etc., can all be distinguished by suitable use of torsion angle constraints. Because most structures crystallize in space groups that demand the presence of the enantiomeric molecule, i.e. the inverse of that reported in the paper, the CSD software will use both enantiomers (via a coordinate inversion procedure) in performing tests of torsion angles.

Descriptors of chirality, as defined by the rules of Cahn, Ingold and Prelog [27] and various updates and extensions [28], are valuable additional attributes for storage in the CSD. These rules are, however, particularly difficult to apply within a systematic algorithmic framework for computational applications. In the past year, we have completed software that will detect chirality on the basis of ranked connectivity trees emanating from atomic centres of maximum valency 4, and will assign R, S descriptors on the basis of the necessary torsional test. Extensions of this procedure to more complex cases are being developed, and results will gradually be made available in future releases of the CSD system. The topics of torsional constraints, and of chirality, are discussed further in Chapter 4.

3.8 Conclusion

This chapter provides an introductory overview of both the information content of the Cambridge Structural Database, and of the associated software systems. Special considerations in the use of the CSD system are discussed throughout the text, and specifically in Section 3.7. We hope that some of the finer points and potential problem areas are identified and that some misconceptions are explained. The CSD is, however, a developing entity, and areas in which information content and software facilities will be improved and extended are also identified in this chapter. Thus, within the effective lifetime of this volume, considerable further upgrading is envisaged. These facilities will be announced through the CSD User Manuals, newsletters and, where appropriate, through articles in the open literature. Indeed, many of these upgrades result from active research projects in which the CSD is used as the source of experimental data, and which suggest the addition of valuable new search fields to the database. Thus, search and research are intimately connected, not only in the development of the CSD, but also in the development of all of the crystallographic databases.

On a more general note, we hope that readers may be encouraged to learn more about computer methods for the storage, manipulation and searching of chemical databases. These topics are rarely covered at undergraduate or graduate level, yet databases are now vital information sources wherever chemistry is studied. This is particularly true in the chemical industry, where large volumes of in-house information are routinely maintained. One problem is that the literature on the subject is widely spread. However, a perusal of articles and leading references contained in the *Journal of Chemical Information and Computer Sciences* (published by the American Chemical Society) is a good starting point. A number of relevant texts, e.g. those by Gray [29], Warr [30] and Willett [17, 18], are now available in this general area, whilst Wilson [31] and Harary [32] provide formal coverage of the necessary graph-theoretical concepts.

References

[1] Watson, D. G., in: *Crystallographic Databases*, Allen, F. H., Bergerhoff, G., Sievers, R. (eds.), International Union of Crystallography, Chester, **1987**

[2] Pauling, L., *The Nature of the Chemical Bond*, Cornell University Press, Ithaca, **1939**

[3] Sutton, L. E. (ed.) *Tables of Interatomic Distances and Configuration in Molecules and Ions*, The Chemical Society, London, **1958** and **1965**

[4] Allen, F. H., Davies, J. E., Galloy, J. J., Johnson, O. Kennard, O., Macrae, C. F., Mitchell, E. M., Mitchell, G. F., Smith, J. M., Watson, D. G., *J. Chem. Inf. Comput. Sci.*, **1991**, *31*, 187–204

[5] Allen, F. H., Bergerhoff, G., Sievers, R. (eds.), *Crystallographic Databases*, International Union of Crystallography, Chester, **1987**

[6] Bergerhoff, G., Hundt, R., Sievers, R., Brown, I. D., *J. Chem. Inf. Comput. Sci.* **1983**, *23*, 66–69

[7] Bernstein, F. C., Koetzle, T. F., Williams, G. J. B., Meyer, E. F., Brice, M. D., Rodgers, J. R., Kennard, O., Shimanouchi, T., Tasumi, M., *J. Mol. Biol.* **1977**, *122*, 535–542.

[8] Islam, S. A., Sternberg, M. J. E., *Protein Engineering* **1987**, *2*, 431–442

[9] Thornton, J. M., Gardner, S. P., *Trends in Biochemical Science* **1989**, *14*, 300–304

[10] Hall, S. R., Allen, F. H., Brown, I. D., *Acta Cryst.* **1991**, *A 47*, 655–685

[11] Wipke, W. T., Dyott, T. M., *J. Chem. Inf. Comput. Sci.* **1975**, *15*, 140–151

[12] Murray-Rust, P., Raftery, J., *J. Mol. Graph.* **1985**, *3*, 50–59

[13] Murray-Rust, P., Raftery, J., *J. Mol. Graph.* **1985**, *3*, 60–68

[14] Motherwell, W. D. S., *PLUTO: a program for the preparation of crystal and molecular diagrams*, Crytallographic Data Centre, Cambridge, **1978**

[15] *Cambridge Structural Database User Manuals*, Crystallographic Data Centre, Cambridge, **1991**

[16] Willett, P., Winterman, V., Bawden, D., *J. Chem. Inf. Comput. Sci.* **1986**, *26*, 36–41

[17] Willett, P., *Similarity and Clustering in Chemical Information Systems*, Wiley/Research Studies Press, Chichester, **1987**

[18] Willett, P., *Three Dimensional Chemical Structure Handling*, Wiley/Research Studies Press, Chichester, **1991**

[19] Morgan, H. L., *J. Chem. Docum.* **1965**, *5*, 107–113

[20] Allen, F. H., Kennard, O., Watson, D. G., Brammer, L., Orpen, G. A., Taylor, R., *J. Chem. Soc. Perkin II* **1987**, S 1–S 19

[21] Orpen, A. G., Brammer, L., Allen, F. H., Kennard, O., Watson, D. G., Taylor, R., *J. Chem. Soc. Dalton* **1989**, S 1–S 83

[22] Altona, C., Sundaralingam, M., *J. Am. Chem. Soc.* **1972**, *94*, 8205–8212

[23] Cremer, D., Pople, J. A., *J. Am. Chem. Soc.* **1975**, *97*, 1354–1358

[24] Kroon, J., Kanters, J. A., van Duijneveldt-van de Rijt, J. G. M. C., van Duijneveldt, F. B., Vliegenhardt, J. A., *J. Mol. Struct.* **1975**, *24*, 109–129

[25] Taylor, R., Kennard, O., *Acc. Chem. Res.* **1984**, *17*, 320–326

[26] Allen, F. H., Doyle, M. J., *Acta Cryst.* **1987**, *A 44*, C 291

[27] Cahn, R. S., Ingold, C. K., Prelog, V., *Angew. Chem.* **1966**, *78*, 413–427

[28] Prelog, V., Helmchen, G., *Angew. Chem.* **1982**, *21*, 567–583

[29] Gray, N. A. B., *Computer-Assisted Structure Elucidation*, Wiley, Chichester, **1986**

[30] Warr, W. A. (ed), *Chemical Structures: The International Language of Chemistry*, Springer, London, **1988**

[31] Wilson, R. J., *Introduction to Graph Theory*, Longman, London, **1985**

[32] Harary, F., *Graph Theory*, Addison-Wesley, London, **1972**

4 Statistical and Numerical Methods of Data Analysis

Robin Taylor and Frank H. Allen

4.1 Introduction and Objectives

In an earlier paper on the systematic analysis of structural data [1] we wrote:

"There are now (1. 1. 82) more than 30,000 organo-carbon structures in the literature, a number which is likely to double in the next 5 – 7 years. Unfortunately, this wealth of information has not been exploited very much: detailed discussions of individual structures are commonplace, but systematic studies of large numbers of related structures are rarely undertaken. The labour involved in locating and extracting relevant information from the literature has probably been a major barrier to such analyses. However, this barrier has now been virtually eliminated by the establishment of crystallographic databases."

By 1991, this situation had changed very considerably. The size of the Cambridge Structural Database (CSD) did indeed double, to 64,604 entries, by the 1st January 1987. Its current size as of 1st August 1991 stands at 92,644 entries, a threefold increase in less than ten years. Further, the crystallographic databases are now well established, as described in the preceding chapter, and software for the location and retrieval of structural data continues to improve. Most importantly, scientists from a wide range of disciplines are now using this accumulated data in a variety of imaginative ways. The number of systematic studies and structure correlations appearing in the literature is growing rapidly; more than 150 such studies have been carried out using the CSD, and applications of the other crystallographic databases are also increasing. Indeed, it is this growing interest that has provided the impetus for the preparation of this current volume.

Given the continued rapid growth in crystallographic output, a chemical substructure of interest may well occur in several hundreds, if not thousands, of crystal structures. Further, a significant number of parameters (internal coordinates) may be required to provide a suitable definition of the substructure geometry. A key feature of recent systematic studies has been the developing use of statistical and numerical methods to summarize these large datasets in forms that can be readily assimilated, understood and interpreted. These techniques can be used to survey, organize and view the raw data matrix, to recognize patterns and relationships that may occur within it, and to give statistical confidence in the results obtained.

It should be remembered, however, that statistical methods are mathematical manipulations of the input data. These methods are used in many domains and for many purposes and they cannot, of themselves, provide an interpretation of the results. This interactive role is reserved for the investigator, using his or her understanding of the data and its origins. Rather, the importance of statistical methods lies in their ability to help us to think clearly about a problem, to comprehend what the data are telling us, and to provide quantitative guidance in making inferences and deductions. This role is identified in Figure 4.1, which constitutes a

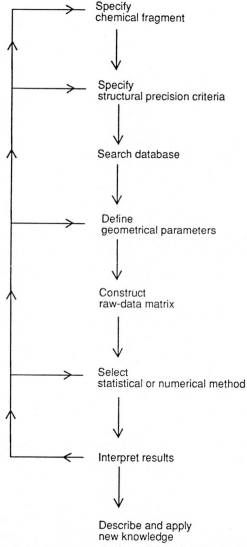

Fig. 4.1. Stages in the acquisition of knowledge from a crystallographic database

procedural "flowchart" for structure correlation studies. The statistical and numerical methods provide the essential link between the raw data extracted from a structural database (the subject matter of Chapter 3), and the knowledge that may be acquired through an interpretation of these data in chemical terms (the subject matter of Chapter 5 and of other chapters in this volume).

To any casual reader who scans a textbook on statistical methods, there appears to be a myriad of techniques that can be used in data analysis. Which ones are likely to prove useful in the chemical context? The answer is that the choice of technique is dictated entirely by the type of data we wish to analyse and the type of problem we wish to solve. A wide variety of statistical methods has already been applied to the analysis of structural data and has generated results of some significance. In this chapter, we discuss these methods under three broad headings:

(a) *Univariate methods.* These methods deal with a single variable, e.g. a specific bond length, torsion angle, etc., for which we have N_f observations, one from each of the chemical fragments of interest. Here we are interested in the distribution of these observations, the modality of the distribution, and a variety of descriptive statistics, such as the mean value, variance, etc., that summarize the distribution.

(b) *Comparison of two distributions.* Here we deal with the significance of the difference between the two mean values, the analysis of variance, simple linear regression, correlation analyses, etc.

(c) *Multivariate statistics.* These methods are used to analyze a complete data matrix, in which each of the N_f fragments is described by N_p parameters, i.e. where each fragment may be regarded as a point in an N_p-dimensional parameter space. Here we deal with methods for dimension reduction, and with methods designed to locate clusters of closely related observations within that space.

We conclude with a brief review of available software systems for statistical analysis, and with some remarks on future possibilities for knowledge acquisition from a large database of chemical structures.

Our approach to this chapter is best described as experimental rather than mathematical. There are many excellent formal texts, some of which will be cited. Here, we wish to summarize the fundamental purpose of each technique, to explore its possible applications and limitations, and to illustrate (or provide references to) its use in systematic studies of molecular structure. Most of the examples have been drawn from the Cambridge Structural Database [2], using the methods of search and retrieval detailed in the preceding chapter. We preface the statistical content with two more general chemical sections: in the first we discuss the selection of geometric parameters that are most appropriate for certain types of analysis, whilst in the second we discuss the possible sources of variation in crystallographic structural data.

4.2 Choice of Parameters for Statistical Analysis

4.2.1 The Coordinate Basis

The primary data recorded in a crystallographic database such as the CSD are the $3N$ fractional coordinates for the N atoms of the crystal chemical unit (see Chapter 3), referred to the unit-cell axes and origin. The opening chapter of this book shows how this primary data can be transformed to Cartesian frames. The resulting external coordinates are generally unsuitable for use in a direct comparison of N_f occurrences of a chemical fragment, since the fragments will have orientational and translational differences, one with another, as well as differences due to chemical or structural effects. However, the orientational and translational differences can be removed (Chapter 1) by analysis of Cartesian coordinates referred to molecular (inertial) axes, and having a common origin at the centre of mass of the fragment. A multivariate data matrix of dimensions $N_f \times 3N$ of these inertial axis coordinates can then be generated. The matrix can be used to generate a graphical superposition of the N_f fragments, and to calculate the similarity of one fragment to another.

There are, however, considerable problems involved in the generation and interpretation of numerical results derived from inertial axis coordinate sets. Firstly, we must use $3N$ descriptors, more than is strictly necessary, to define each N-atom fragment. Secondly, and most important, it is difficult (if not impossible) to interpret differences in the external coordinates of superimposed fragments in basic chemical terms. Essentially, then, it is the arbitrariness and lack of chemical significance of most axial frames that preclude the use of external coordinates for systematic studies: external coordinates are only of use when the axial frame is directly related to the structure of the fragment.

4.2.2 Internal Coordinate Axes

We may increase the chemical relevance of the parameters by describing each fragment in terms of a (constant) set of internal (geometric) coordinates. Intuitively, this is only reasonable: if we wish to acquire chemical knowledge from an analysis, then we must use basic descriptors that are chemically meaningful. In contrast to external coordinates, we may select for analysis only those internal coordinates that describe the structural features of interest; molecular dimensions, atomic configuration, fragment conformation, etc. By choosing N_p internal coordinates in this way, we now represent each fragment as a point in an N_p-dimensional parameter space. Since the same set of N_p parameters are used to describe each of the N_f fragments, the axial frame is the same for each fragment and each axis represents some structural feature of interest. All of the N_f fragments are directly comparable, one with another, and the description is ideally suited to statistical and numerical analysis. We may now (a) study the distribution of values along individual axes using univariate methods, (b) compare the distributions of values along pairs of axes using hypothesis testing, correlation analysis, etc., or (c) examine the full distribution of N_f points in N_p-space by use of multivariate techniques.

The full set of bond lengths, valence angles and torsion angles is regarded as defining the standard internal coordinate axes. These quantities are intimately related to concepts of chemical connectivity and form a natural starting point for systematic studies. With this standard system, we can identify coordinate subsets that are directly associated with specific aspects of 3 D molecular structure. Further, a wide range of "non-standard" internal coordinate axes can be generated from the primary crystallographic data. These can be simple analogues of the standard set, but involving non-bonded vectors in their definition. They can also be more complex quantities, often linear combinations of simpler parameters, such as symmetry-adapted deformation coordinates (Chapter 2), ring puckering coordinates [3], and distances and angles derived from mean plane calculations.

4.2.3 Internal Coordinates and Chemical Fragments

The key question that now remains is: which particular subset of internal coordinates should we select for a particular analysis? Obviously this selection is intimately related to the nature of the chemical fragment (see Chapter 1.1), and to the type of analysis we wish to perform. Here we must be guided by our prior chemical and structural knowledge, and also by our preconceptions of the type of results that we hope to obtain. In practice, we should also be guided by intermediate stages of analysis, which may indicate modifications to the initial selection of parameters, or changes to the specification of the chemical fragment.

In simple cases, for example the derivation of a mean value for a particular bond length, the question of parameter selection does not arise. Instead, the initial results may indicate that it is the chemical environment of the bond that should be modified or varied, by the introduction of chemical constraints or specific substituent types into the basic chemical search. For example, in a study of C(alkyl) − O − C ether and ester systems [4] it was found that the C(alkyl) − O bond length varied over a broad range, from 1.418 Å to 1.475 Å. It was possible to subdivide this distribution by modification of the initial chemical definition of the fragment. Thus, by specifying the second carbon atom to be (a) alkyl, (b) aryl, (c) vinyl, and (d) carbonyl, it was shown that bond lengthening could be correlated with increased electron withdrawing ability of the variable oxygen substituent. Further subdivision on the basis of C(alkyl) as methyl, or primary, secondary or tertiary alkyl, showed that the bond length also increased with increasing degree of substitution.

The relationship between chemical and structural search specifications and parameter selection is an important one. Chapter 3 shows that it is possible to impose geometrical, as well as chemical, constraints on the fragments retrieved in a search.

Thus, in a study of a conjugated system, such as $-C=C-C=O$, we obtain two peaks in the distribution of the central $C-C$ bond length. This can be resolved by constraining the torsion angle (τ) about the central bond (a) to be close to $0°$ or $180°$, and (b) to lie in the range (say) $20 < \tau < 160°$. This use of geometric constraints is explicit: it limits the variation of one of the initially chosen parameters so that we may concentrate on the other(s). More subtly, the purely chemical specifications of a search fragment often place implicit limitations on the variability of some of the standard internal coordinates. We can make use of this to reduce the number of parameters used in the analysis. Thus, if we define a seven-membered ring of $C(sp^3)$ atoms connected by single bonds, then we immediately limit the range of values that can be adopted by the seven bond lengths and the seven valence angles. We may proceed with an analysis of the conformational variations in the rings by using the seven torsion angles as parameters, assuming that geometrical variance in bond lengths and valence angles is minimal.

There is a considerable advantage in trying to minimize the number of internal coordinate descriptors that are used in any analysis. Each additional descriptor adds a further dimension to the coordinate system, and the human brain has difficulty in assimilating and visualizing information in more than three dimensions. The explicit or chemically implied application of constraints, noted above, helps to reduce the number of parameters. Another way is to use some non-standard parameter(s) to replace a greater number of "standard" parameters. Consider the cyclopropyl-carbonyl fragment of Figure 4.2a. Here we are interested in the conformational relationship beween the three-atom ring frame and the carbonyl substituent. This can be measured by the two standard torsion angles $C2-C1-C4-O5$ and $C3-C1-C4-O5$. Instead, it is easier and more informative to use the non-standard torsion angle $X1-C1-C4-O5$, where X1 is the mid-point of the $C2-C3$ bond. In assessing the planarity of three-coordinate N atoms, we could employ all three of the $X-N-X$ valence angles, already a three-dimensional problem. Instead, we may calculate the sum (or mean) of these angles to reduce the assessment to a single variable. These are simple cases, but many analyses are conducted on simple fragments. If we can reduce these to univariate or bivariate problems then we simplify the process of structure correlation considerably.

Despite the indications given above, it will often be impossible to reduce the number of standard parameters significantly, or to visualize suitable linear combinations of parameters that may help in this aim. We are then left with a truly multivariate problem. However, one of the aims of multivariate analysis is the reduction of dimensionality, i.e. the detection of a subset of parameters or, more often, of linear combinations thereof, that best describe the total variance in the data set. In effect, the techniques are trying to detect automatically the set of axes in parameter space that are most useful for visualizing the data. We return to this topic in some depth in Section 4.6.3.

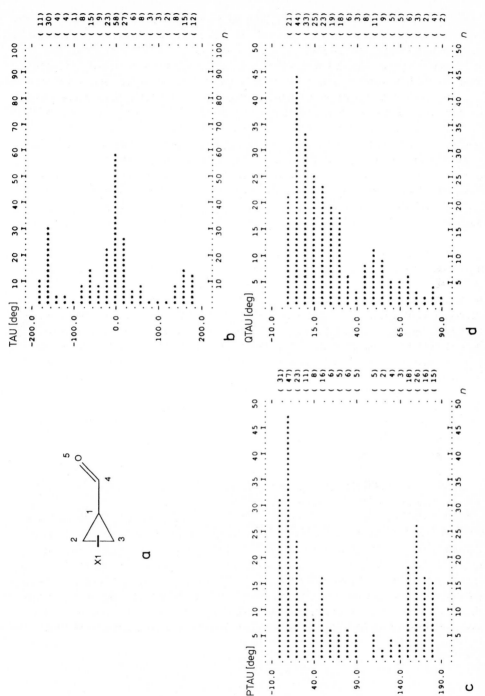

Fig. 4.2. a: Use of non-standard torsion angle TAU = X1−C1−C4 = O5 to define conformation of −C=O with respect to cyclopropane ring. b: Histogram of TAU over full numerical range of −180° to +180°. c: Histogram of |τ| = PTAU. d: Histogram of QTAU in which *cis/trans* conformations are equivalenced to yield a 0−90° angular range

4.2.4 Internal Coordinates and Types of Analysis

The previous section has made general points about the choice of parameters, and how this choice is often closely connected with the chemical specification of the fragment. Here we briefly summarize some parameter sets that are commonly used for certain specific types of analysis.

4.2.4.1 Bond Length Studies

The object of such studies (e.g. [5, 6], see also Appendix A to this volume) is usually to focus on one particular type of bond in a specific chemical environment (see Section 4.2.3). In practice, a rather general fragment is often specified first and some preliminary analysis is required to establish possible chemical and structural sub-classifications of the bond. In these preliminary studies, it is wise to examine structural parameters that are associated with the environment of the bond of interest: lengths of direct substituent bonds, valence angles at the two atoms defining the bond, etc. Most importantly, when there is freedom of rotation about the bond in question, then the relevant torsion angle(s) should also be examined. This list will be reduced as the analysis proceeds.

4.2.4.2 Studies of Coordination Geometry at Atomic Centres

Studies of this type are frequently performed to investigate configurational variations, e.g. at metal centres. The obvious initial parameters are the relevant valence angles $L-M-L$, where L represents a ligating atom. Such studies may then be broken down chemically according to the nature of L. However, it is often more informative to study the deviations of observed coordination geometries from some idealized symmetric form [7, 8, 9], as described in Chapter 2. This requires use of symmetry-adapted deformation coordinates which can readily be calculated, using the CSD System, for instance, as simple linear combinations of standard internal coordinates.

4.2.4.3 Conformational Studies

These are normally carried out using a basis set of torsion angles to define the fragment conformation. For ring systems, the Cremer-Pople puckering parameters [3], generally applicable to any ring size, or those of Altona and Sundaralingam [10] for five-membered rings, may prove useful. Where only a single torsion angle is involved, the use of the absolute numerical value is often sufficient, if we are only interested in the magnitudes of deviations from local symmetry and not in the direction of the deviation. This is exemplified in Figure 4.2c for the cyclopropyl-carbonyl fragment. In some cases, we may even wish to ignore differences between *cis* and *trans* conformations (see Figure 4.2 d) and a further reduction to the $0-90°$ quadrant is then sufficient. When using torsion angles, it is important to remember the phase change at $\pm 180°$ in the Klyne and Prelog definition [11]. This definition has been coded into the CSD software, but other statistical packages may well handle circular measure within a $0-360°$ frame. Finally, when using sets of torsion angles to define fragment shape in a multivariate analysis, we have the option to work in a chiral or in an achiral environment (equivalencing of enantiomers). We return to this topic in Section 4.6.2.

4.2.4.4 Studies of Hydrogen Bonding

In studies of hydrogen bonding, we are interested not only in the dimensions and angles involved in the hydrogen bonded interaction, but also in the direction of approach of the donor, $X-H$, to the lone pairs of the acceptor group. Two additional and important parameters are the elevation of the $X-H$ vector from the putative lone-pair (θ) and the direction of approach of $X-H$ to the lone pair in that plane (ϕ). These parameters are defined [12, 13] using spherical polar constructs as shown in Figure 4.3 for sp^2 hybridized lone pairs. These parameters may be calculated by the CSD software package (Chapter 3).

4.3 Sources of Variation in Crystallographic Structural Data

Most crystallographers have been asked the following question: how do you know that the conformation you observe in the crystal structure is the same as the solution conformation? Although the question is obviously flawed (there is no single solution

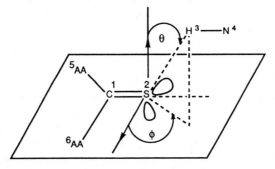

Fig. 4.3. The spherical polar angles θ, ϕ used to define the direction of approach of an $N-H$ donor to the lone pairs on an $(AA)_2-C=S$ acceptor group

conformation for anything other than a completely rigid molecule), it serves to highlight an important fact: in any given crystal structure, the geometry of each molecular entity is perturbed by the forces exerted on it by surrounding molecules in the crystal. These crystal-packing forces have traditionally been regarded as something of a nuisance, since they generally preclude direct observation of global minimum-energy geometries. However, the structure-correlation approach requires us to look at crystal-packing forces in a new light: far from being a nuisance, they are distinctly useful. This is because they enable us to measure the geometries of a given molecular fragment in a wide variety of environments. If we make the reasonable, though qualitative [14], assumption that the resulting observations tend to congregate in the low-energy regions of conformational space, we can see that crystal-packing forces allow us to deduce information about a variety of low-lying points on the potential-energy surface, rather than just the global minimum.

The influence of environmental effects can be formalized as follows: Suppose that we measure the value of a particular molecular parameter (say, the $O-N-O$ valence angle in nitro groups) in a series of crystal structures. The values observed in the ith structure, x_i, can be expressed as:

$$x_i = \mu_i + \varepsilon_i \tag{4.1}$$

where μ_i is the true value of the bond angle in this structure and ε_i is the experimental error in our measurement. Importantly,

$$\mu_i \neq \mu_j , \quad i \neq j \tag{4.2}$$

because the environment of the nitro group will differ in the ith and jth structures. Variation in the x_i is therefore due to two major effects: experimental errors, and crystal-field environmental effects [15]. The statistical consequences are explored in the next Section.

In Equation (4.1), ε_i represents the *random* experimental error, which is nominally represented by the e.s.d derived from the least-squares covariance matrix. However, there are a number of *systematic* effects that may also influence the observed x_i. These are usually ignored in structure correlation and related studies, but it is as well at least to bear them in mind. Firstly, thermal motion produces systematic shortening of bond lengths, which can be as large as 0.04 Å, even at 23 K [16]. Secondly, crystal-packing forces can introduce a systematic bias into crystallographic measurements. For example, the $O \ldots O$ hydrogen-bond distance in ice I is 2.75 Å [17], compared with an equilibrium $O \ldots O$ separation of 2.98 Å in the gas-phase water dimer [18]. The difference is due to the compressive effect of the crystal lattice and to cooperative effects that occur in the solid but not in the gas phase. As another example, it has been shown [19] that nearly planar biphenyl fragments occur more often in the crystalline state than would be expected from the torsional potential around the $C-C$ single bond of biphenyl *in vacuo*. Presumably, this is because planar molecules pack favourably. Thirdly, we cannot always assume that the available set of x_i constitutes a random sample, since crystal structures are not determined at random. For example, many structures have been determined by neutron diffraction because they contain short, possibly symmetrical $O-H\ldots O$ hydrogen bonds. An estimate of the mean $H\ldots O$ distance based on all available neutron data is therefore likely to be biased towards short values.

4.4 Mean Values and Other Simple Descriptive Statistics

4.4.1 Characteristics of Distributions

Figure 4.4 shows a histogram of $C-C$ bond lengths in $C(ar)-CN$ fragments [5]. The histogram conveys three types of information. Firstly, it tells us something about the central tendency of the distribution (the bond lengths are centred around a value of about 1.445 Å). Secondly, it conveys information about spread, or dispersion (individual observations vary between about 1.42 and 1.46 Å). Thirdly, it shows the shape of the distribution (approximately bell-shaped, but perceptibly skewed). In this section, we discuss how these three different types of information can be described numerically.

4.4.2 Means of Normal or Near-Normal Distributions

The usual estimate of central tendency is the arithmetic mean. Extensive tabulations of mean bond lengths are available ([5, 6], see also Appendix A to this volume). These are useful for model building, and have a great deal of inherent interest as well. It is therefore worth considering how a mean molecular dimension may best be estimated. Taylor and Kennard [20], using standard statistical methodology [21], have shown that the best estimate of a mean dimension depends crucially on the relative importance of experimental errors and crystal-field environmental effects.

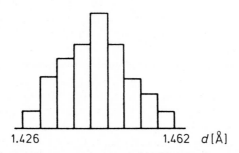

Fig. 4.4. Histogram of C(aromatic)−CN bond lengths after outlier removal

This can be illustrated by a pair of hypothetical examples [22]. Suppose that we have four observations of a covalent bond length; let them have e.s.d.'s of 0.002, 0.008, 0.009 and 0.010 Å, respectively, and assume, for the moment, that these e.s.d.'s are reliable. Now, most bond lengths are "hard" parameters, i.e. considerable energy is required to distort them from their equilibrium values. They are therefore relatively insensitive to changes in the crystal-field environment. Thus, it is likely that the true value of the ith bond length (i.e. its value in the ith crystal structure, if we could measure it without experimental error; μ_i in Equation 4.1), is virtually identical to that of the jth. Differences between the μ_i may therefore be assumed negligible compared with experimental errors, and (4.1) may be simplified to:

$$x_i = \mu + \varepsilon_i \tag{4.3}$$

where μ is now considered constant across all crystal structures. If the experimental errors, ε_i, are normally distributed, there is a 99% chance that the true value of the quantity, μ, lies within about 3 standard deviations of each observation, i.e. between ± 0.006 Å of the first, ± 0.024 Å of the second, etc. These confidence ranges suggest that the first observation gives us more information about the value of μ than all the rest put together. Accordingly, the mean is best estimated by weighting each observation by the reciprocal of its estimated variance:

$$\bar{x}_w = \sum_{i=1}^{n} [x_i/\sigma_i^2]/ \sum [1/\sigma_i^2] \tag{4.4}$$

where $\bar{x}_w$ = the weighted mean, σ_i = e.s.d. of x_i, n = number of observations. Note that in our hypothetical example the value of $\bar{x}_w$ is almost completely determined by the first, most precise, observation.

Now suppose that the molecular dimension is not a covalent bond length but a hydrogen-bond distance. This is a "soft" parameter, i.e. easily distorted by environmental effects. Thus, differences between the μ_i are large compared with the ε_i, and (4.1) cannot be simplified to (4.3). This is illustrated in Figure 4.5. The solid curve

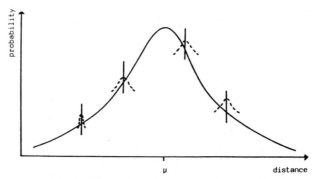

Fig. 4.5. Schematic drawing of environmental (solid curve) and experimental (broken curves) distributions for the hypothetical example discussed in the text

represents the probability distribution of the hydrogen-bond distance over all possible environments, the four vertical lines indicate individual observations, and the broken curves show the experimental uncertainties associated with these observations. Effectively, we now seek to estimate the average value of the hydrogen-bond distance over all possible environments, i.e. the mean of all possible μ_i. Since the broken curves are much narrower than the solid curve, the differences in experimental precision are unimportant: the information content of the least precise observation is almost as high as that of the most precise. If we were to use the weighted mean (4.4), we would simply be throwing away approximately three quarters of our data set. In this situation, an unweighted mean is preferable:

$$\bar{x}_u = \sum_{i=1}^{n} x_i/n \ .$$
(4.5)

In summary, the best estimate of the mean depends on the importance of environmental effects compared with experimental uncertainties.

In principle, the importance of environmental effects can be inferred by computing the quantity:

$$\chi^2 = \sum_{i=1}^{n} [x_i - \bar{x}_w]^2/\sigma(x_i)^2 \ .$$
(4.6)

This should follow a χ^2 distribution with $(n-1)$ degrees of freedom if environmental effects are negligible [21]. A significantly large χ^2 value therefore suggests that environmental effects are important and that the unweighted mean should be used. The test assumes that the x_i are normally distributed, and this assumption could be invalidated by gross departures from normality.

A practical difficulty arises in the use of formulae such as (4.4) and (4.6): both involve the e.s.d.'s of the x_i, which are, of course, derived from the least-squares covariance matrix. The accuracy of these estimates is therefore a matter of some importance. Ideally, we would like $\sigma(x_i)$ to be an unbiased estimate of the standard deviation of ε_i in (4.1). Unfortunately, there is a wealth of evidence to suggest that this is not the case. A study by the International Union of Crystallography, involving the collection of seventeen independent X-ray data sets from tartaric acid, indicated that atomic-coordinate e.s.d.'s are typically too small by a factor of about 1.4 [23, 24, 25]. Substantially the same conclusion was reached in a study [26] of one hundred crystal structures, each of which had been determined independently by two different research groups. A pragmatic approach is to increase the estimated $\sigma(x_i)$ by 50% and use these "corrected" estimates in (4.4) and (4.6).

However the mean is calculated, it must be remembered that it is only an *estimate* based on a limited number of observations. The uncertainty of $\bar{x}_u$ (i.e. the "standard error" of $\bar{x}_u$) is given by:

$$\sigma(\bar{x}_u) = \left\{ \sum_{i=1}^{n} [x_i - \bar{x}_u]^2/n(n-1) \right\}^{1/2} \ .$$
(4.7)

The standard error of $\bar{x}_w$ is more problematic, since it depends on the e.s.d.'s of the x_i, which, as already noted, tend to be of dubious reliability. In consequence, standard errors of weighted means can be grossly over-optimistic [27] and it is undesirable to use weighted means in hypothesis testing.

So far, the discussion has been confined to the estimation of means for well-behaved (i.e. normal or near-normal) distributions. Difficulties arise for more complex distributions, and these are considered in Section 4.4.5.

4.4.3 Dispersion

The simplest measure of the "spread" of a distribution is the range, which is the difference between the largest and smallest observations. In practice, this statistic is almost never used, mainly because it is very vulnerable to the presence of outliers. The usual measure is the sample standard deviation:

$$\sigma_{sample} = \left\{ \sum_{i=1}^{n} [x_i - \bar{x}_u]^2 / n - 1 \right\}^{1/2} . \tag{4.8}$$

Roughly 95% of the observations in a normal distribution lie within $2\sigma_{sample}$ of the mean.

It is worth emphasising the difference between (4.8) and (4.7): the quantity given by (4.8) measures the amount of variation between the observations in the sample; that given by (4.7) is an estimate of how precisely the mean of the distribution has been estimated. For very big samples, i.e. large n, it is possible to estimate the mean with great precision, even though the sample standard deviation may be large. In consequence, if enough data are available, it may be possible to show that the means of two distributions are significantly different, even though the distributions themselves overlap appreciably. We return to this point in Section 4.5.

4.4.4 Departures from Normality

Distributions of molecular dimensions show three common types of departure from normality. The first is illustrated by Figure 4.6, which shows the observed distribution of B−F bond lengths in tetrafluoroborate ions [5]. The histogram is clearly asymmetric, or "skewed". This is presumably a thermal-motion effect, but skewed distributions may also arise for reasons of structural chemistry. For example, the distribution of H...O hydrogen-bond distances in a sample of about 1300 N−H ...O=C bonds was found [28] to be skewed in the opposite direction to that of Figure 4.6. This is presumably due to the fact that the energy of a hydrogen bond

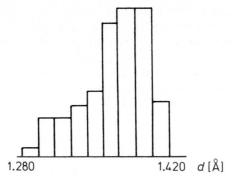

Fig. 4.6. Histogram of B−F bond lengths in BF_4^- ions

varies asymmetrically with distance. The energy increases steeply at short H...O separations because of steric repulsion between the donor and acceptor moieties. In contrast, the donor-acceptor interaction approaches zero asymptotically as the H...O separation is increased. This asymmetry in the potential energy vs. distance curve is reflected in the skewed distribution of H...O distances.

The second common departure from normality is the presence of outliers. Thus, the distribution shown in Figure 4.4 has been doctored somewhat; the original histogram was as shown in Figure 4.7, and clearly contains a rogue observation. Such outliers may arise from gross experimental error (quite common, if one of the structures in the data set is of much lower precision than the rest) or for chemical reasons.

The third common type of non-normality is multimodality, i.e. the presence of two or more peaks in the distribution. This commonly occurs in a distribution of torsion angles, since torsional potential-energy profiles frequently contain several minima (e.g. corresponding to *gauche* and *anti* conformations).

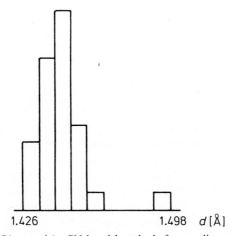

Fig. 4.7. Histogram of C(aromatic)−CN bond lengths before outlier removal

A number of statistical techniques exist for assessing whether a given distribution departs from normality. Historically, the most commonly used method in crystallography is the normal probability plot [29, 30]. In this procedure, the n observations in a sample are normalised to zero mean and unit variance, and then ranked. The resulting ranked quantities are plotted against the order statistics expected for a random sample of size n taken from the standard normal distribution. Departures from a straight-line plot indicate non-normality. Various type of departure (e.g. bowed lines or S-shaped curves) are characteristic of particular sorts of deviations from normality.

Although commonly used, normal probability plots are of limited value because they do not give a numerical estimate of the significance of the observed non-normality of a sample. Such an estimate can be obtained from a variety of hypothesis tests that are generally referred to as "goodness-of-fit" tests. Some of these are used to examine particular types of departure from normality. For example, calculation of the third moment about the mean gives a statistic which can be used to test whether a distribution is significantly skewed [31]. This statistic, the coefficient of skewness, has an expectation value of zero for a normal (or, indeed, for any symmetric) distribution. In the hydrogen-bond example quoted above, the coefficient was 0.80, which conclusively established skewness.

Other goodness-of-fit tests are "general purpose", being designed to detect any sort of departure from normality. The most commonly used tests are the Kolmogorov-Smirnov and the χ^2 tests [32]. The Kolmogorov-Smirnov test involves determination of the largest absolute deviation between the observed cumulative distribution and the cumulative distribution that would theoretically be obtained under the assumption of normality. Obviously, large values of the Kolmogorov-Smirnov statistic indicate significant non-normality. In the χ^2 test, a count is made of the number of observations falling in various ranges (e.g. for the $O-N-O$ valence angle in nitro groups, we might choose the ranges: $<118; 118-119; 119-120; 120-121;$ $>121°$). The number expected in each range is then computed, assuming normality. The χ^2 statistic is then:

$$\chi^2 = \sum_{i=1}^{r} [O_i - E_i]^2 / E_i \tag{4.9}$$

where r = number of ranges, O_i = number of observations in ith range, E_i = number expected in ih range, and there are $r-1$ degrees of freedrom. Again, large values of the statistic indicate non-normality.

In fact, these tests are more versatile than the above discussion suggests, because they can be used to compare an observed sample with any sort of mathematical distribution, not just the normal distribution. For example, the χ^2 test has been used to show that the distribution of hydrogen bonds around carbonyl oxygen atoms in organic crystal structures is significantly non-uniform [33]; in fact, the hydrogen bonds tend to fall along, or close to, the directions of the acceptor oxygen lone pairs. Here, therefore, the theoretical distribution used in the test was uniform rather than normal. In a similar vein, a binomial test [32] was used to show that the nitrogen atom of a nitro group prefers to form short non-bonded contacts to electronegative rather than electropositive atoms [34].

Automatic detection of outliers is extremely difficult. In principle, an observation that lies more than, say, three or four standard deviations from the mean of a normal or near-normal distribution is likely to be an outlier (i.e. not a *bona fide* member of the population to which the other observations belong). Unfortunately, the standard deviation is usually estimated from the sample, and the presence of the rogue observation(s) tends to inflate the estimate. For example, the sample standard deviation of the data set shown in Figure 4.7 is 0.012 Å; this drops to 0.008 Å if the outlier is removed. Outliers may therefore appear to lie within three standard deviations of the mean simply because the value being used for the standard deviation is too large. A variety of techniques has been suggested for outlier detection [35, 36], but, in structure-correlation studies, there is probably no substitute for visual inspection of histograms. In any case, outliers in crystal-structure data are often interesting and are therefore worth seeking out. Many of the same comments apply to multimodality.

4.4.5 Estimates of Central Tendency for Non-normal Distributions

Arithmetic means are not very suitable as estimates of the central location of highly skewed distributions. They are, of course, biased in the direction of the long tail and do not occur in the position that seems natural to most people. The sample median is the most commonly used alternative; this is the value that divides the sample into two halves, 50% of the observations lying above it and 50% below. In a similar vein, the dispersion of skewed distributions can usefully be expressed by the upper and lower quartiles. These are defined such that 25% of the observations in a sample lie below its lower quartile and 25% above its upper quartile. Typically, the upper and lower quartiles will not be symmetrically disposed about the mean if a distribution is highly skewed. Medians are also useful for defining the central location of a sample containing outliers, being much less sensitive to their presence than the mean. Another useful alternative is the trimmed mean. This is calculated by rejecting the *m* largest and *m* smallest observations, and averaging the rest. The value of *m* is under the user's control. Again, the trimmed mean is much less affected by outliers than the mean of the complete sample.

Statistics such as the median and the trimmed mean are variously described as robust (i.e. suitable for use with a wide variety of population types) and/or resistant to outliers. Traditionally, robust and resistant statistics have been unpopular in classical statistics because it is often impossible to derive an analytical expression for the precision with which they can be estimated (i.e. formulae analogous to (4.7) above). This made it difficult to use the estimates in hypothesis tests. However, the advent of fast computers has radically altered the situation, since estimates of the precision of almost any *ad hoc* statistic can now be obtained by simulation techniques [37]. Consequently, robust and resistant statistics such as the trimmed mean may be expected to become increasingly popular.

Finally in this section, we turn to a problem of particular importance to structural chemists – the estimation of the mean of a torsion-angle distribution. It has already been mentioned that these distributions are often multimodal, which inevitably causes difficulties. Moreover, an additional problem exists, even for unimodal torsion-angle distributions. As is well known, torsion angles are conventionally expressed in the range $-180°$ to $+180°$ [11]. The phase discontinuity at $180°$ must obviously be handled properly: to crystallographers, the mean of $-179°$ and $+179°$ is $180°$, not zero! Allen and Johnson [38] have devised an algorithm which deals with this problem by temporarily re-defining the angular range in which the torsion angles are expressed. Thus, if the conventional range $-180°$ to $+180°$ is replaced by $0°$ to $360°$, then the two torsions mentioned above $(-179°, +179°)$ become $(+181°, +179°)$ and can be averaged correctly. The result is then re-expressed in the conventional range.

Torsion angles are, however, properly represented as a circular distribution and can be treated by the established formalisms of circular statistics. This is a branch of statistics much used for examining the spatial distribution of features in landscapes [39]. In this formalism, each torsion angle in a sample is regarded as a vector. Thus, for example, a torsion angle of $+90°$ would be represented by a unit vector pointing to 3 o'clock, and a torsion angle of $180°$ would be represented by a unit vector at 6 o'clock. A circular mean, or "preferred direction", can then be obtained by calculating the direction in which the vector sum of the torsion angles lies. The magnitude of this sum gives information about the variance of the torsion angles: for samples of the same size, the magnitude will tend to be larger for torsion-angle distributions with small variance. Some sampling theory exists which enables confidence limits to be estimated for the circular mean from this magnitude [39].

4.5 Comparison of Distributions

4.5.1 Introduction

Frequently, the structural chemist will wish to test hypotheses about the differences between, or the inter-relationships of, two or more different molecular parameters. For example: do amides form stronger (i.e. shorter) hydrogen bonds than ketones? Are the $C-N$ and $C-O$ bond lengths in ureas inversely correlated? Are the internal valence angles of benzene rings systematically affected by the nature of exocyclic substituents? Are acidic $C-H$ groups significantly more likely to form short non-bonded contacts to oxygen atoms than less acidic $C-H$ groups? These questions are generally handled by the simple, classical methods of statistics: t-tests, calculation of correlation coefficients, simple linear regressions, etc. Some examples will be given below, and many standard texts may be found on the subject; the exposition by Snedecor and Cochran [31] is particularly lucid.

4.5.2 Significance of Differences between Means

Perhaps the most common requirement is to determine whether the means of two samples are significantly different. For example, theoretical considerations suggest that there is a cooperative phenomenon in hydrogen bonding, such that the individual hydrogen bonds in long ... O−H ... O−H ... O−H ... chains tend to be stronger than isolated O−H ... O bonds [40]. This phenomenon was investigated [41] by dividing a set of 74 neutron diffraction observations of hydrogen bonds into two samples, one comprising O−H ... O bonds that formed part of a chain, and the other comprising isolated O−H ... O bonds. Calculation of the mean O ... H distances of these samples gave results of 1.805 and 1.869 Å, respectively. The statistical significance of the difference between these means was estimated by using the unpaired t-test [31], which confirmed the theoretical hypothesis at the 99.5% confidence level.

The unpaired t-test is an example of a parametric method, which means that it is based on the assumption that the two samples are taken from normal, or approximately normal distributions. Generally, parametric tests should be used where possible because they are more "powerful" (effectively, more sensitive) than the alternative non-parametric methods [32]. However, significance levels obtained from parametric tests may be inaccurate, and the true power of the test may decrease, if the assumption of normality is poor. The non-parametric alternative to the unpaired t-test is the Mann-Whitney test [32]. In this test, a rank is assigned to each observation (1 = smallest, 2 = next smallest, etc.), and the test statistic is computed from these ranks. Obviously, the test is less sensitive to departures from normality, such as the presence of outliers, since, for example, the rank assigned to the smallest observation will always be 1, no matter how small that observation is.

Tests such as the t-test are useful because they can give the research worker confidence that a real difference in means exists, even if two samples overlap appreciably, as happens frequently. For example, Desiraju [42] calculated that the mean C ... O distance for 105 C−H ... O hydrogen bonds involving alkyne donors was 3.46 Å, compared with a mean of 3.64 Å for a sample of 622 alkene C−H ... O interactions. The difference in means is in the direction expected, given that the C−H ... O interaction is primarily electrostatic (H[δ+] ... O[δ−]) and alkyne C−H groups are more acidic than alkene C−H ones. However, the actual C ... O distributions, when plotted together as histograms, show very considerable overlap. A hypothesis test here gives extra confidence that the difference in means is not only as expected, but is genuine. Significant results can be obtained in this situation because there is a large amount of data available (i.e. n is large in (4.7)), so the mean of each sample can be estimated very precisely.

In the studies described immediately above, the samples are unpaired. This means that there is not a one-to-one correspondence between the observations in one sample and those in the other. Occasionally, such a pairing exists. For example, we might search the CSD for crystal structures that have been determined both by X-ray and neutron diffraction [43]. This will give rise to paired samples, since each observation from an X-ray study will have a counterpart from the corresponding neutron study. In comparing paired samples, the tests mentioned above (unpaired t-test, Mann-Whitney) can be replaced by their paired equivalents (paired t-test, see [31]; Wilcoxon test, see [32]). When applicable, paired tests are more effective than unpaired tests, provided that the pairs of observations (x_i, y_i) are positively correlated (i.e. x_i tends to be large when y_i is large, and *vice versa*). This makes the variance of $(x_i - y_i)$ less than the sum of their individual variances, and paired tests exploit this property.

Occasionally, we may wish to compare several different mean values. For example, a sample of $N-H \ldots O=C$ hydrogen bonds was divided [28] into various sub-groups, depending on the precise nature of the donor and acceptor. The mean $H \ldots O$ distance of each sub-group was then computed: for example, this mean is 1.916 Å when the donor is an unsubstituted ammonium ion and the acceptor is an unionized carboxylic acid. The question arises: do the mean $H \ldots O$ distances vary significantly across the sub-groups? The appropriate statistical technique is analysis of variance. Essentially, this partitions the variance of the complete sample into two parts, the variance that occurs within the sub-groups, and the variance between sub-groups. The larger the latter is relative to the former, the more significant is the difference between the means. In the case cited above, the variance between sub-groups was very significant compared to that within sub-groups, implying that hydrogen-bond distances are dependent on the precise nature of the donor and acceptor groups.

In principle, an alternative way of handling the above problem would be to conduct all possible pairwise comparisons between the sub-groups, using the t-test. This is generally held to be a less desirable method, because it involves performing a large number of hypothesis tests, rather than just one. Apart from the extra computational overhead, the gratuitous proliferation of tests is bad practice. Remember that a value of t that is significant at the 95% confidence level has a one in twenty probability of occurring by chance (i.e. in the absence of any underlying physical cause). The more tests that are performed, the more likely it is that spurious "significant" results will be obtained. There has been some attention given to how a set of genuinely independent tests of significance can be combined to give a single overall probability [44, 45, 46].

4.5.3 Covariance, Correlation and Regression

The extent to which two parameters are related can be measured in several ways. The covariance:

$$\text{Cov}(x,y) = \sum_{i=1}^{n} [(x_i - \bar{x}_u)(y_i - \bar{y}_u)]/n \tag{4.10}$$

is the product moment of the two variables about their respective means. If two variables are linearly related, then Cov (x,y) will have a positive or negative value, depending on the direction of slope of the x,y-relationship. The actual size of the coefficient, however, is difficult to interpret since it depends upon the units in which the two variables are expressed. For this reason, the covariance is little used in descriptive statistics and Cov (x,y) is usually normalized, through x and y, to yield the correlation coefficient:

$$r = \frac{\sum_{i=1}^{n} [(x_i - \bar{x}_u)(y_i - \bar{y}_u)]}{\left\{ \left[\sum_{i=1}^{n} (x_i - \bar{x}_u)^2 \right] \left[\sum_{i=1}^{n} (y_i - \bar{y}_u)^2 \right] \right\}^{1/2}} \tag{4.11}$$

which varies between -1 and $+1$, these values indicating perfect negative and positive correlation, respectively. Its distribution under random sampling is well known, so it can be used to determine whether two parameters are significantly correlated. For example, Blessing [47] calculated that the value of r for the $C-N$ and $C-O$ bond lengths in 114 ureas and ureido rings was -0.83, which shows that these two parameters are significantly and inversely correlated. The non-parametric alternative to r is the Spearman correlation coefficient, which is based on ranks [32].

A distinction should be made between a significant and a strong correlation. In the study of ureas mentioned above, Blessing also calculated the value of r for the $C-O$ and $C-N$ bond lengths in a sample of 78 peptides collated by Benedetti [48]. This time, the correlation coefficient was only -0.28. Whereas a scatterplot of the urea $C-N$ and $C-O$ data showed an obvious linear relationship, that of the peptide data looked more like a Jackson Pollock painting. Nevertheless, a value of $r = -0.28$ is still highly significant for a sample of size 78. Basically, this means that we can be confident that there is some underlying relationship between $C-N$ and $C-O$ bond lengths in peptides, but the relationship is not strong enough to be of any real value in estimating one from the other.

Estimation of one parameter (the dependent variable) from another with which it is strongly correlated (the independent variable) is commonly done by use of simple linear regression. Thus, Blessing derived the predictive relationship $y = 2.869 - 1.206x$ for estimating the $C-O$ distance in ureas and ureido rings (y) from the $C-N$ distance (x). Amongst other examples of the use of linear regression in structural chemistry, a particularly interesting one focussed on the prediction of valence-angle deformations in benzene rings as a result of various types of substitution [49].

Linear regression becomes increasingly difficult as the number of independent variables increases. Thus, difficulties often arise in determining which independent variables are worth including in the predictive equation, and which not. Methods such as stepwise multiple regression are designed to search for the best subset, but they are prone to produce chance correlations because of the large number of ways in which subsets can typically be chosen. This problem is notorious in other areas of research, notably in the study of quantitative structure-activity relationships [50, 51].

4.5.4 Comparison of Ratios

Finally in this section, we note that a number of tests exist for estimating the significance of the difference between ratios. These include the χ^2 and Fisher tests, both of which are described in standard texts. They have occasional use in structure correlation studies. For example, a χ^2 test was used to show that the distribution of non-bonded contacts around nitro groups in organic crystal structures is significantly anisotropic [34].

4.6 Multivariate Statistics

4.6.1 Introduction

It is frequently necessary to use three or more parameters to define the geometry of a chemical fragment. Obvious examples are the n intra-annular torsion angles used to define the conformation of an n-membered ring, or the $n(n-1)/2$ valence angles used to define the coordination geometry about an atom of ligancy n. In the general case, we retrieve N_f examples of the fragment from the CSD, and we define the same N_p parameters to describe the geometry of each example. Hence, we define a matrix of N_f rows and N_p columns, referred to here as the data matrix $G(N_f, N_p)$. This matrix may correspond exactly to a geometrical tabulation generated by the CSD system (see Chapter 3). Alternatively, the contents of the G-matrix can be derived from a more extensive tabulation by selection of those N_f rows and N_p columns which are to be considered in a specific analysis. This selection procedure is available in the CSD software.

Of course, many of the analyses described in Sections 4.4 and 4.5 were performed on geometry tables containing more than the one or the two parameters discussed in those sections. These parameters were chosen as the focus of specific interest in a particular fragment, perhaps because we knew *a priori* that they were the only significant variables. The need for multivariate analysis arises in those frequent cases where we do not have that *a priori* knowledge. The set of parameters chosen to describe the fragments are then said to "arise on an equal footing", as descriptors of the fragment. A variety of techniques are available to analyze the G-matrix as a whole, rather than through selection of individual parameters or pairs of parameters. The purpose of multivariate techniques is to explore and simplify the data by use of two basically different approaches. The *variable-directed* approach is used to analyze and report any interdependence between the variables and to examine whether the variances in the data can be better explained on the basis of a smaller set of new variables. The primary aim of *object(fragment)-directed* approaches is that of classification and pattern recognition through the identification of clusters of points that are in close mutual proximity.

The fth row vector of $G(N_f, N_p)$ defines the position of the fth fragment in an N_p-dimensional parameter space (in the same way that the x, y, z-coordinates define the position of an atom in 3-space). Conversely, the column vectors can be regarded as points in an N_f-dimensional fragment space. The interdependence between the N_p variables can be studied through analysis of their covariances or correlations (see Section 4.5.3), expressed as $(N_p \times N_p)$ square symmetric matrices $Cov(N_p, N_p)$ or $R(N_p, N_p)$. The standardized correlations are normally used (see Section 4.5.3), and it can be shown that an individual correlation coefficient (r, Equation 4.11) is the cosine of the angle between two standardized, N_f-dimensional column vectors. In general, these column vectors will be non-orthogonal, indicating some degree of correlation between pairs of parameters. The most common variable-directed technique, principal component analysis, attempts to describe the N_f fragments in terms of a new set of uncorrelated variables or "principal components", as described in Section 4.6.3.

We may study the relationship between fragments in N_p-space by comparing the row vectors that define the point positions of each fragment. Here we calculate a "distance" or "dissimilarity" between each pair of fragments as a measure of mutual proximity in N_p-space. Hence, we may generate a square $(N_f \times N_f)$ matrix that contains all possible pairwise distances. This matrix, $D(N_f, N_f)$ is, of course, symmetric about a zero diagonal. The object(fragment)-directed techniques, based on analysis of $D(N_f, N_f)$, are designed to examine the structure of N_p-space in terms of its configuration, or in terms of cluster formation amongst the N_f fragments. The methods of multidimensional scaling and cluster analysis are described in Section 4.6.4.

The methods to be described below are mutually complementary. Results from principal component analyses and from multidimensional scaling are usually presented graphically whilst, although the process of clustering can be followed graphically, the results of a cluster analysis are presented numerically. These numerical results can be used to derive mean geometrical parameters for each subset of fragments that are found to have "similar" geometry, i.e. to fall within the same cluster. Some considerations involved in the calculation and reporting of mean fragment geometries are discussed in Section 4.6.5.

We begin the discussion of multivariate methods, however, by returning to the topic of symmetry in the chemical fragment, introduced in Chapter 2. The presence of symmetry in the 2D topological representation of the fragment has implications for the structure of the data matrix $G(N_f, N_p)$ that are related to the symmetry of the underlying parameter space, e.g. a valence-angle space or a conformational space. Fragment symmetry must be recognized and handled correctly throughout any multivariate analysis. Further, in 3D we must also exercise care when using parameters such as torsion angle sets, which define a particular enantiomorph of the fragment.

4.6.2 Fragment Symmetry and Chirality

Relatively few molecular structures in the CSD exhibit molecular symmetry. However, many substructural fragments of interest in structure correlation studies are small and symmetric. This fact is recognized in Chapter 2, and various aspects of fragment symmetry are discussed there. We now examine the consequences of fragment symmetry on the search process itself and, hence, on the relative ordering of the N_p geometrical parameters recorded for each fragment in the multivariate data matrix $G(N_f, N_p)$.

Consider the cyclohexane fragment of Figure 4.8. It is specified as a 2D chemical connectivity query (Chapter 3) with the six carbon atoms enumerated in one way or another, perhaps cyclically, as shown. The 2D fragment has plane symmetry D_{6h}. Substructure searching in any chemical database system proceeds by an atom-by-atom, bond-by-bond matching of the query graph against a succession of graphs representing complete molecules (targets). Mathematically we are looking for subgraph isomorphisms within these targets to establish database hits. For cyclohexane, atom 1 of the query may be mapped (equivalently) onto any of the atoms of a cyclohexane ring in a given target. If atom 1 (query) maps to atom 3 (target), then atom 2 of the query may map either to atom 2 or to atom 4 of the target. This gives rise to 12 independent, and totally equivalent, mappings of the query to the target. The fragment search routines in the CSD System will arbitrarily choose one of these. Essentially, then, there are 12 possible, and topologically equivalent, enumerations of the atoms in the fragment: there is no way of imposing a unique fragment enumeration, as there always is when dealing with an asymmetric fragment.

Imagine now that we wish to define the conformation of each cyclohexane ring using the cyclically ordered set of torsion angles ($\tau 1$ to $\tau 6$) of Figure 4.8. We now examine the effect of choosing at random one of the 12 possible atomic enumerations of the fragment on the ordering of these torsion angles. Remember that each fragment is selected by mapping 2D chemical connectivity representations of the query onto the targets. However, for each fragment located in a target there are a set of associated 3D coordinates derived from the X-ray analysis. Only in the special case of a planar ring (all torsion angles equal to zero) is the D_{6h} symmetry preserved, in which case no problems arise. In more general cases, the 3D symmetry is lower than D_{6h}, e.g. C_{2v} (boats), D_{3d} (chairs), etc. Here the alternative enumerations obviously give rise to alternative orderings of the cyclic sequence of torsion angles. A small section of a raw data matrix for boat conformations [52] is shown in Table 4.1a, and illustrates the misalignments that result from alternative atomic numberings of the fragment.

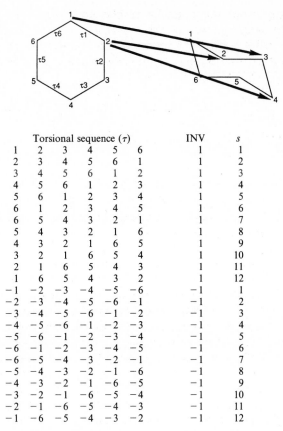

Torsional sequence (τ)						INV	s
1	2	3	4	5	6	1	1
2	3	4	5	6	1	1	2
3	4	5	6	1	2	1	3
4	5	6	1	2	3	1	4
5	6	1	2	3	4	1	5
6	1	2	3	4	5	1	6
6	5	4	3	2	1	1	7
5	4	3	2	1	6	1	8
4	3	2	1	6	5	1	9
3	2	1	6	5	4	1	10
2	1	6	5	4	3	1	11
1	6	5	4	3	2	1	12
-1	-2	-3	-4	-5	-6	-1	1
-2	-3	-4	-5	-6	-1	-1	2
-3	-4	-5	-6	-1	-2	-1	3
-4	-5	-6	-1	-2	-3	-1	4
-5	-6	-1	-2	-3	-4	-1	5
-6	-1	-2	-3	-4	-5	-1	6
-6	-5	-4	-3	-2	-1	-1	7
-5	-4	-3	-2	-1	-6	-1	8
-4	-3	-2	-1	-6	-5	-1	9
-3	-2	-1	-6	-5	-4	-1	10
-2	-1	-6	-5	-4	-3	-1	11
-1	-6	-5	-4	-3	-2	-1	12

Fig. 4.8. Multiple mappings of search fragment due to 2D topological symmetry

In 3D, we must also remember that each fragment will also have an enantiomorph of equal interest. If we are using torsion angles as conformational descriptors, then these quantities are enantiomorph-sensitive: a change in enantiomorph results in a reversal of signs of the torsion angles for each of the 12 possible enumerations. This raises the number of possible equivalent torsion angle sequences to 24 for the cyclohexane fragment. Since we are usually unsure as to which enantiomorph is represented by the 3D coordinates in the CSD, then examples of alternative sign sequences will also occur at random in the raw data matrix, as shown in Table 4.1 a, b.

We need to consider enantiomorphs in most multivariate analyses which use crystallographic data. Most of the structures are either (a) non-chiral, or (b) chiral molecules for which the absolute chirality has not been determined. In some cases, however, the 3D coordinate sets in the CSD do reflect absolute chirality: either the absolute stereochemistry has been established explicitly by the X-ray experiment, or it is so well known that we expect the coordinates to represent the correct enantiomorph as, for example, in the case of the pyranose sugars. In these instances we might wish to ignore the enantiomorphs.

Table 4.1. Representative torsional sequences generated by CSD Software System for six-membered carbocycles

Fragment	$\tau 1$	$\tau 2$	$\tau 3$	$\tau 4$	$\tau 5$	$\tau 6$
a: Boats						
63	1.7	70.7	−73.6	2.2	70.2	−72.0
69	−1.2	−71.7	71.3	1.7	−70.8	70.4
114	−78.1	72.9	0.2	−69.8	64.4	5.2
121	70.1	−1.0	−72.5	72.8	−4.7	−65.8
131	−70.5	−0.6	71.1	−68.7	−2.5	72.7
134	68.2	−73.2	1.8	70.9	−75.6	4.7
b: Chairs						
1	−60.5	59.7	−61.1	63.4	−60.4	58.9
9	56.1	−56.0	56.4	−53.9	54.7	−57.2

The derivation of the equivalent atomic permutations resulting from C_{3v} symmetry is shown in Figure 4.9 and is relevant to the example of Section 4.6.3.3. The handling of fragment symmetry, in terms of these atomic permutation sequences, within the principal component analysis and cluster analysis routines of the CSD System are discussed in the next two sections. At present the permutations must be entered manually, but software for their direct generation from the 2D fragment description is nearing completion. It is unlikely that external statistical packages will take internal account of this specialist symmetry requirement. However, methods by which the data matrix can be pre-processed to include symmetry information prior to such external analyses are also indicated below.

4.6.3 Principal Component Analysis (PCA)

Principal component analysis (PCA: see e.g. Chatfield and Collins [53]) interprets the overall variance in the multivariate dataset in terms of a new set of uncorrelated variables: the principal components or PC's. Because they are uncorrelated, the PC axes are mutually orthogonal. Most importantly, the PC's are generated in decreasing order of importance, i.e. in decreasing order of the percentage total variance that is explained by each PC. The hope is that a very large proportion (say, >95%) of the total variance in the dataset will be explained by the first N_c principal components, where $N_c \ll N_p$. Thus, the underlying purpose of the technique is one of dimension reduction. The number of parameters required to describe the fragments is reduced to a minimum, and it is now possible to view the complete dataset in an N_c-dimensional PC-space, rather than the original more complex N_p-dimensional parameter space.

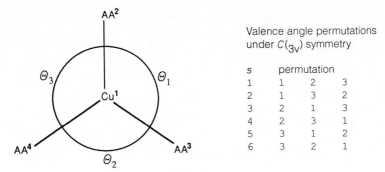

Valence angle permutations
under $C_{(3v)}$ symmetry

s	permutation		
1	1	2	3
2	1	3	2
3	2	1	3
4	2	3	1
5	3	1	2
6	3	2	1

Fig. 4.9. Permutations of valence angles under C_{3v} symmetry

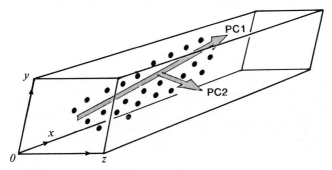

Fig. 4.10. Relationship of principal component axes to original axial frame in the hypothetical example discussed in the text

The basic principles of the technique are illustrated in Figure 4.10. Here, each data point is described by three parameters, but the plot shows that the points all lie on a plane in this 3-space. PCA transforms the coordinate system so that the data points are described by two new PC axes in that plane; when taken together, the two PC's explain 100% of the total variance. Hence, the dimensionality has been reduced from $N_p = 3$ to $N_c = 2$. Obviously, it is now much simpler to visualize the dataset when it is plotted as a 2D scattergram based on the PC axes. In the general case, we can hope to visualize a more complex dataset through the generation of the $N_c(N_c-1)/2$ unique 2D scatterplots, a significant reduction of the $N_p(N_p-1)/2$ 2D scatterplots required to view the original parameter space. The PC scattergrams are used to detect clusters of similar fragments within the dataset, deformations from some standard geometry, pathways that may be interpreted as conformational or configurational interconversions and, importantly, any outliers that may exist in the dataset.

Mathematically, the technique operates through an eigenanalysis of either the covariance matrix $Cov(N_p, N_p)$, or of the correlation matrix $R(N_p, N_p)$, see Sections 4.5.3 and 4.6.1. Full formal details of the derivation of PC's are given in most texts on multivariate analysis (see [53]). Mathematical treatments are also given by

Murray-Rust and Bland [15] and by Auf der Heyde [54]. In summary, the principal components are the eigenvectors of the covariance (or correlation) matrix. The PC's appear as linear combinations of the original variables $[x_i\,(i = 1, 2, \ldots, N_p)]$ in the form:

$$(PC)_j = a_{1j}x_1 + a_{2j}x_2 + a_{3j}x_3 + \ldots a_{N_p j}x_{N_p} \tag{4.12}$$

where the coefficients a_{ij} are referred to as "loadings" that indicate the relative contribution of each of the original variables to $(PC)_j$. The eigenvalues of the matrix, $\lambda_j\,(j = 1, 2, \ldots, N_p)$, represent the proportion of the total variance (or correlation) in the dataset that is accounted for by each of the PC's. Finally, the technique generates the principal component "scores" for each fragment, i.e. the coordinates of each fragment in the N_c-dimensional PC-space. These are the coordinates that are used in the generation of the PC scatterplots. In the original literature, the technique is sometimes referred to as "factor analysis" and the PC's are often termed as "factors". Formally, the factor analysis technique is closely related to PCA, but differs in the underlying statistical model. Modern texts make this distinction very clearly (see e.g. [53]).

Principal component analysis has been used to simplify and explore multivariate datasets that arise in many areas of human activity: social sciences, biostatistics, economics, etc. It has also been used effectively in analytical chemistry and in chemometrics [55]. PCA was introduced as a valuable tool in structure correlation work by Murray-Rust and Bland [15]. The first application [56] involved the analysis and classification of the conformations adopted in crystal structures by the β-1'-aminoribofuranosyl fragment. Since then, PCA has been applied to studies of substituent-induced deformations of benzene rings [57], configurational distortions from ideal symmetry at metal centres [7, 8, 9 and references therein], and to analyses of ring conformations in terms of archetypal forms and their interconversions (see e.g. [58]).

In those cases where PCA is successful in providing a simplification of the dataset, there is a natural tendency to take the analysis to its logical conclusion: the interpretation of the PC's in chemical terms. The PC's are, after all, new geometrical parameters derived from existing parameters of known chemical meaning. The N_c equations of type 4.12 that emerge from a PCA do contain some obvious information. If a single coefficient, a_{ij}, is dominant, then the associated x_i is closely aligned with $(PC)_j$. If one or more of the a_{ij}'s are close to zero, then their associated x_i's contribute little to $(PC)_j$. In the general case, however, the equations of type 4.12 may be difficult, even impossible, to rationalize in terms of the underlying chemistry. Methods do exist for the rotation of axes (see e.g. Harman, [59]), and for the grouping of associated variables, in a search for functional relationships between the PC's and the original parameter set, but they must be used with very great care.

Within the CSD System, it is now possible to include the PC scores within an extended data matrix. This matrix can also include linear combinations of standard parameters set up by the investigator (see Chapter 3) as putative equivalents to the PC's. The extended data matrix can then be used to examine the correlations between the PC's and any of the wide range of standard and non-standard geometrical quantities that can be included in a CSD data matrix. The treatment of PC's in this way is beginning to reveal standard chemical interpretations for specific types of fragments; some examples will be described below. Consideration of fragment symmetry, and of the symmetry of the underlying configurational or conformational spaces, have been crucial to these interpretations.

Despite these successes, the main attraction of the PCA method lies in its ability to reduce the dimensionality of a complex problem to manageable proportions. Even when it fails, i.e. the number of significant PC's is equal to the number of original parameters, the method is still providing the information that the original parameters are essentially uncorrelated and already represent a minimum dimensionality. Indeed, this lack of reduction of dimensionality in a PCA can be used to check possible interpretations of the PC's derived from a higher dimensional dataset. We illustrate the use of PCA with three examples.

4.6.3.1 An Asymmetric Example: β-1′-Aminoribofuranosides

This fragment (Figure 4.11) was studied by Murray-Rust and Motherwell [56], using 110 examples retrieved from the CSD. The thirteen torsion angles, $\tau 1$ to $\tau 13$, of Figure 4.11 were used to define the conformation of each fragment. Variations in

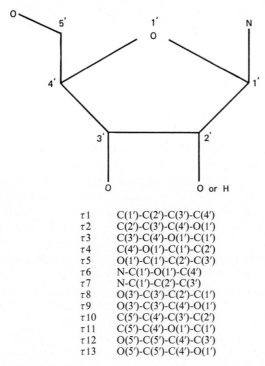

$\tau 1$	C(1′)-C(2′)-C(3′)-C(4′)
$\tau 2$	C(2′)-C(3′)-C(4′)-O(1′)
$\tau 3$	C(3′)-C(4′)-O(1′)-C(1′)
$\tau 4$	C(4′)-O(1′)-C(1′)-C(2′)
$\tau 5$	O(1′)-C(1′)-C(2′)-C(3′)
$\tau 6$	N-C(1′)-O(1′)-C(4′)
$\tau 7$	N-C(1′)-C(2′)-C(3′)
$\tau 8$	O(3′)-C(3′)-C(2′)-C(1′)
$\tau 9$	O(3′)-C(3′)-C(4′)-O(1′)
$\tau 10$	C(5′)-C(4′)-C(3′)-C(2′)
$\tau 11$	C(5′)-C(4′)-O(1′)-C(1′)
$\tau 12$	O(5′)-C(5′)-C(4′)-C(3′)
$\tau 13$	O(5′)-C(5′)-C(4′)-O(1′)

Fig. 4.11. The β-1′-aminoribofuranoside fragment and the thirteen torsion angles used to define its conformation

Table 4.2. Principal component analysis (on correlation matrix) of β-1'-aminoribofuranoside fragment [56, 60]

a: Eigenvalues for PC's together with percentage variance explained by each

PC No.	Eigenvalue	% Variance
1	7.98	61.4
2	3.03	23.3
3	1.95	15.0
4	0.02	0.1
5 – 13	<0.01	–

b: PC loadings for the torsion angles $\tau_1 - \tau_{13}$

	Loadings		
Torsion angle	PC 1	PC 2	PC 3
$\tau 1$	− 1.00	0.06	− 0.01
$\tau 2$	0.97	0.22	0.04
$\tau 3$	− 0.66	− 0.73	0.13
$\tau 4$	− 0.53	0.84	− 0.15
$\tau 5$	0.93	− 0.35	0.06
$\tau 6$	− 0.48	0.86	− 0.13
$\tau 7$	0.93	− 0.36	0.06
$\tau 8$	− 1.00	0.04	0.01
$\tau 9$	0.97	0.23	− 0.60
$\tau 10$	0.95	0.20	0.04
$\tau 11$	− 0.67	− 0.72	− 0.13
$\tau 12$	0.03	0.26	0.96
$\tau 13$	0.03	0.26	0.97

bond lengths and angles were assumed to be minimal due to the chemical consistency of the examples (this aspect was checked, for bond angles at least, in the original analysis). The original data were later reviewed and reworked by Taylor [60], and the summary given here is distilled from both sources.

It is obvious that the 13 torsion angles cannot all be independent. This was confirmed by applying PCA to the correlation matrix $R(13, 13)$. Results are summarized in Table 4.2. It can be seen that only three PC's account for 99.7% of the total variance in the dataset, a very significant reduction in dimensionality. The three 2D scatterplots of the principal component scores [60] are shown in Figure 4.12a, b, c. Scatterplot (b) is the most informative: it reveals five clearly defined groupings respectively containing, in decreasing order of occupation, 33, 30, 12, 8 and 6 fragments. There are also a number of minor groupings and singleton outliers.

Visual examination of the dataset and the PC loadings led to a chemical interpretation of the PC's themselves. The loadings of Table 4.2 clearly show that $\tau 12$ and

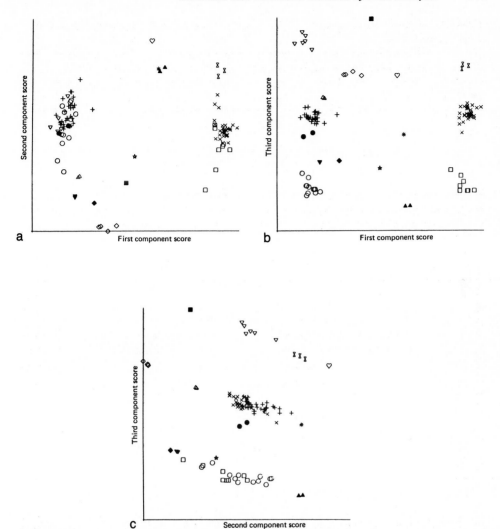

Fig. 4.12. Principal component plots for the *β*-1′-aminoribofuranoside fragment. Observations adopting the same conformation are represented by the same symbol

τ13 are the sole contributors to PC3, i.e. this component is related entirely to the conformation of the side chain at C5′. Conversely, these torsion angles make negligible contributions to PC1 or PC2, which are dominated by the torsional description of ring conformation. Examination of relevant fragments from the highly populated subgroups indicated that two conformations of the five-membered ring were predominant. Since a five-membered ring has two degrees of freedom for out-of-plane deformations [61], a given conformation is defined by just two parameters which can be chosen in several ways [3, 10, 62, 63]. Murray-Rust and Motherwell [56] showed that PC1 and PC2 in the furanose analysis could be directly correlated with the Altona-Sundaralingam parameters P (the angle of pseudorotation) and τ(max) (a torsional description of the maximum out-of-plane displacement in the ring). Taylor [60] tabulates the values of τ(max), P and $\tau 13$ for all 110 fragments in sorted order. Typical values for the five major subgroups identified in the 2D scattergram of Figure 4.12b were readily identified.

4.6.3.2 A Symmetrical Example: Conformations of Six-membered Rings

We return now to the problems that occur in the analysis of multivariate datasets derived for symmetric fragments (Section 4.6.2). The example given there, that of six-membered rings, has been studied using PCA [58] based on the six intra-annular torsion angles defined in Figure 4.8. Two fragment subsets were selected for analysis. The first, termed 6-C(1), consisted of 71 rings known to contain a wide variety of conformations (chair, boat, envelope, half-chair, planar, etc). The second, termed 6-C(2), was a much larger subset of 952 rings chosen at random but, inevitably, containing a similar conformation mix. Full details of the search and selection procedures are given by Allen et al. [58].

In order to provide a framework for the interpretation of the PCA results, and to illustrate the effects of fragment symmetry on the analysis, we first summarize the features of the 3D conformational space for six-membered rings [64]. The conformation of a given ring can be represented (Figure 4.13) by the spherical coordinates Q (the total puckering amplitude), θ, and ϕ. These coordinates are simply related to the three degrees of freedom q_2, ϕ_2, q_3 [3]. Thus, the conformations of n six-membered rings can be represented by n concentric spheres, of radii Q_1 to Q_n, since each ring may have a different puckering amplitude. Special symmetrical conformations exist on these spherical surfaces (Figure 4.13). One chair form occupies the north pole ($\theta = 0°$), with its inverse at the south pole ($\theta = 180°$). The equator at

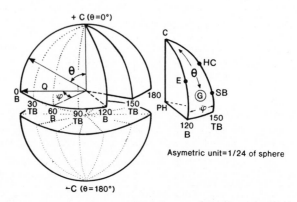

Fig. 4.13. Conformational space for six-membered rings: C = chair, B = boat, PH = planar, TB = twist-boat, HC = half-chair, SB = screw-boat, E = envelope

$\theta = 90°$ represents a pseudorotation pathway that connects six equivalent boat forms at $\phi = 0, 60, \ldots 300°$ and six equivalent twist-boats at $\phi = 30, 90, \ldots 330°$. Six equivalent envelope conformations exist in the northern hemisphere, each lying at $\theta = 55°$ on one of the six arcs connecting the chair and boat forms; six inverse envelopes lie in the southern hemisphere at $\theta = 125°$. Similarly, six half-chairs ($\theta = 50°$) and six screw-boats ($\theta = 68°$) exist on arcs joining the chair and twist-boat conformations, with their inverses at $\theta = 130°$ and $112°$ respectively on similar arcs in the southern hemisphere. The planar hexagon ($Q = 0$ and θ, ϕ indeterminate) occupies the centre of the family of spheres.

The PCA results for the larger (random) dataset 6-C(2) are given in Table 4.3. They show that all the variance in the 6-dimensional torsional dataset is accounted for by three PC's, as expected. Further, the symmetry of the PC loadings can be identified as corresponding (almost) to the symmetries of perfect chair (PC1), boat (PC2) and twist-boat (PC3) forms. The scatterplot of PC2 vs. PC3 (Figure 4.14a) shows a central strong peak around the origin, surrounded by six peaks at approximately 60° intervals. The axis connecting two of these peaks through the origin is slightly offset from the PC2 axis. The heights of the six peaks are not identical, and there are a small number of fragments which lie between these peaks in this projection of PC-space.

In the next PCA, the smaller (selected) dataset, 6-C(1), was used. However, in this case, the 23 additional sets of torsion angles generated by the atomic permutations and enantiomorphic inversions (Section 4.6.2, Figure 4.8) were included in the raw data matrix for each of the 71 fragments. The dimensions of this matrix now rise from G (71,6) to G(1704,6). The PCA results derived from this "symmetry-expanded" matrix are also given in Table 4.3. Again the first three PC's account for all of the variance in the dataset, but now PC2 and PC3 each account for 25% of the

Table 4.3. Principal component analysis (on covariance matrix) of six-membered carbocycles [58] for datasets 6-C (1), 6-C (2) as defined in the text. PC loadings for each torsion angle $\tau_1 - \tau_6$ and % variance accounted for by each PC

	PC1		PC2		PC3	
	6-C (2)	6-C (1)	6-C (2)	6-C (1)	6-C (2)	6-C (1)
% Variance	75.4	49.4	14.5	25.3	10.1	25.3
τ_1	38.7	33.7	2.0	5.4	− 19.7	33.6
τ_2	− 38.5	− 33.7	20.2	26.4	11.4	− 21.5
τ_3	38.8	33.7	− 22.2	− 31.8	8.7	− 12.1
τ_4	− 39.8	− 33.7	1.7	5.4	− 20.2	33.6
τ_5	40.2	33.7	20.7	26.4	11.3	− 21.5
τ_6	− 39.4	− 33.7	− 22.2	− 31.8	8.4	− 12.1
Symmetry of PC	Chair	Chair	Boat	Boat	Twist Boat	Twist Boat

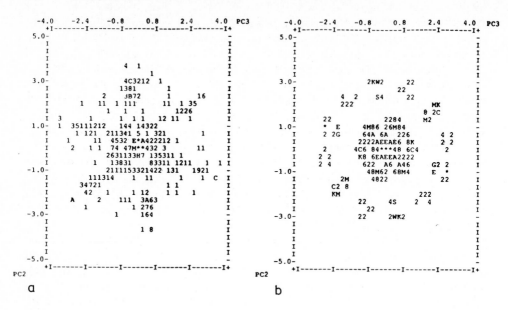

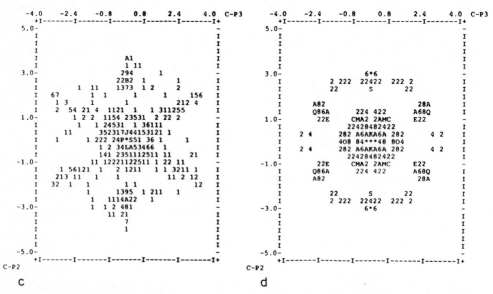

Fig. 4.14. a, b: Principal component plots derived for six-membered ring torsional datasets 6-C (1) and 6-C (2) respectively. c, d: Plots of Cremer-Pople puckering parameters for six-membered ring datasets 6-C (1), 6-C (2) respectively

variance, with 50% being associated with PC1. Further, the PC loadings can now be correlated to the symmetries of the chair, twist-boat, and boat forms respectively. Most importantly, the PC2 vs. PC3 scatterplot (Figure 4.14b) now shows the large central peak to be surrounded by (a) six large peaks distributed as in Figure 4.14 a, but now of equal height, and (b) six smaller peaks, again of equal height, midway between the peaks of type (a).

Interpretation of the scatterplots of Figure 4.14 a, b in terms of the conformational sphere of Figure 4.13 is now clear. Both represent a view of conformational space along the chair-chair (vertical) axis. In this projection the boat – twist-boat pseudorotational pathway should appear as a circle of twelve peaks separated by intervals of 30°. The relative heights of adjacent boat and twist-boat peaks will, of course depend upon the relative frequencies of each conformer in the dataset. The centre of the projection will contain the chairs and planar rings, with envelopes and half-chairs closely surrounding the central peak. Half-chairs predominate in this sample dataset, and the outline of the central area is hexagonal, the points of the hexagon (half-chairs) being directed from the origin towards the small twist-boat peaks.

In this case, and indeed for all ring systems, the chemical meaning of the PC's now becomes obvious. The spherical conformational space of Figure 4.13 is related directly to the three Cremer-Pole puckering parameters q_2, ϕ_2, q_3 [3]. Figure 4.14 c, d shows the scatterplots generated directly from values of $q_2 \sin \phi_2$ ($= C$-$P\,2$) and $q_2 \cos \phi_2$ ($= C$-$P\,3$) calculated for each fragment. Although there are small axial rotations relating Figure 4.14 a and c, and Figure 4.14 b and d, it is clear that the functional forms of the PC space and that of the Cremer-Pole (C-P) space are identical. The axial rotations occur since PC 2 and PC 3 are degenerate and the eigenanalysis will choose an arbitrary pair of orthogonal vectors in the PC 2, PC 3 plane. In view of the results of Murray-Rust and Motherwell [56], it is no surprise that the PCA of a symmetry-expanded torsional dataset for five-membered rings [58] exactly maps the classic circular pseudorotation itinerary identified for these rings [61].

4.6.3.3 Symmetrical Examples: Coordinate Geometries at Metal Centres

PCA has been successfully applied to the mapping of valence angle deformations at metals and other atomic centres. For example, Murray-Rust [65] has studied the deformations from ideal symmetry in PO_4 tetrahedra, whilst Auf der Heyde and Bürgi [7, 8, 9] have used PCA to study the Berry [66] pseudorotational interconversion of trigonal bipyramidal and square planar five-coordinate metal centres. The use of symmetry-adapted deformation coordinates (see Chapter 2) is now well established for this kind of work, and the chemical interpretation of results is covered in detail in Chapter 5. In our final example [67], we examine deformations at three-coordinate copper centres to show how a PCA based on the L-Cu-L valence angles leads naturally to an interpretation in terms of the relevant symmetry-adapted angular deformation coordinates. This is an example of the analogy between normal coordinate analysis and PCA, as noted by Murray-Rust [65].

A total of 116 entries containing three-coordinate Cu atoms, retrieved from the CSD, yielded 192 unique fragments for analysis. The three valence angles and the symmetry permutations of Figure 4.9 were used to generate a raw data matrix

Table 4.4. Principal component analysis (on covariance matrix) of three-coordinate copper centers [67]. % variance accounted for by each PC and PC loadings for each valence angle

	PC1	PC2	PC3
% Variance	48.6	48.6	0.6
θ_1	11.12	-6.42	0.99
θ_2	-11.12	-6.42	0.99
θ_3	0.0	12.84	0.99

$G(1152,3)$. Results of the PCA, summarized in Table 4.4, show that the first two factors account (equally) for a total of 97.2% of the sample variance. A third factor accounts for a further 0.6%. Scattergrams of PC1 vs. PC2 and of PC1 vs. PC3 are shown in Figure 4.15a, b respectively. Normally, one might be tempted to ignore PC3 and say that variance in the dataset is adequately explained by PC1 and PC2 alone. However, in this case, the factor loadings are of considerable interest since they may be interpreted in terms of linear combinations of the original valence angles that have the functional forms:

$$PC1 = \quad f(\theta_1)-f(\theta_2)$$
$$PC2 = -f(\theta_1)-f(\theta_2)+2f(\theta_3)$$
$$PC3 = \quad f(\theta_1)+f(\theta_2)+f(\theta_3)$$

The relevant symmetry-adapted angular coordinates are of exactly this form:

$$SA1 = \quad \delta(\theta_1)-\delta(\theta_2)$$
$$SA2 = -\delta(\theta_1)-\delta(\theta_2)+2\delta(\theta_3)$$
$$SA3 = \quad \delta(\theta_1)+\delta(\theta_2)+\delta(\theta_3)$$

where $\delta(\theta_{1,2,3}) = \theta_{1,2,3}-120°$, i.e. the angular deformations from ideal trigonal planar geometry.

The coordinates SA1,2,3 were calculated and included for each fragment in an extended data matrix (see Section 4.6.2), together with the PC scores. Correlation analysis of pairs of parameters in this extended matrix showed coefficients $r(PC1,SA1)$, $r(PC2, SA2)$ and $r(PC3, SA3)$ of exactly unity. Scattergrams of SA1 vs. SA2 and of SA1 vs. SA3 (Figure 4.15c, d) show a functional identity to the corresponding PC plots (Figure 4.15a, b). The highly populated centres of Figures 4.15a, c represent the (predominant) near-perfect trigonal planar arrangements, with in-plane angular deformations being mapped by the PC1 and PC2 axes. The perpendicular view, represented by Figure 4.15b, d, shows the out-of-plane movement of Cu (maximum 0.61 Å in this dataset) towards a trigonal pyramidal arrangement. It might be said that this interpretation of the geometry of 3-coordination in terms of angular deformations and out-of-plane movement of the central atom is rather obvious. What is important about this analysis is that the PCA has pointed the way towards this interpretation purely on the basis of valence angle

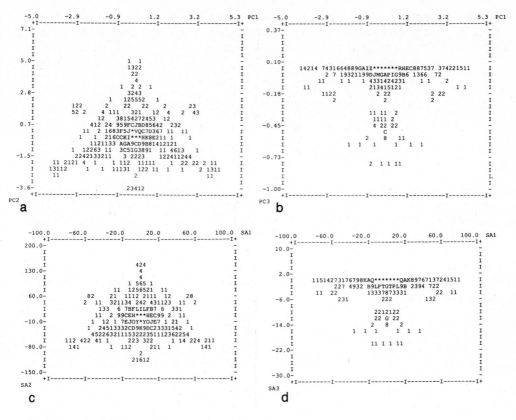

Fig. 4.15. a, b: Principal component plots derived for CuL₃ angular datasets. **c, d:** Plots of symmetry-adapted angular deformation coordinates for CuL_3 system

input. The dimensionality of the problem is not reduced (in this simple three-parameter case); instead the interpretation of the PC's is the vital part of the analysis.

4.6.4 Methods Based on Dissimilarity Matrices

4.6.4.1 Introduction

We now turn to methods that are based on analysis of the $(N_f \times N_f)$ dissimilarity, or distance, matrix. In structural chemistry, most applications of such techniques have been directed towards the automatic classification of molecular conformations. Murray-Rust and Raftery [68] distinguish three common situations: (a) all crystallographic observations of a molecular fragment show essentially the same geometry; (b) the molecular fragment has several distinct conformations, i.e. the crystallographic observations form clusters in the multidimensional conformational space; (c) there are large, continuous variations between widely different geometries – this occurs if the molecular fragment has a potential-energy surface with extensive flat regions. Of these situations, (a) is easy to deal with, since there is only one conformation and the question of classification does not arise. In contrast, (c) is extremely difficult to analyze objectively, and attempts to distinguish distinct conformers may not even be meaningful. Consequently, our attention must focus on the intermediate, and arguably the most interesting, situation (b).

Classification techniques fall into two general categories: graphical methods, where the aim is to produce a pictorial representation of the data set, such that clusters of observations can be detected by visual inspection; and non-graphical methods, where the aim is to assign each observation automatically (i.e. without user-intervention) to a cluster, so that similar observations belong to the same cluster and dissimilar observations fall in different clusters. PCA is a graphical technique based on the correlation matrix $R(N_p, N_p)$. However, many methods also use the dissimilarity matrix, $D(N_f, N_f)$ as their starting point, and we shall begin by considering how this matrix may be calculated.

4.6.4.2 Measures of Dissimilarity

The dissimilarity between two fragments, r and s, each described by N_p parameters, may be estimated by the Euclidian distance between them in N_p-space:

$$D(r,s) = \left\{ \sum_{i=1}^{N_p} [G(r,i) - G(s,i)]^2 \right\}^{1/2} \tag{4.13}$$

where $G(r,i)$, $G(s,i)$ are values of the ith parameter for fragments r and s in the raw data matrix. A more general definition of dissimilarity is the Minkowski metric:

$$D(r,s) = \left\{ \sum_{i=1}^{N_p} [G(r,i) - G(s,i)]^n \right\}^{1/n} \tag{4.14}$$

where the user-specified integer n is usually chosen as 1 (the "city-block" metric) or 2 (when Eq. 4.14 reduces to Eq. 4.13: the Euclidian metric). Some authors prefer to express dissimilarity by the Mahalanobis distance, which aims to incorporate the effects of parameter correlation by including the inverse of the covariance matrix in the distance expression [54, 69]. Most published work, however, has been based on the Euclidian or city-block measures. There is no clear indication that one is better than the other; in principle, the city-block metric should be less affected by outliers, but at least one case has been found [60] where use of the Euclidian metric resulted in a more complete identification of the conformational minima.

Two minor complications may occur when using expressions such as (4.14) in structural chemistry research. Firstly, if the fragment geometries are described by torsion angles, it is necessary to allow for the phase discontinuity at 180°. This is achieved by defining the difference between two torsion angles, $\tau(1)$ and $\tau(2)$, as the lesser of $|\tau(1)-\tau(2)|$ and $360-|\tau(1)-\tau(2)|$. The second complication arises if the parameters used to describe the fragment geometries are not all on the same scale (e.g. if there is a mixture of torsion angles and valence angles). The usual remedy is to normalize each variable to unit variance – however, it must be emphasized that this is a sensible, but arbitrary, operation. Fortunately, the problem rarely occurs in conformational work, where the variables are usually all torsion angles or all Cartesian coordinates. Also, calculation of the variance of the torsion angles is complicated by the phase discontinuity at $\pm 180°$, as discussed in Section 4.4.5. Major complications arise in calculating the dissimilarity of symmetrical fragments; these are considered further in the Section on cluster analysis (4.6.4.4).

4.6.4.3 Multidimensional Scaling

Multidimensional scaling [70] is a method for obtaining the "best" low-dimensional representation of a high-dimensional data set. Normally, a two- or three-dimensional representation is required, since it can then be plotted and inspected visually for clusters. In the classical scaling technique, the low-dimensional representation is obtained by extracting the eigenvectors of the $(N_f \times N_f)$ dissimilarity matrix. However, it can be shown that this operation is equivalent to a PCA of the $(N_p \times N_p)$ covariance matrix, provided that the distances in the dissimilarity matrix are Euclidian or near-Euclidian [53]. Since the covariance matrix is invariably of smaller order than the dissimilarity matrix, PCA is to be preferred on computational grounds. The only exception is if the dissimilarity matrix is available but the covariance matrix is not, a circumstance that rarely arises in structural chemistry work.

Ordinal multidimensional scaling is an empirical technique for iteratively improving a low-dimensional representation of a high-dimensional dataset [53, 70]. The technique is non-parametric, because it is based on the ranks of the dissimilarities rather than their actual values. The iterative procedure aims to minimize a quantity called "stress". Small values of stress are obtained if the ranks of the Euclidian distances between points on the low-dimensional plot are similar to the ranks of the dissimilarities of the corresponding observations in the original multidimensional space.

The technique was used [60] to improve the scatterplot of the first and second PC scores of the aminoribofuranoside dataset (Figure 4.12a). After optimization of stress, the resulting representation was as shown in Figure 4.16. On this plot, as on the original one, observations associated with the same conformation are represented by the same symbol. It can be seen that the ordinal scaling has considerably improved the original plot; for example, it is now possible to distinguish clearly the five major conformational clusters. Note, however, that one member of the cluster represented by squares has become detached from the rest. This sort of discrepancy may be expected from an iterative technique that might find a local rather than the global minimum value of the stress function.

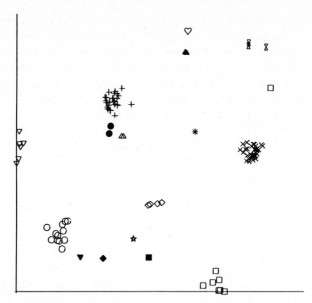

Fig. 4.16. Result of application of ordinal multidimensional scaling to principal component scatterplot of Figure 4.12a. The axes are arbitrary

4.6.4.4 Cluster Analysis

A variety of cluster analysis algorithms has been proposed [69, 71]. Virtually all examples of the use of the technique for classifying molecular conformations have employed hierarchical methods. Here, each observation is initially regarded as a separate cluster. A stepwise procedure then ensues (see [52, 72] for a detailed account): at each step, the two most similar clusters are merged, so that the total number of clusters is reduced by one. If taken to its ultimate conclusion, the algorithm ends with all observations in the same cluster. However, it is usual to interrupt the procedure once a "natural" clustering pattern has been found. Different hierarchical techniques use different definitions of the dissimilarity between clusters. In the single-linkage (nearest-neighbour) method, the dissimilarity coefficient of two clusters is set equal to the dissimilarity of their nearest members. Conversely, in the complete-linkage (furthest-neighbour) algorithm, the dissimilarity of two clusters is defined as the distance between their furthest members. In Ward's method, cluster dissimilarity is measured by the distance between cluster centroids.

Cluster analysis is not a robust technique: in a classic review, Cormack [73] wrote:

> "The availability of computer packages of classification techniques has led to the waste of more valuable scientific time than any other 'statistical' innovation (with the possible exception of multiple-regression techniques)".

Different algorithms may produce radically different results when applied to the same dataset. For example, the single-linkage method has difficulty in recognizing two separate clusters linked by a chain of intervening observations (a situation particularly likely to occur with conformational data, where the two clusters could represent different energy minima and the chain would represent a connecting valley in the potential-energy hypersurface). Ward's method is prone to a different problem, the tendency to find spherically-shaped clusters, even if this means splitting a single, elongated cluster into two, more symmetrical subsets. Moreover, we must remember that clustering is, in any case, an inherently ill-defined process, there being no generally accepted mathematical criterion of an "optimum" clustering pattern for any given data set. Thus, we must expect different algorithms to produce discrepant results, even in the absence of additional technical problems such as those mentioned above.

Despite these difficulties, cluster analysis has been applied with considerable success to a variety of molecular fragments. One of the earliest studies [68] used Ward's method to partition a sample of tripeptide β-loops into subsets. It was found that type II turns form a clearly defined class on their own, but type I and III turns have very similar geometries and are not clearly distinguished. Both single-linkage and complete-linkage analysis were applied [60] to the aminoribofuranoside data discussed above, and substantially reproduced the manual classification of conformations made on the basis of the graphical PCA method. Another clustering algorithm, the Jarvis-Patrick method [74], was used to cluster a sample of 6-ring carbocycles [72]. The major conformers (chairs, boats, etc.) were easily identified. This particular algorithm is somewhat different from the hierarchical techniques mentioned above, because it is based on an analysis of the "nearest-neighbour table". The ith row of this table contains the m fragments in the data set that are most similar to fragment i, where m is a parameter chosen by the user. In Jarvis-Patrick analysis, two observations (i,j) are placed in the same cluster if: (a) i is one of the m nearest neighbours of j; (b) j is one of the m nearest neighbours of i; (c) in addition, i and j share at least s nearest neighbours in common, where s is a user-chosen "threshold", and must obviously lie between zero and $m-1$. The algorithm has been shown to be particularly suitable for clustering entries in chemical databases on the basis of chemical connectivity (i.e. two-dimensional structure [75]). The study of ref. [72] suggests that it works well for molecular conformations as well.

In the Jarvis-Patrick technique, the number of clusters found depends on the value chosen for the threshold parameter s. Clearly, large values of s make it difficult for observations to be assigned to the same cluster, and result in a large number of small clusters. Conversely, small values of s produce a small number of large clusters. The ability of the user to control the results in this way is, at first sight, frighteningly arbitrary. However, we emphasize again that clustering is inherently an ill-defined process, and there is rarely a clear-cut "right" answer.

In hierarchical methods, such as the single-linkage algorithm, the number of clusters found depends solely on the step at which the agglomerative procedure is interrupted. A number of techniques have been suggested for identifying a suitable stopping point. Most simply, the dissimilarity of the clusters being combined can be plotted against the step in the analysis. Figure 4.17 shows this plot for the single-linkage analysis of aminoribofuranosides [60]. It can be seen that the curve takes a sudden upwards turn at about step 96, implying that, at this stage, the algorithm is beginning to merge clusters that are, in fact, rather dissimilar. Step 96 is therefore

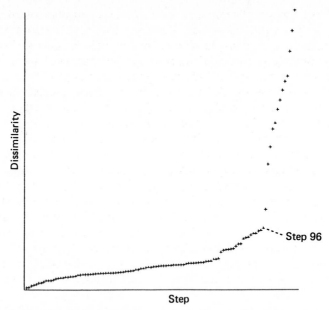

Fig. 4.17. Plot of cluster-fusion dissimilarity vs. step number in single-linkage clustering, indicating that step 96 is a suitable stop point (see text)

an appropriate point at which to interrupt the process. Unfortunately, the stopping point is not always so clearly signposted: [54] gives a brief discussion. In preference to the simple plot of Figure 4.17, Nørskov-Lauritsen and Bürgi [76] plotted a cubic clustering criterion [77] against step number. Allen et al. [52] plotted step number against the difference between the dissimilarity of the clusters fused at step i and the dissimilarity of the clusters fused at step $i+1$ (essentially, the derivative of the quantity plotted along the vertical axis in Figure 4.17).

4.6.4.5 Cluster Analysis of Symmetrical Fragments

Difficulties arise when cluster analysis is used on fragments with topological symmetry, due to the problems discussed in Section 4.6.2. An early attempt to overcome these difficulties was made [76] in a cluster analysis of 62 observations of the fragment $M(PPh_3)_2$ (M = metal). Here, the conformation of an individual fragment can be described by eight torsion angles, two $(\tau1, \tau2)$ describing the orientation around the M-P bonds, three $(\tau11, \tau12, \tau13)$ describing the orientation of the three phenyl rings in one of the triphenylphosphine groups, and a further three $(\tau21, \tau22, \tau23)$ for the other PPh_3 group. Clearly, $\tau1$ and $\tau2$ can be interchanged, since it is arbitrary which PPh_3 group is numbered 1 and which 2; similar arguments apply to the phenyl ring torsions within each group. Nørskov-Lauritsen and Bürgi solved the problem by numbering each fragment in all possible ways, hence expanding the original data set by a factor of 36, and generating a data set with the correct inherent symmetry. Ward's method was then used to cluster the expanded data set, with successful results. The study suggested a model for conformational interconversion in this system, which involves a gearing motion of the two PC_3 fragments, alternating with stepwise interconversions of the helicities of the PPh_3 propellers.

One difficulty with the above "symmetry-expansion" approach is that it leads to a considerable increase in the size of the data set, with consequent computational overheads. Allen et al. [52, 72, 78, 79] avoid this difficulty by an alternative stratagem. In estimating the dissimilarity of each pair of fragments in their data set (six-membered carbocycles), they tried all possible ways of overlaying the atoms of the second fragment on those of the first. The lowest dissimilarity coefficient thus obtained was accepted. These dissimilarity coefficients were then used as the basis for the subsequent cluster analyses, in which slightly modified versions of the single-linkage, complete-linkage and Jarvis-Patrick methods were employed. The algorithms produced clusters in which the different fragments were optimally overlaid on one another, and also placed the different clusters into a single asymmetric unit of conformational space. After the cluster analysis had been performed, the asymmetric unit of clusters could be expanded by application of the appropriate symmetry operations, giving a data set with the full, correct symmetry.

4.6.4.6 Symmetry-Modified Clustering: An Example

The symmetry-modified clustering algorithms described in the previous section were developed using a test dataset of 222 six-membered carbocycles [52, 72, 78], each described by the six intraannular torsion angles $\tau1$ to $\tau6$ of Figure 4.8. The dataset was selected to contain examples of a variety of conformers: chair, boat, half-chair, envelope, planar, etc. The symmetry transformations involved in the clustering process are illustrated in Figure 4.18 for some general conformer (g), which will occur 24 times in the relevant conformational space (see Figure 4.8 and 4.13). In Figure 4.18a, we show the effect of overlaying all possible symmetry variants of fragments g_1 and g_3 to g_6 onto a randomly selected fragment g_2, which now becomes the "root" of the developing cluster. Figure 4.18b shows the possible result of this process for general (g), chair (c) and boat (b) conformers: fragments with the same conformation have been correctly assigned to the same cluster, but the three clusters lie in different asymmetric units of conformational space, depending on the initial random choice of their respective "root" fragments. These clusters may now be brought into their closest mutual proximity (Figure 4.18c) by applying symmetry-modified single-linkage clustering to the centroids of the clusters. The agglomerative process is allowed to go to completion, which automatically finds the optimum overlay of the cluster centroids, one upon another.

Figure 4.19 shows (a) overlay plots of all members of the four major clusters and (b) the most representative fragment (see Section 4.6.5.2) from each of the top 9 clusters as obtained by the single-linkage clustering algorithm. The mean torsion angles for the top 10 clusters are given in Table 4.5a. These data reveal a further problem: many of the mean torsion-angle sequences, $\tau1$ to $\tau6$, show marginal distortions from some ideal symmetrical form, i.e. the cluster centroid lies close to a special posi-

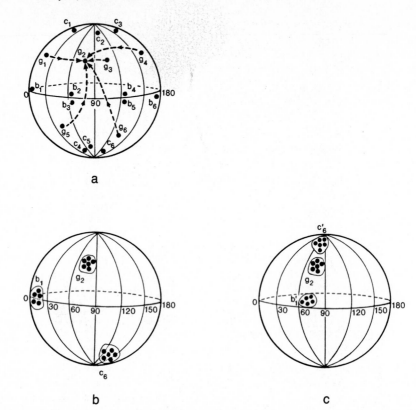

Fig. 4.18. a: Representative random conformations of six-membered rings in the c(hair), b(oat) and g(eneral) conformers. The arrows indicate symmetry transformations performed during cluster formation in which fragment g_2 is arbitrarily chosen as the cluster root. b: Cluster formed around randomly chosen cluster roots g_2, b_1 and c_6. c: Symmetry transformation of clusters around b_1 and c_6 to new positions (b_1' and c_6') which are in closest mutual proximity to g_2

tion in conformational space. A simple, general solution to this problem has been proposed [79] based upon the calculation of the dissimilarities $D(c, c')$, where c is the centroid of a given cluster, and c' denotes the centroid(s) of its symmetry-related counterpart(s). If $D(c, c')$ is less than a test value, based upon the "spread" of geometries within the cluster centred on c, then the clusters centred on c and c' may be coalesced. The results of this process on the initial asymmetric clusters of Table 4.5a are illustrated in Table 4.5b. Here $(D_c)_{max}$, the maximum centroid-to-fragment distance in a cluster, represents the "spread" of that cluster, whilst O_c indicates the number of symmetry variants that have been coalesced to yield the mean torsion angles of Table 4.5b.

Further examples of symmetry-modified clustering, for the azacycloheptane ring and for the C_{17} side chain in cholesterol and related steroids, have been presented [78].

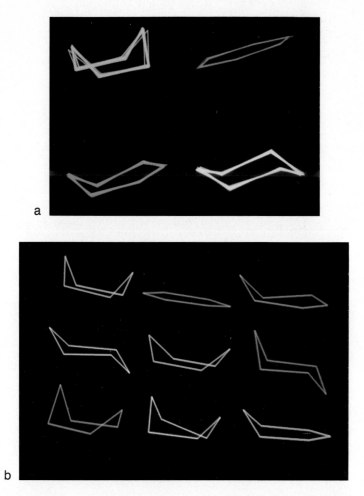

Fig. 4.19. a: Overlay plots of all members of the four major clusters identified by symmetry-modified single-linkage analysis. Top: norbornane boats (34 fragments), planar hexagons (35 fragments); bottom: half-chairs (26 fragments), chairs (51 fragments). b: Most representative fragments from each of the top 9 clusters. Top (l to r) norbornane boats, planar hexagons, half-chairs; Middle (l to r) chairs, normal (flattened) boats, distorted chairs; Bottom (l to r) two forms of distorted boats, and envelopes

4.6.5 Mean Geometries for Complete Fragments

The object of the various cluster analysis algorithms, then, is to perform a numerical dissection of the multivariate dataset $G(N_f, N_p)$, such that each individual cluster contains fragments that have closely similar geometries. The numerical results are best summarized in terms of an archetypal or mean fragment geometry for each ma-

Table 4.5. Mean torsional angles [deg] for the ten major clusters (population $N_p > 5$) obtained with the symmetry-modified Jarvis-Patrick algorithm for a trial dataset of 222 six-membered rings [79]. Column headings are N_c = cluster number, $\tau 1$ to $\tau 6$ are torsion angles, $(D_c)_{max}$ is the maximum "city-block" centroid-fragment distance [deg], O_c is the "order" of symmetry (coalesced) clusters (see text). Conformation descriptions are PH = planar, C = chair, B = boat, HC = half-chair, SB = screw-boat (1,3-diplanar), TB = twist boat

Class	N_c	N_p	$\tau 1$	$\tau 2$	$\tau 3$	$\tau 4$	$\tau 5$	$\tau 6$	$(D_c)_{max}$	O_c
a: Asymmetric clusters
PH	1	35	−1.0	0.5	1.2	−2.4	2.0	−0.3	16.1	1
C	2	46	52.1	−50.8	53.3	−57.7	58.7	−55.3	36.2	1
	3	9	37.5	−41.3	56.2	−66.1	62.3	−48.5	27.8	1
B	4	30	−72.2	70.2	0.8	−70.3	68.2	2.4	32.4	1
	5	14	−56.2	59.4	−4.4	−52.6	56.7	−2.2	27.3	1
	6	9	−55.3	71.6	−6.5	−66.2	83.8	−18.8	38.0	1
	7	6	−50.8	58.2	15.2	−84.3	93.8	−27.1	53.2	1
HC	8	29	9.2	1.2	19.2	−48.7	60.6	−39.9	44.1	1
SB	9	9	−3.0	15.4	7.3	−40.3	52.7	−31.4	55.1	1
TB	10	9	−36.6	73.3	−22.2	−50.5	89.1	−36.5	50.0	1
b: Values for coalesced clusters
PH	1	840	0.0	0.0	0.0	0.0	0.0	0.0	22.4	24
C	2	552	54.6	−54.6	54.6	−54.6	54.6	−54.6	36.2	12
	3	18	37.5	−44.9	59.2	−66.1	59.2	−44.9	37.2	2
B	4	120	−70.3	70.3	0.0	−70.3	70.3	0.0	34.0	4
	5	28	−54.4	58.0	−3.3	−54.4	58.0	−3.3	30.9	2
	6	9	−55.3	71.6	−6.5	−66.2	83.8	−18.8	38.0	1
	7	6	−50.8	58.2	15.3	−84.3	93.8	−27.1	53.2	1
HC	8	58	14.2	1.2	14.2	−44.3	60.6	−44.3	63.0	2
SB	9	18	2.2	15.4	2.2	−35.8	52.7	−35.8	62.9	2
TB	10	9	−36.6	73.3	−22.2	−50.5	89.1	−36.5	50.0	1

jor cluster. Obviously, the univariate statistical methods of Section 4.4 can be used to obtain separate mean values for each of the N_p parameters (e.g. Table 4.5). However, these mean values may not be self-consistent and, if we try to derive orthogonal 3D coordinate sets from them, e.g. for graphical display purposes, then problems will occur in ensuring, for example, that ring systems form a closed pathway. Further, for symmetrical fragments, we must be aware of certain special positions in the relevant parameter space as discussed in Section 4.6.4.6. We examine these problems and some simple solutions below (see also Chapter 1).

4.6.5.1 Least-Squares Superposition Methods

The necessary mathematical background to the generation of mean Cartesian coordinates referred to inertial axes of the fragment is covered in Chapter 1. This method has been followed in the derivation of mean rigid-body coordinates for subsets of chair-form pyranose rings [80]. These coordinates were then used to generate archetypal conformations, in torsion angle terms, for rings occurring in the various stereoisomers of the pyranose sugars. An alternative approach, involving least-squares minimization based on internal coordinates, was used to derive Cartesian coordinates for standard nucleic acid base units [81]. It was assumed that these residues were precisely planar, hence the expression minimized was:

$$\sum_i (l_i - \bar{l}_i)^2/\sigma_i^2 + \sum_j (\theta_j - \bar{\theta}_j)^2/\sigma_j^2 \tag{4.15}$$

where the summations are over all i bond lengths l_i, and over all j valence angles θ_j. The quantities $\bar{l}_i$ and $\bar{\theta}_j$ are the respective mean values, σ_i and σ_j are the standard errors of these means. This procedure yields dimensions that are as close as possible to the average values, subject to the necessity for ring closure.

4.6.5.2 Averaging Clusters: The Most Representative Fragment

It is a simple matter to generate a univariate statistical summary for each parameter employed in the clustering process. These statistics should include (at least) the mean value, its standard error and the standard deviation of the sample [52]. Now, instead of entering a least-squares minimization procedure, it is reasonable to choose the archetypal fragment as that fragment whose N_p individual geometrical parameters are closest to the N_p mean values determined for the cluster. This is termed the most representative fragment (MRF). For a cluster of population N_c, the MRF is located by calculating the N_c values $D(m, q)$, i.e. the dissimilarity between the cluster mean (m) and the $q = 1, 2, \ldots, N_c$ individual fragments. The fragment corresponding to the smallest $D(m, q)$ is taken as the MRF. This simple method obviates any problems due to ring-closure constraints. Essentially, the MRF is closest to the cluster centroid, and its coordinates and corresponding geometry are perfectly adequate for most purposes, e.g. graphical display and molecular modelling. The simple method for the location of the MRF described here must be modified for fragments which exhibit topological symmetry [78].

4.6.6 Miscellaneous Graphical Methods

We conclude this section on multivariate methods by mentioning a number of additional techniques for obtaining graphic representations of multidimensional data sets. A good general review may be found in Everitt [82]. Amongst the techniques described there is the Andrews plot [83]. For the rth observation in a multivariate data set, the Andrews function is defined as:

$$\phi_r(\theta) = x_{r1}/\sqrt{2} + x_{r2} \sin \theta + x_{r3} \cos \theta + x_{r4} \sin 2\theta + \ldots \qquad (4.16)$$

where $x_{r1}, x_{r2}, \ldots$ are the coordinates of the observation in the multidimensional data space. The function is periodic in the range $-180° < \theta \le 180°$. Observations that are close together in data space tend to have similar Andrews functions. Consequently, if the Andrews functions of several molecular fragments are plotted together on the same diagram, it should be possible to distinguish between fragments with different conformations. The method is limited to relatively small samples, and gives more weight to the variables involved in the lower-order terms of Expression (4.16). Taylor [60] gives an example of its use in structural chemistry.

A number of other graphical techniques could, in principle, be applied to structural data but (to the authors' knowledge) have not yet been used for this purpose. These include projection pursuit [84, 85], which aims to find informative projections of a multivariate dataset by iteratively improving a coefficient of "interestingness". This coefficient is a measure of the deviation of the projection from the multivariate normal distribution. Projections in which the various observations form clusters tend to deviate appreciably from normality, and are therefore regarded as interesting. Livingstone et al. [86] have recently described the use of neural networks for the display of multivariate data, whilst Chernoff [87] used pictorial representations of the human face to describe multivariate data points. Different features of the face can be altered according to the position of an observation in the data space. Those coordinates that are deemed the most important are used to determine the obvious features of the face (e.g. whether it smiles or frowns), whilst less important features determine the minutiae (length of left ear lobe, etc).

4.7 Statistical Software

Most of the statistical methods discussed above are standard. Many of them may be accessed directly through the CSD System for application to raw data matrices generated by the programs GSTAT or QUEST 3D. Taylor [88] has written a library of statistical subroutines (CAMAL) which is now being blended into the CSD System software. Adaptations of standard statistical algorithms to problems involving fragment symmetry are also being introduced into distributed software associated with the CSD.

Since they are standard, much of the statistical functionality required for structure correlation work will also be found in any of several large, multi-functional statistics programs. Examples are SAS (SAS Institute, Cary, NC 27512, USA), SPSS [89] and GENSTAT (Numerical Algorithms Group, Oxford, UK). One or two packages are specifically designed for pattern-recognition work, the best known probably being ARTHUR (Infometrix Inc., Seattle, WA 98121, USA).

Although the use of formal statistical methods is valuable, it is always advisable to examine data sets manually as well. This is facilitated by general data-handling and spreadsheet programs such as LOTUS (Lotus Development Corporation, Cambridge, MA 02142, USA) and RS/1 (BBN Software Products, Staines, UK). Recently, plotting and spreadsheet functionality has been made available as part of large molecular modeling packages such as SYBYL (Tripos Associates, St. Louis, MO 63117, USA). The close integration of these data-manipulation facilities with routines for calculating and displaying molecular geometries can be particularly useful.

4.8 Conclusion

Virtually all our examples of the use of statistical and numerical methods for data analysis have been drawn from the small molecule arena. However, the same or similar techniques also find wide application to the systematic analysis of features of macromolecular structure (see e.g. [90, 91, 92] and references therein). The sum result of all of this activity is a rapidly expanding literature which reports the acquisition of new chemical, biochemical and crystallographic knowledge from the reservoir of basic structural facts that have been accumulated in the crystallographic databases. Further, these studies suggest that modifications to existing methodologies for data analysis will lead to a greater automation of this knowledge acquisition process.

Most existing analyses of structural data have been driven by a need for some specific type(s) of knowledge that is required for some specific area of application. In particular, systematic knowledge of crystal and molecular structure is vital in the determination and interpretation of new structures, in model building as part of rational drug and pesticide design programs, and in studies of molecular recognition processes. However, the ability to recall, manipulate and interpret this growing body of knowledge depends crucially upon the mental agility of the individual researcher. There is, at present, no formal mechanism for the representation and storage of knowledge acquired from the crystallographic databases so that it may be searched, manipulated and used for problem solving. Recently, a semantic network model has been proposed for the long-term memory of crystal and molecular knowledge [93, 94]. A knowledge base of this type could then be accessed through an artificial intelligence infrastructure, to establish further knowledge (in the form of rules and algorithms) and to assist and guide individual researchers in their problem-solving activities.

We would predict, then, that the next decade will see a significant increase in the development of knowledge acquisition methodologies applied to crystallographic databases. Methods for the representation, storage and application of this knowledge will soon become a necessity and then a reality.

References

[1] Allen, F. H., Kennard, O., Taylor, R., *Acc. Chem. Res.* **1983**, *16*, 146–153
[2] Allen, F. H., Davies, J. E., Galloy, J. J., Johnson, O., Kennard, O., Macrae, C. F., Mitchell, E. M., Mitchell, G. F., Smith, J. M., Watson, D. G., *J. Chem. Inf. Comp. Sci.* **1991**, *31*, 187–204
[3] Cremer, D., Pople, J. A., *J. Am. Chem. Soc.* **1975**, *97*, 1354–1358
[4] Allen, F. H., Kirby, A. J., *J. Am. Chem. Soc.* **1984**, *106*, 6197–6202
[5] Allen, F. H., Kennard, O., Watson, D. G., Brammer, L., Orpen, A. G., Taylor, R., *J. Chem. Soc. Perkin II* **1987**, S1–S19
[6] Orpen, A. G., Brammer, L., Allen, F. H., Kennard, O., Watson, D. G., Taylor, R., *J. Chem. Soc. Dalton* **1989**, S1–S83
[7] Auf der Heyde, T. P. E., Bürgi, H.-B., *Inorg. Chem.* **1989**, *28*, 3960–3969
[8] Auf der Heyde, T. P. E., Bürgi, H.-B., *Inorg. Chem.* **1989**, *28*, 3970–3981
[9] Auf der Heyde, T. P. E., Bürgi, H.-B., *Inorg. Chem.* **1989**, *28*, 3982–3989
[10] Altona, C., Sundaralingam, M., *J. Am. Chem. Soc.* **1972**, *94*, 8205–8212
[11] Klyne, W., Prelog, V., *Experientia* **1960**, *16*, 521–528
[12] Kroon, J., Kanters, J. A., van Duijneveldt-van de Rijt, J. G. C. M., van Duijneveldt, F. B., Vliegenthart, J. A., *J. Mol. Struct.* **1975**, *24*, 109–129
[13] Taylor, R., Kennard, O., *Acc. Chem. Res.* **1984**, *17*, 320–326
[14] Bürgi, H.-B., Dunitz, J. D., *Acta Cryst.* **1988**, *B44*, 445–448
[15] Murray-Rust, P., Bland, R., *Acta Cryst.* **1978**, *B34*, 2527–2533
[16] Jeffrey, G. A., Ruble, J. R., McMullan, R. K., DeFrees, D. J., Pople, J. A., *Acta Cryst.* **1981**, *B37*, 1885–1890
[17] Kamb, B., in: *Structural Chemistry and Molecular Biology:* Rich, A., Davidson, N., (eds.). Freeman, San Francisco, **1968**, pp. 509–542
[18] Dyke, T. R., Mack, K. M., Muenter, J. S., *J. Chem. Phys.* **1977**, *66*, 498–510
[19] Brock, C. P., Minton, R. P., *J. Am. Chem. Soc.* **1989**, *111*, 4586–4593
[20] Taylor, R., Kennard, O., *Acta Cryst.* **1983**, *B39*, 517–525
[21] Cochran, W. G., *Biometrics* **1954**, *10*, 101–129
[22] Taylor, R., Kennard, O., *J. Chem. Inf. Comput. Sci.* **1986**, *26*, 28–32
[23] Abrahams, S. C., Hamilton, W. C., Mathieson, A. McL., *Acta Cryst.* **1970**, *A26*, 1–18
[24] Hamilton, W. C., Abrahams, S. C., *Acta Cryst.* **1970**, *A26*, 18–24
[25] Mackenzie, J. K., *Acta Cryst.* **1974**, *A30*, 607–616
[26] Taylor, R., Kennard, O., *Acta Cryst.* **1986**, *B42*, 112–120
[27] Taylor, R., Kennard, O., *Acta Cryst.* **1985**, *A41*, 85–89
[28] Taylor, R., Kennard, O., Versichel, W., *Acta Cryst.* **1984**, *B40*, 280–288
[29] Abrahams, S. C., Keve, E. T., *Acta Cryst.* **1971**, *A27*, 157–165
[30] Hamilton, W. C., Abrahams, S. C., *Acta Cryst.* **1972**, *A28*, 215–218
[31] Snedecor, G. W., Cochran, W. G., *Statistical Methods, 7th edn.*, Iowa State University Press, Ames, **1980**
[32] Siegel, S., *Nonparametric Statistics for the Behavioral Sciences*, McGraw-Hill, Tokyo, **1956**
[33] Taylor, R., Kennard, O., Versichel, W., *J. Am. Chem. Soc.* **1983**, *105*, 5761–5766
[34] Taylor, R., Mullaley, A., Mullier, G. W., *Pestic. Sci.* **1990**, *29*, 197–213
[35] Grubbs, F. E., *Technometrics* **1969**, *11*, 1–21
[36] Barnett, V., Lewis, T., *Outliers in Statistical Data*, Wiley, Chichester, **1978**
[37] Efron, B., Tibshirani, R., *Science* **1991**, *253*, 390–395
[38] Allen, F. H., Johnson, O., *Acta Cryst.* **1991**, *B47*, 62–67
[39] Upton, G. J. G., Fingleton, B., *Spatial Data Analysis by Example, Vol. 2, Categorical and Directional Data*, Wiley, Chichester, **1989**
[40] Newton, M. D., *Acta Cryst.* **1983**, *B39*, 104–113
[41] Ceccarelli, C., Jeffrey, G. A., Taylor, R., *J. Mol. Struct.* **1981**, *70*, 255–271
[42] Desiraju, G. R., *J. Chem. Soc. Chem. Comm.* **1990**, 454–455
[43] Allen, F. H., *Acta Cryst.* **1986**, *B42*, 515–522
[44] Birnbaum, A., *J. Am. Stat. Assoc.* **1954**, *49*, 559–574
[45] Lancaster, H. O., *Biometrika* **1949**, *36*, 370–382

[46] Pearson, E.S., *Biometrika* **1950**, *37*, 383–398
[47] Blessing, R.H., *J. Am. Chem. Soc.* **1983**, *105*, 2776–2783
[48] Benedetti, E., in: *Peptides, Proceedings of the Fifth American Peptide Symposium:* Goodman, M., Meienhofer, J. (eds.). Wiley, New York, **1977**, pp. 257–273
[49] Domenicano, A., Murray-Rust, P., *Tetrahedron. Lett.* **1979**, 2283–2286
[50] Topliss, J.G., Edwards, R.P., *J. Med. Chem.* **1979**, *22*, 1238–1244
[51] Cramer, R.D. III, Bunce, J.D., Patterson, D.E., Frank, I.E., *Quant.-Struct. Act. Relat.* **1988**, *7*, 18–25
[52] Allen, F.H., Doyle, M.J., Taylor, R., *Acta Cryst.* **1991**, *B47*, 29–40
[53] Chatfield, C., Collins, A.J., *Introduction to Multivariate Analysis*, Chapman and Hall, London, **1980**
[54] Auf der Heyde, T.P.E., *J. Chem. Educ.* **1990**, *67*, 461–469
[55] Malinowski, E.R., Howery, D.G., *Factor Analysis in Chemistry*, Wiley, New York, **1980**
[56] Murray-Rust, P., Motherwell, W.D.S., *Acta Cryst.* **1978**, *B34*, 2534–2546
[57] Domenicano, A., Murray-Rust, P., Vaciago, A., *Acta Cryst.* **1983**, *B39*, 457–463
[58] Allen, F.H., Doyle, M.J., Auf der Heyde, T.P.E., *Acta Cryst.* **1991**, *B47*, 412–424
[59] Harman, H.H., *Modern Factor Analysis*, University of Chicago Press, Chicago, **1967**
[60] Taylor, R., *J. Mol. Graph.* **1986**, *4*, 123–131
[61] Kilpatrick, J.E., Pitzer, K.S., Spitzer, R., *J. Am. Chem. Soc.* **1947**, *69*, 2483–2488
[62] Altona, C., Geise, H.J., Romers, C., *Tetrahedron* **1968**, *24*, 13–32
[63] Lifson, S., Warshel, A., *J. Chem. Phys.* **1968**, *49*, 5116–5129
[64] Pickett, H.M., Strauss, H.L., *J. Am. Chem. Soc.* **1970**, *92*, 7281–7288
[65] Murray-Rust, P., *Acta Cryst.* **1982**, *B38*, 2765–2771
[66] Berry, R.S., *J. Chem. Phys.* **1960**, *32*, 933–941
[67] Howard, J.A.K., Pitchford, N.A., Allen, F.H., in preparation
[68] Murray-Rust, P., Raftery, J., *J. Mol. Graph.* **1985**, *3*, 50–59
[69] Everitt, B., *Cluster Analysis, 2nd edn.*, Wiley, New York, **1980**
[70] Davison, M.L., *Multidimensional Scaling*, Wiley, New York, **1983**
[71] Massart, D.L., Kaufman, L., *The Interpretation of Analytical Chemical Data by the Use of Cluster Analysis, Chemical Analysis Monographs, Vol. 65*, Wiley, New York, **1983**
[72] Allen, F.H., Doyle, M.J., Taylor, R., *Acta Cryst.* **1991**, *B47*, 41–49
[73] Cormack, R.M., *J. R. Statist. Soc.* **1971**, *A 134*, 321–367
[74] Jarvis, R.A., Patrick, E.A., *IEEE Trans. Comput.* **1973**, *C22*, 1025–1034
[75] Willett, P., Winterman, V., Bawden, D., *J. Chem. Inf. Comput. Sci.* **1986**, *26*, 109–118
[76] Nørskov-Lauritsen, L., Bürgi, H.B., *J. Comput. Chem.* **1985**, *6*, 216–228
[77] Searle, W.S., *SAS Technical Report A-108*, SAS Institute Inc., Cary, NC, **1983**
[78] Allen, F.H., Doyle, M.J., Taylor, R., *Acta Cryst.* **1991**, *B47*, 50–61
[79] Allen, F.H., Taylor, R., *Acta Cryst.* **1991**, *B47*, 404–412
[80] Sheldrick, B., Akrigg, D., *Acta Cryst.* **1980**, *B36*, 1615–1620
[81] Taylor, R., Kennard, O., *J. Am. Chem. Soc.* **1982**, *104*, 3209–3212
[82] Everitt, B.S., *Graphical Techniques for Multivariate Data*, North-Holland, New York, **1978**
[83] Andrews, D.F., *Biometrics* **1972**, *28*, 125–136
[84] Friedman, J.H., Tukey, J.W., *IEEE Trans. Comput.* **1974**, *C23*, 881–890
[85] Jones, M.C., Sibson, R., *J. R. Statist. Soc.* **1987**, *A 150*, 1–36
[86] Livingstone, D.J., Hesketh, G., Clayworth, D., *J. Mol. Graph.* **1991**, *9*, 115–118
[87] Chernoff, H., *J. Am. Stat. Ass.* **1973**, *68*, 361–368
[88] Taylor, R., *J. Applied Cryst.* **1986**, *19*, 90
[89] Nie, N.H., Hull, C.H., Jenkins, J.G., Steinbrenner, K., Bent, D.H., *Statistical Package for the Social Sciences, 2 edn.*, McGraw-Hill, New York, **1975**
[90] Sali, A., Blundell, T.L., *J. Mol. Biol.* **1990**, *212*, 403–428
[91] Singh, J., Thornton, J.M., *J. Mol. Biol.* **1990**, *211*, 595–615
[92] Engh, R.A., Huber, R., *Acta Cryst.* **1991**, *A 47*, 392–398
[93] Glasgow, J.I., Fortier, S., Allen, F.H., *Proc. 7th IEEE Conference on Artificial Intelligence Applications*, Miami, Florida, *1991*
[94] Allen, F.H., Rowland, R.S., Fortier, S., Glasgow, J.I., *Tetrahedron Computer Methodology*, **1990**, *3*, 757–774

5 Structure Correlation; the Chemical Point of View

Hans Beat Bürgi and Jack D. Dunitz

5.1 Introduction

In the preceding chapters we have seen how to define and describe the structure of a molecule or molecular fragment (Chapters 1 and 2), how to retrieve the relevant information from electronic databases (Chapter 3) and how to analyze this information with a variety of statistical methods (Chapter 4). Fragment definition, data retrieval and statistical analysis in structure correlation have their analogs in ordinary chemical experiments: First a problem is identified, then the data are acquired from an appropriate experiment, filtered to distinguish the relevant from the irrelevant, and processed into a clear summary form. However, this is not the final step. The data still need to be understood.

This chapter describes ways to interpret in chemical or physical terms correlations among structural parameters and also results of more sophisticated statistical analyses. We start with qualitative arguments relating interatomic distances to chemical reactivity (Section 5.2). Then we discuss in some detail the concept of the molecular potential energy surface (Section 5.3), which has already been alluded to several times in the preceding chapters. A consideration of the symmetry properties of the energy surface can tell us which types of molecular distortions are expected to be coupled, and which form the couplings may take. In certain cases we can determine relative magnitudes and ratios of force constants. Such considerations lead us to the principle of structure correlation (Section 5.4). Finally, we present a principle of structure-energy correlation, which combines the structural similarities and differences of related molecules with equilibrium and activation energies of their reactions (see Section 5.5) and also provides a common viewing point for structure-structure, structure-energy and energy-energy correlations.

5.2 Structural Probes of Reactivity, Non-Bonded Distances

By the late thirties, Pauling and others had laid the basis for systematic structural chemistry in tables of covalent, ionic, van der Waals, metallic and other radii [1]. These quantities were found to depend only little on atomic environment. Combination of structural information with thermodynamic data showed that, in general, interatomic distance correlates with bonding energy. Thus, covalent distances are typically $1-2$ Å shorter than van der Waals distances, and covalent bond energies are typically $10-100$ times larger than van der Waals energies. As a result of such correlations, interatomic distance can be used as an easily available, rough and ready measure of relative stability.

As reliable structural data accumulated, an increasing number of interatomic distances were found to deviate from standard values. Early examples are hydrogen bonds between proton donors DH and proton acceptors A (see Chapter 12 and [2]). A typical structural feature of such bonds is an approximately linear arrangement $D-H...A$. The distances $H...A$ and $D...A$ may be as much as $0.5-1.0$ Å shorter than the respective van der Waals distances, but longer by about the same amount than the respective covalent ones. Short hydrogen bond distances imply strong interactions. Hydrogen bonding may be considered as a structural expression of Brönsted acidity and basicity, terms which describe aspects of proton transfer, i.e. a chemical reaction. In fact, given a choice, proton acids surround themselves in the crystal by the strongest available bases [3].

Hydrogen bonds are only one, though probably the most widespread, of many analogous interactions found primarily between main group elements. In 1968 Bent reviewed the then available data under the heading of "Donor-Acceptor Interactions" [4] and made the comparison with hydrogen bonding. The salient features of donor-acceptor interactions may be illustrated by a single example, that of the planar sheets in crystalline iodine (Figure 5.1) [5]. The I_2-molecules are arranged in a herring-bone pattern. There are three nearest neighbor distances: 2.70 Å between the two bonded atoms, 3.54 Å between two non-bonded atoms collinear with a bonded one, and 4.07 Å between non-bonded atoms of parallel molecules. The bond distance is close to the $I-I$ distance in the gas phase (2.66 Å), the long non-bonded distance is close to the van der Waals separation (4.3 Å). The intermediate distance at 3.54 Å is almost midway ($+0.88$ Å and -0.76 Å) between these two limits! Bent's interpretation of these distances implies that I_2 is a Lewis base (electron donor) perpendicular to its molecular axis and a Lewis acid (electron acceptor) along this axis.

Bent also mentions the packing of triketoindane [6], which shows short intermolecular distances between carbonyl oxygen and carbon atoms (Figure 5.2). In the language of organic chemistry, this is an interaction between a nucleophilic oxygen and an electrophilic carbon atom. A later study of the environments of divalent sulfur $(Y-S-Z)$ in crystals showed the shortest non-bonded contacts between sulfur

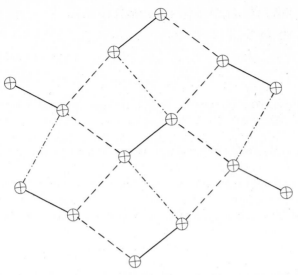

Fig. 5.1. A planar sheet of atoms in the crystal structure of I_2 (solid line: covalent bond, 2.10 Å; dashed line: secondary bond, 3.54 Å; dash-dotted line: Van der Waals contact, 4.07 Å)

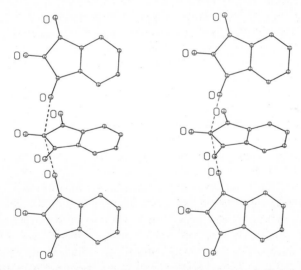

Fig. 5.2. Stereo-drawing of packing motif from the crystal structure of triketoindane. The O . . . C intermolecular contacts (2.83 Å) are indicated by dashed lines

and electrophiles roughly 20° from the perpendicular to the plane Y–S–Z, whereas nucleophiles approach sulfur roughly along the extension of the Y–S or Z–S bond [7].

There is no point in extending the list of examples, as they all follow the same pattern: an electron donor and an electron acceptor in close proximity. The distance between them is longer than a normal bond but short enough to qualify it as a "secondary bond" [8]. Such bonds, like normal bonds, have specific directional preferences.

These observations call for a reconsideration of the concept of van der Waals radii, which implies that distances between non-bonded atoms are independent of direction and equal to the sums of corresponding radii. Besides the sulfur example already mentioned, other studies have been concerned with mapping the directionality of non-bonded contacts. Nyburg and coworkers [9] have analyzed X ... X contacts among $C-C\equiv N$, $(C)_2C=O$, $C=S$, $C=Se$, $C-F$, $C-Cl$, $C-Br$ and $C-I$ fragments in terms of orientation-dependent van der Waals surfaces. For O and N such surfaces are virtually spherical, but for S, Cl, Br, I the shape is that of a rotation ellipsoid with its short radius along the $C-X$-bond vector (Figure 5.3). In a similar study, Murray-Rust and Motherwell [10] showed that the probability of finding a short non-bonded contact is also orientation dependent, small when $C-I...I$ (or O) forms a right angle and larger when the arrangements are linear. A natural extension of these approaches is to map the surface and associated probabilities of entire functional groups. This has been done, for example, for the quaternary ammonium group, $RN(CH_3)_3^+$ [11].

The loss of simplicity associated with direction-dependent contact analysis is compensated by the gain in information about the nature of non-bonded interactions, the very stuff of supramolecular chemistry. This new complexity has hardly begun to be included in force fields for atom-atom potential calculations, which adhere, for the most part, to spherical atom models as far as the non-bonded interactions are concerned. Other aspects of intermolecular distances are taken up in Chapter 12 on Hydrogen Bonding and in Chapter 13 on Molecular Packing.

5.3 Conceptual Framework: Energy Surfaces

So far, an intuitively appealing analogy between non-bonded interatomic distances and chemical reactivity has been given. As a more general basis for relating structure and chemical properties, we need the concept of the molecular potential energy surface: each point in configuration space is associated with a definite potential energy. In the following sections we shall discuss ways to relate features of statistical correlations with the general topology of energy minima and maxima on such surfaces.

Strictly speaking, each molecule is characterized by a specific energy surface, different from those of other molecules. It can be expected, however, that energy surfaces of similar molecules are themselves similar, at least as far as their low-energy regions are concerned. We shall describe two approaches to the structure-energy problem:

(1) A qualitative one, which concentrates on general low-energy features shared by a family of molecules with a common fragment.

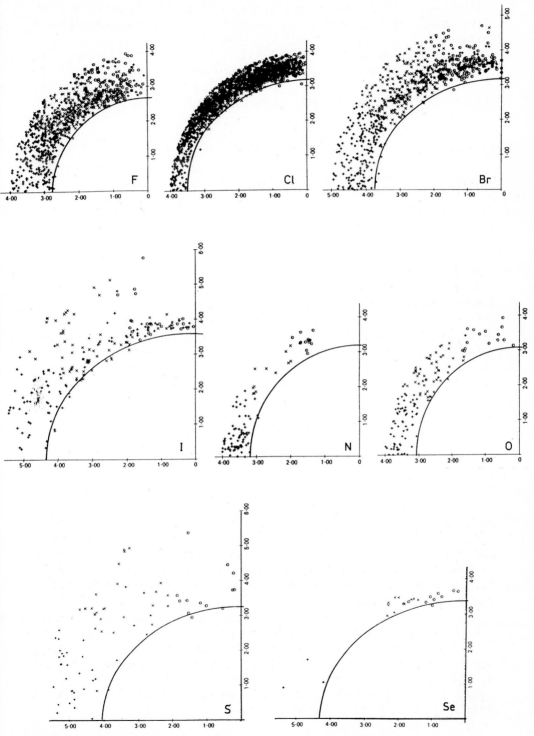

Fig. 5.3. Scatter-plots of X . . . X distances vs. C−X . . . X angles, linear arrangements along the vertical. The significance of the circles, crosses, and arrows is irrelevant here

(2) A quantitative one, which concentrates on the systematic aspects of the energy differences between surfaces of related molecules on the one hand, and the correlation between such differences and structure, on the other.

5.3.1 Energy Minima, Force Constants and Structure Correlation

A convenient way to represent an energy surface in the neighborhood of an arbitrary point D_0 of configuration space, is in terms of a Taylor expansion about this point.

$$
\begin{aligned}
V(D_0,D) = V(D_0) &+ \sum_i (\partial V/\partial d_i)\, d_i \\
&+ \sum_{ij} \sum (\partial^2 V/\partial d_i \partial d_j)\, d_i d_j/2 \\
&+ \sum_{ijk} \sum \sum (\partial^3 V/\partial d_i\, \partial d_j\, \partial d_k)\, d_i\, d_j\, d_k/3!
\end{aligned}
\tag{5.1}
$$

$$+ \text{higher order terms}$$

where the d_i's are deformations of the reference structure D_0 along the displacement vectors p_i (Chapter 1, Section 1.4.3; Chapter 2, Section 2.3). If D_0 corresponds to an extremum of V the first derivatives are zero, the second derivatives are the force constants f_{ij} of the reference structure, the terms $(\partial^3 V/\partial d_i\, \partial d_j\, \partial d_k)/3!$ and analogous higher order derivatives are the so-called anharmonicity constants f_{ijk}, etc. The mixed elements, $f_{ij}\, d_i\, d_j$, imply that coupled deformations d_i, d_j and $d_i, -d_j$ do not require the same energy. The principal deformation directions are given by the eigenvectors q_i and eigenvalues f_i of the matrix of force constants f. The eigenvector associated with the smallest eigenvalue of f points in the direction of slowest increase in V, i.e. the direction of easiest deformability. In quadratic approximation

$$
V = V(D_0) + \sum_i f_i q_i^2
\tag{5.2}
$$

For a fixed deformation energy $V(D_0, Dq) - V(D_0)$, this equation represents an equipotential hyperellipsoidal surface with principal axes along the eigenvectors q_i.

As an example, consider the bond-stretching deformations of a symmetric I_3 fragment with distance deformations d_1 and d_2.

$$
V = V(D_0) + f_{11} d_1^2/2 + f_{22} d_2^2/2 + f_{12} d_1 d_2
\tag{5.3}
$$

The choice of a symmetric reference structure D_0 implies $f_{11} = f_{22}$. The mixed element $f_{12} d_1 d_2$ says that changing both bond lengths in the same sense does not require the same deformation energy as increasing one and decreasing the other. In matrix representation we obtain

$$V = V(d_0) + (1/2)[d_1 d_2] \begin{bmatrix} f_{11} & f_{12} \\ f_{12} & f_{22} \end{bmatrix} \begin{bmatrix} d_1 \\ d_2 \end{bmatrix} . \tag{5.4}$$

The principal directions are

$$\begin{bmatrix} q_1 \\ q_2 \end{bmatrix} = \begin{bmatrix} 1/\sqrt{2} & 1/\sqrt{2} \\ 1/\sqrt{2} & -1/\sqrt{2} \end{bmatrix} \begin{bmatrix} d_1 \\ d_2 \end{bmatrix} = \begin{bmatrix} (d_1+d_2)/\sqrt{2} \\ (d_1-d_2)/\sqrt{2} \end{bmatrix} \tag{5.5}$$

The associated force constant matrix is

$$\begin{bmatrix} f_1 & 0 \\ 0 & f_2 \end{bmatrix} = \begin{bmatrix} f_{11}+f_{12} & 0 \\ 0 & f_{11}-f_{12} \end{bmatrix} \tag{5.6}$$

For a positive value of f_{12} a symmetric stretch $(d_1+d_2)/\sqrt{2}$ is more costly in energy than an antisymmetric stretch $(d_1-d_2)/\sqrt{2}$. Or, in other words, for a given amount of deformation energy, larger deformations are possible along q_2 than along q_1. Thus, if a molecular fragment, here I_3, is forced by its environment to deviate from the reference structure, it is likely to do so along the direction of least resistance, here along q_2, the antisymmetric stretch coordinate. This is confirmed, at least for I_3, by Figure 5.4 (similar to Figure 1.5) which shows the central part of the distribution of sample points extending along $(d_1-d_2)/\sqrt{2}$.

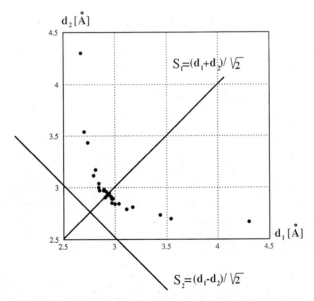

Fig. 5.4. Correlation plot of interatomic distances in linear triiodide anions. The directions of the symmetric and antisymmetric stretch coordinates $q_1 = S_1 = (d_1+d_2)/\sqrt{2}$ and $q_2 = S_2 = (d_1-d_2)/\sqrt{2}$ are indicated

The I_3 example thus illustrates how signs of force constants and their ordering by magnitude can be obtained from structural scatter plots. Under certain conditions, it may even be possible to deduce ratios of force constants from structural data. Again, this is best illustrated by a simple example. Suppose that the relevant part of V for a structural fragment takes the form

$$V = V(D_0)+f_{11}d_1^2/2+f_{22}d_2^2/2+f_{12}d_1d_2 \qquad (5.7)$$

If one of the two structural parameters, d_1 say, is constrained to assume a specific value by the fragment environment, then the other, here d_2, will assume a value which minimizes V, i.e.

$$dV/dd_2 = 0 = f_{22}d_2+f_{12}d_1 \qquad (5.8)$$

Thus $f_{12}/f_{22} = -d_2/d_1$, and the ratio of force constants is obtained from the experimental values of d_1 and d_2. This type of argument has been applied to strained cyclic polyenes [12], where the energy involving small out-of-plane deformations of the double bond can be expressed in simplified form as

$$V = V(\omega_0)+f_{11}\omega^2(CC=CC)/2$$

$$+f_{22}\{\omega^2(H_1C=CC)+\omega^2(CC=CH_2)\}/2$$

$$+f_{12}\{\omega(CC=CC)[\omega(H_1C = CC)+\omega(CC=CH_2)]\} \ . \qquad (5.9)$$

The ω's are the three independent torsion angles about the $C=C$ double bond. With the usually observed approximate local C_2-symmetry $\omega(H_1C=CC)$ $\approx\omega(CC=CH_2) = \omega(H)$. If the torsion angle $\omega(CC=CC)$ is now assumed to be largely determined by ring closure contraints, as in small and medium ring systems, then minimization of V with respect to $\omega(H)$ leads to

$$f_{12}/f_{22} = -\omega(H)/\omega(CC=CC) \ . \qquad (5.10)$$

The ratio of force constants estimated in this way from structural data, or rather the dependence of this ratio on double-bond character, is matched by corresponding ratios from vibrational spectroscopy [12a].

5.3.2 Energy Minima, Symmetry Force Constants
and Structure Correlation

It is sometimes convenient to transform to principal directions of f in steps, by going first to symmetry coordinate S_i. With the notation of Chapter 2, the displacement vector from the symmetric reference structure is

$$D = \sum_i D_i S_i \; . \tag{5.11}$$

In quadratic approximation the deformation energy may be written as

$$V = V(\boldsymbol{D}_0) + \sum_{kij} \sum \sum F_{ij}(\Gamma_k) D_i(\Gamma_k) D_j(\Gamma_k) \; . \tag{5.12}$$

The terms are now grouped according to the irreducible representations Γ_k of D_i and D_j. Note that there are no terms involving mixed irreducible representations, i.e. $D_i(\Gamma_k) D_j(\Gamma_l)$. This comes about because the energy V is a structure invariant, that is, its value cannot change when a coordinate transformation is made. For a given symmetry transformation, the D's remain the same or change sign. If D_i and D_j belong to the same irreducible representation their product transforms as the totally symmetric representation, that is, it does not change, not even in sign, for any symmetry operation of the group. If, however, they would belong to different irreducible representations, then their product must change sign for at least one symmetry operation of the group.

For the I_3 example, the symmetry coordinates are

$$S_1 = (d_1 + d_2)/\sqrt{2}$$
$$S_2 = (d_1 - d_2)/\sqrt{2} \; . \tag{5.13}$$

The energy is

$$V = V(\boldsymbol{D}_0) + F_{11} D_1^2/2 + F_{22} D_2^2/2 \; . \tag{5.14}$$

Here, there is no term $F_{12} D_1 D_2$, since S_1 is symmetric and S_2 antisymmetric with respect to the inversion symmetry of the reference structure (or the mirror line in configuration space, Figure 5.4). In this simple example, the symmetry coordinates coincide with the principal directions q_1 and q_2.

For a more interesting example, we consider a tetrahedral MX_4 system. As discussed in Chapter 2 (Table 2.4), the symmetry coordinates transform as $2A_1$, E, and $2T_2$. The two sets of triply degenerate deformation coordinates are S_3, in the distances r_i, and S_4, in the angles θ_{ij}. In quadratic approximation, the potential energy for each of these has spherical symmetry, but there is a cross-term. The T_2-part of the potential energy is then

$$V(T_2) = V(\boldsymbol{D}_0, T_2) + F_{33}(D_{3a}^2 + D_{3b}^2 + D_{3c}^2)/2$$

$$+ F_{44}(D_{4a}^2 + D_{4b}^2 + D_{4c}^2)/2$$

$$+ F_{34}(D_{3a}D_{4a} + D_{3b}D_{4b} + D_{3c}D_{4c}) \tag{5.15}$$

We can discuss the sign of F_{34} in the same way we discussed f_{12} for the I_3 example in Section 5.3.1. For positive F_{34}, V is smaller in the second and fourth quadrants of the S_{3a}, S_{4a}-subspace than in the first and third quadrants: distortions along S_{3a} should then be associated with distortions along $-S_{4a}$ (i.e. lengthening two distances and shortening the other two while closing the angle between the distances that become longer and opening the angle between those that become shorter). This is what is observed (Figure 5.5) [13], so we conclude that $F_{34} > 0$. We can even proceed a little further. If we have reason to believe that the principal directions of the distribution in Figure 5.5 (see Chapter 1.2.4) map the principal deformation directions $q_i(T_2)$ of $F(T_2)$, then the S_{3a}- and S_{4a}-vector components of q_{3a} and q_{4a} are obtained directly from the experimental distribution. In matrix notation

$$\begin{bmatrix} q_{3a} \\ q_{4a} \end{bmatrix} = \begin{bmatrix} B_1 & B_2 \\ -B_2 & B_1 \end{bmatrix} \begin{bmatrix} S_{3a} \\ S_{4a} \end{bmatrix} = BS \qquad (5.16)$$

with $B_1^2 + B_2^2 = 1$. By definition, B diagonalizes F, i.e. $BFB^T = F'$ with $F'_{12} = 0 = F_{34}(B_1^2 - B_2^2) - B_1 B_2 (F_{33} - F_{44})$. This leads to another relationship between a ratio of force constants and a ratio of structure-derived quantities [12b]:

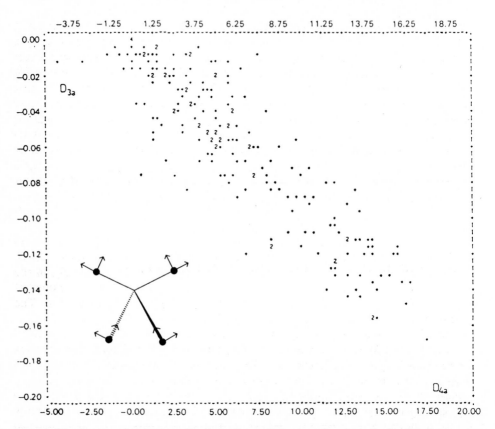

Fig. 5.5. Distribution of distortions D_{3a} along S_{3a} vs. distortions D_{4a} along S_{4a} for PO_4-fragments. The correlation says that as two distances lengthen and two shorten, the angle between the longer distances becomes smaller while the angle between the shorter distances becomes wider

$$\frac{F_{34}}{F_{33}-F_{44}} = \frac{B_1 B_2}{B_1^2 - B_2^2} . \qquad (5.17)$$

For a distribution in n dimensions the corresponding $(n \times n)$ matrix $\mathbf{B}$ leads to $n(n-1)/2$ conditions between the elements of $\mathbf{F}$. The limitations of this approach are discussed in detail in Section 5.4.

The real usefulness of expanding energy in terms of symmetry coordinates becomes apparent when we consider third-order terms [12b]. The condition for third-order terms to occur is similar to that for second-order terms, namely that F_{ijk} is non-zero only if the product $D_i(\Gamma_l)D_j(\Gamma_m)D_k(\Gamma_n)$ transforms as the totally symmetric representation. This condition reduces the number of allowable F_{ijk}'s substantially.

For the I_3 example, the energy expansion in symmetry coordinates contains the following third-order terms.

$$F_{111}D_1^3 + F_{222}D_2^3 + F_{112}D_1^2 D_2 + F_{122}D_1 D_2^2 . \qquad (5.18)$$

Since the inversion operation changes the sign of D_2 but leaves D_1 unchanged, it follows that both F_{222} and F_{112} must be zero. If F_{111} were negative, the increase in energy along $-S_1$ would be steeper than along $+S_1$; in other words, it would cost more energy to contract the molecule than to expand it. This is what we expect for bonds following a Morse-like potential, although Figure 5.4 in itself contains nothing that would lead to this conclusion. On the other hand, Figure 5.4 does have something to say about F_{122}. Suppose it were negative; then the term $F_{122}D_1 D_2^2$ would lower the energy symmetrically in the $+S_1$, $\pm S_2$ quadrants but would increase it in the $-S_1$, $\pm S_2$ quadrants. Thus the shallowest direction of the potential, which was shown above to correspond to S_2 in quadratic approximation, now curves around and takes up increasing positive components along S_1. This is just what is observed, so we conclude that F_{122} is indeed negative.

5.3.3 Reaction Profiles and Structure Correlation

Energy surfaces are also used to describe chemical reactions, albeit usually in vastly simplified form, that of the energy reaction-coordinate diagram (or reaction profile for short), examples of which are to be found in every general chemistry textbook (Figure 5.6). Reaction profiles are a convenient way of summarizing many kinds of experimental data and for discussing reactions from a molecular viewpoint.

Figure 5.6 shows two minima representing reactant and product and a maximum representing the transition state of the reaction. The reaction coordinate is taken to represent all changes of interatomic distances and angles during the reaction. The energy difference between the minima indicates relative thermodynamic stability, while the difference between minima and maximum is related to (kinetic) lability.

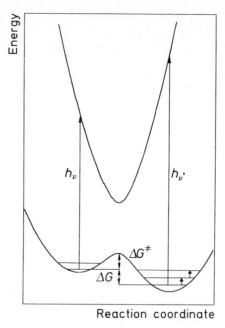

Fig. 5.6. Schematic representation of a reaction profile summarizing structural, spectroscopic, thermodynamic and kinetic experimental data

The curvature about the minima is an expression of molecular flexibility, which can be ascertained, in principle at least, from knowledge of the molecular vibrations. From a more general viewpoint, the reaction profile must be regarded as a more or less complicated curve on a multi-dimensional energy hyper-surface.

One might ask how crystal-structure studies can contribute to the understanding of chemical reactivity beyond the characterization of the reactants, products and possibly intermediates of a series of chemical transformations, i.e. beyond confirming the expected, or elucidating the unexpected, outcome of a reaction. In terms of Figure 5.6 such studies merely fix the structures associated with the starting and end points of reaction, but say nothing about what happens in between, at least not directly. However, as described in Section 5.2 for a few representative examples, unusual structural features can be suggestive of the way a reaction might go. A natural next step is, therefore, to compile and compare unusual structures, to search for correlations among their geometric parameters, and to look for similarities with the structural changes expected to occur during chemical reactions.

For an example we return to the scatterplot Figure 5.4 of the two distances in the I_3 fragment, an early version of which was given by Bent in his review of donor-acceptor interactions [4]. Following earlier suggestions by the Slaters [14], Bent noted a reciprocal relationship between the two distances, namely a "lengthening of an intramolecular interaction with a shortening of a *trans* intermolecular interaction (until finally, perhaps, the two interactions become indistinguishable)". His conclusion: "The hyperbolic-like curve may be presumed to show, approximately, the changes that occur in the distances between nearest neighbors in the linear exchange reaction $I_1 + I_2I_3 = I_1I_2 + I_3$" [4]. This statement shows unambiguously that Bent drew the parallels between his correlation curve and a chemical reaction.

This approach for delineating the structural course of a chemical reaction in configuration space has been applied by now to many organic and inorganic transformations, including bond-breaking/bond-making processes as well as conformational interconversion. Many examples are discussed in Chapters 6 to 9 of this book.

5.4 The Principle of Structure-(Structure) Correlation

The approach outlined in Section 5.3.3 has been described in the following way:

"Although direct observation of a molecule along the reaction pathways does not seem feasible, its visualization at least does. According to the structure correlation hypothesis, the gradual distortion or static deformation that a molecular fragment of interest manifests collectively over a large variety of crystalline frameworks may be assumed to mirror the distortion which that fragment would undergo along a given reaction coordinate. The various crystal or molecular structures are considered to constitute a series of 'frozen-in' points, or snapshots, taken along the reaction pathway, which, when viewed in the correct order, yield a cinematic film of the reaction" [15].

More generally, the relationship of structure correlations to features of the energy surface is expressed in the principle of structure correlation [16a]:

"If a correlation can be found between two or more independent parameters describing the structure of a given structural fragment in a variety of environments, then the correlation function maps a minimum energy path in the corresponding parameter space";
or, more circumspectly:
"Observed structures tend to concentrate in low lying regions of the potential energy surface" [16b].

This means that the regions in Figures 5.4 and 5.5 that are populated by sample points correspond to energetically favorable atomic arrangements.

It would be tempting to conclude that the density of points in a scatterplot can be taken as a *quantitative* measure of relative energy, high density corresponding to low energy. Indeed, several authors have succumbed to this temptation (see below). Consider, however, the following argument: Suppose, for example, that we want to study PO_4-tetrahedra but are unaware of the possibility that there might have been a special concentration of interest in, say, the structures of diesters of phosphoric acid. We take a sample of PO_4 tetrahedra, including anions, esters — whatever we can find in the Cambridge Structural Database. Diesters will tend to contain PO_4 groups with twofold symmetry, as opposed to other types of PO_4 groups with threefold symmetry. Thus, even if we find an accumulation of points imaging groups with twofold symmetry, it would be wrong to attribute this to a special energetic preference. In general, we do not know the reasons for variations in the density of scatterpoints.

Nevertheless, inspection of the data base may sometimes give us a feeling for the presence or absence of bias. The structural data for I_3 in Figure 5.4 [17], for example, come from gaseous and solid I_2, from salts of I_3^-, I_5^-, I_7^- and I_8^{2-} with various counterions, such as NH_4^+, $N(CH_2CH_3)_4^+$, $As(C_6H_5)_4^+$, Cs^+, and $[Fe(C_5H_5)_2]^+$, etc. The anions and cations differ substantially in size and polarizability. There would not seem to be too much of an obvious bias.

Several studies have attempted to quantify differences in the density of scatterpoints. All of them make explicit or implicit use of Boltzmann distributions. One form of the argument pertains to molecules which may assume two (or more) different structures, e.g. the C2'-*endo* and C3'-*endo* conformations of furanose [18]. The ratio of the probabilities P_i for finding one or the other structures is *assumed* to follow a Boltzmann-like distribution

$$P_1/P_2 = \exp\left(-\Delta G_{12}/RT_c\right) \qquad (5.19)$$

where RT_c has been interpreted as "the mean energy of deformation due to intermolecular interactions in crystals..." [18].

A modified form of the argument addresses scatterplots which may be approximated by a normal distribution of the deformation x with mean $\langle x \rangle$ and variance $\sigma^2(x)$

$$P(x) \approx \exp\left[-(x-\langle x \rangle)^2/2\sigma^2(x)\right] . \qquad (5.20)$$

It is then assumed that the probability of observing a fragment structure with a particular deformation decreases exponentially with the deformation energy ΔG or V, which can be taken, for small deformations at least, as a quadratic function of $(x-\langle x \rangle)$, leading to

$$P'(x) = \exp\left(-f(x-\langle x \rangle)^2/2E_c\right) \qquad (5.21)$$

where E_c is a constant to be determined, not necessarily RT as in the Boltzmann distribution. On the basis of the similarity of P and P' the conclusion is drawn that the variance of x is inversely proportional to the force constant f

$$\sigma^2(x) = E_c/f . \qquad (5.22)$$

E_c has been described as "the average amount of energy available from packing forces to distort a functional group" [19]. Different values of E_c and RT_c have been found from statistical analysis of different types of deformation.

While it is true that large, high-energy deformations are less likely to occur (and be observed) than small, low-energy ones, there is a serious flaw in these arguments. An ensemble of structural parameters obtained from chemically different compounds in a variety of crystal structures does not even remotely resemble a closed system at thermal equilibrium and does not therefore conform to the conditions necessary for the application of the Boltzmann distribution. It is thus misleading to draw an analogy between this distribution and those derived empirically from statistical analysis of observed deformations in crystals [20].

This criticism can be formalized in terms of a simple physical model that furnishes an alternative relationship between $\sigma^2(x)$ and f for a structural parameter x with equilibrium value x_0 in the absence of perturbing forces. The energy increase for small deformations $x-x_0$ is assumed to be quadratic and the crystal environment is supposed to exert some perturbing force a on the system, so that the linearly perturbed energy becomes:

$$E(x) = f(x-x_0)^2/2 + a(x-x_0) . \tag{5.23}$$

The new equilibrium value x_e is thus displaced to

$$x_e = x_0 - a/f . \tag{5.24}$$

Each crystal environment can be expected to exert a different perturbing force, and for a collection of such environments there will be some distribution $P(a)$, which could well be supposed to be normal (since it results from a large number of independent causes). Whatever the type of distribution

$$P(a/f) = P(x_e - x_0) \tag{5.25}$$

that is, the distribution of a determines the distribution of x_e. In particular, for $\langle a \rangle = 0$, $\langle x_e \rangle = \langle x_0 \rangle$. This model leads to the relationship

$$\sigma^2(x_e - x_0) = \sigma^2(a)/f^2 . \tag{5.26}$$

The force constant can only be determined if $\sigma^2(a)$ is known, and this will seldom be the case. Hence the actual deformation energies cannot, in general, be derived from the observed parameter variances. Moreover, in contrast to the Boltzmann-type argument where the parameter variance was proportional to f^{-1} it is now proportional to f^{-2}. Qualitatively there is an inverse relationship for both models.

This discussion can be extended to cover a multi-dimensional distribution involving several different structural parameters (interatomic distances, bond angles, torsion angles, etc.) by rewriting the above equations in matrix form:

$$E(x) = (x-x_0)^T F(x-x_0)/2 + a^T(x-x_0) \tag{5.27}$$

$$P(x_e - x_0) = P(F^{-1}a) = P(Ca) \tag{5.28}$$

$$\langle (x_e - x_0)(x_e - x_0)^T \rangle = C\langle aa^T \rangle C^T . \tag{5.29}$$

The left-hand side of the last equation involves the variance-covariance matrix of the observed distribution of structural parameters, and the right-hand side involves the compliance matrix $C = F^{-1}$ for the system in question as well as $\langle aa^{T}\rangle$, the second-moment matrix of the distribution of perturbing forces. This model provides a conceptual basis for relating observed distributions $P(x_e - x_0)$ to the energy surfaces associated with small x_e, but its actual application is beset with obstacles.

The variance-covariance matrix $\langle (x_e - x_0)(x_e - x_0)^{T}\rangle$ is usually contaminated by contributions from experimental error. Indeed, for small deformations and poor experimental data, it could be dominated by these experimental uncertainties. On the other hand, for very large deformations the quadratic energy dependence cannot be expected to hold. Once the harmonic approximation breaks down, the above equations would need to be replaced by more complicated expressions involving a larger number of unknown (anharmonic) force constants.

Even if these equations are assumed to be valid, the variance-covariance matrix $\langle (x_e - x_0)(x_e - x_0)^{T}\rangle$ cannot, in general, be resolved into C and $\langle aa^{T}\rangle$. However, if sufficient information about the force constants $F(=C^{-1})$ were available, the equation for $E(x)$ could be used to obtain the perturbing forces a for each observed environment separately. For a sufficiently large sample of different environments, $\langle aa^{T}\rangle$ could be calculated and its dependence on crystal environment studied. Conversely, the compliance matrix C could be derived if $\langle aa^{T}\rangle$ were known.

Clearly we cannot possibly be expected to know these quantities. The distribution of forces could be different for different kinds of structural parameter and is likely to depend in a complex way on the types of interaction (van der Waals, hydrogen bond, ionic forces) operative in the selection of crystal structures included in the investigation. The detailed analysis of these forces for a statistically significant sample of structures would be extremely laborious and could hardly be free from assumptions about the natures and magnitudes of these interactions.

After this rather sobering discussion of the limitations of interpreting structural correlations in a quantitative way, one may ask whether we can learn anything at all from scatterplots. We make three points here:

(1) In several cases the results of structure correlations have been compared with results of ab initio or force-field calculations. Invariably, it is found that regions in configuration space populated by data points coincide, approximately, with regions of low energy. We conclude that the presence of data points in a scatterplot indicates regions of low potential energy, whereas their absence does not necessarily imply high potential energy – it may merely mean that the sampling of configuration space with the available data is insufficient.

(2) It is a general experience in structure correlation studies that the definition of a structural fragment influences the appearance of scatterplots: They tend to be sparse with well defined correlations for restrictive fragment definitions, but well populated with fuzzy correlations for fuzzy fragment definition. Fuzziness of correlation is not necessarily a disadvantage because the diversity of perturbations implicit in fuzzy fragment definition will tend to distort a fragment structure in more than one direction. As a result, more can be learned about the shape of low energy regions of an energy surface.

(3) At first sight there seems to be a discrepancy between the logic of the formal treatment given in this section and the qualitative nature of the information derivable from structure correlations. However, the two aspects serve different purposes. From structure correlations we can glean at least some information, which, for many of the examples in this book, would have been difficult to obtain in any other way. The formal treatment, however, provides a framework for analysis and for finding out about the possibilities and limitations of the approach. As we shall show in the following sections, the formal treatment may be extended to analyze correlations between structure and energy.

5.5 Structure-Energy Correlation

The reaction paths described in Section 5.4 and in other chapters of this book have been derived from geometric structural data only. Structure correlation studies make no explicit reference to energy, although their results are usually interpreted in terms of gross, qualitative features of an energy surface, the lowlands of an energy landscape. In the real world the situation is different; it is not the gross, qualitative features of an energy surface common to a family of related molecules that determine the course of a reaction, but rather the characteristics associated with a specific molecule. A number of questions thus arise:

(1) What can be said about the reaction profile of a specific reaction, given the general information derived from structure correlation?

(2) Is it possible to combine qualitative conclusions from structure correlation with specific, quantitative information on an actual reaction, and if yes, how?

(3) Is there a similarity between the reaction profiles of several different but similar compounds undergoing the same type of reaction; how can such a similarity be described?

(4) What are the corresponding transition state structures? What are their similarities and differences?

These questions pose formidable problems. They entail nothing less than the reconstruction and comparison of individual reaction profiles from experimental data. Such reconstructions will be more or less detailed, depending on the type and amount of available experimental data and depending on the assumptions about the energy surface that are necessary for the analysis. In the following sections we indicate how to approach different aspects of this problem. In Section 5.5.1 we concentrate on the reactants and products of a reaction, their structures and relative energies. In Sections 5.5.3 and 5.5.4 we consider the region between, that around the transition state in particular. While the discussion is based on well known thermodynamic and kinetic principles (Section 5.5.2), an attempt is made to be as explicit as possible about the role of structure in describing and understanding reaction profiles.

5.5.1 Equilibria in Crystals

The thermodynamic quantities of a reaction, ΔG, ΔH, etc., and the structures of reactants and products fix the starting and end point of a reaction profile (Figure 5.6). In general these points are well characterized − structurally with the help of standard crystal structure analyses and energetically by equilibrium measurements in solution. In favorable cases, the experimental information necessary to define both reactant and product can be obtained from structure analyses on a single crystal specimen at different temperatures. In such cases the starting material and the product are present in the crystal simultaneously, but in varying proportions, depending on temperature, to give disordered structures.

A simple example is reorientation of C_{60} in its disordered low-temperature crystal structure [21]. The molecule can assume two crystallographically inequivalent orientations related by a rotation of about 60° about the crystallographic threefold axis of C_{60} (or of about 42° about one of its molecular twofold axes, Figure 5.7). The populations of the two orientations are not equal and depend on temperature. Diffraction studies at 110, 153 and 200 K yield $\Delta H = 257(13)\,\text{cal mol}^{-1}$, $\Delta S = 0.0(1)\,\text{cal mol}^{-1}\,\text{K}^{-1}$. The reaction coordinate is the rotation angle of C_{60}. From solid-state NMR the barrier to reorientation has been estimated to be about $4\,\text{kcal mol}^{-1}$ [22].

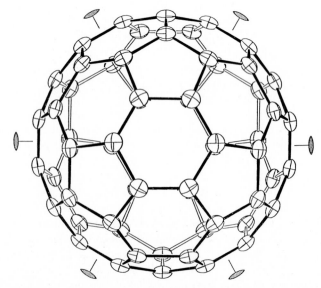

Fig. 5.7. The C_{60} molecule viewed down a threefold rotation-inversion axis. The upper half of the molecule is identified with filled, the lower half with unfilled bond lines. Molecular twofold axes are also shown. The unfilled bond lines also represent the upper half of the second orientation of C_{60} found in the crystal

A chemically more interesting example is proton transfer in the fungal metabolite citrinine (Figure 5.8) [23]. At low temperature only the more stable tautomer is found in the crystal. At higher temperatures the existence of both tautomers in varying proportions is needed to explain the diffraction data ($\Delta H = 1.6(6)$ kcal mol^{-1}, $\Delta S = 4.5(2.2)$ cal mol^{-1} K^{-1} from measurements between 20 and 293 K). Note that these results give no indication on the energy barrier of the reaction, nor on whether the protons change place in a concerted way or in a two-step process.

The Cu(II) complex, [Cu(bpy)$_2$ONO] (NO$_3$) (bpy: 2,2′-bipyridine), has been shown to undergo a ligand rearrangement in the crystal (Figure 5.9) [24]. The short Cu−O bond lengthens by about 0.5 Å, and the Cu−N bond *trans* to it by about 0.1 Å. Simultaneously, the long Cu−O bond and the Cu−N bond *trans* to it shorten correspondingly. For a molecule in an isotropic environment this process would be degenerate, i.e. the two forms would have the same energy. The observed energy difference of 0.22 kcal mol^{-1} in the crystal thus reflects the anisotropy of the crystal environment. The barrier has been estimated from spectroscopic considerations to be about 1.6 kcal mol^{-1}.

Other examples of reversible solid-state "reactions" in crystals have been described, e.g. conformational changes in organic crystals [25] and spin equilibria in

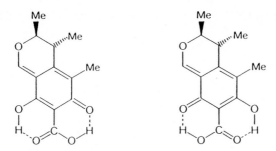

Fig. 5.8. Double proton shift in citrinine (more stable isomer on the left)

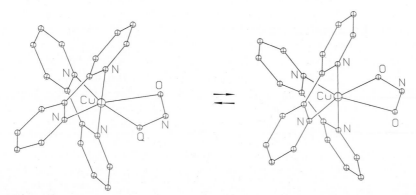

Fig. 5.9. Degenerate rearrangement of nitrite ligand in [Cu(bpy)$_2$ONO]$^+$-complexes (bpy: 2,2′-bipyridine)

various crystalline transition metal complexes [26]. However, variable temperature diffraction studies on such systems are still not numerous, and chemical interpretations of the results are given for only few examples. This is probably because the diffraction technique is capable of assessing the amount of the minor component only if the equilibrium constant is in the range 1 to 10 (or 1 to 100, at best) and because crystallographic disorder is often considered as a nuisance rather than as an opportunity. Molecules in crystals may also undergo photochemical reactions, and, in favorable cases, the sample remains a single crystal during the reaction. Due to high interconversion barriers, reactants and products are usually not in thermodynamic equilibrium and the reaction may proceed sufficiently slowly to allow measurement of several stages of partial conversion corresponding to disordered arrangements of reactant and product molecules. Attempts have been made to measure kinetic parameters for such processes and to relate them to features of the reaction cavity, i.e. the volume available for motions of the reacting groups [27a]. Little can usually be said about the detailed structural changes along the reaction path although some preliminary attempts have been made [27b]. The crystalline environment may impose severe limitations on the molecular motions [28], leading sometimes to highly stereoselective "topochemical" reactions [29].

5.5.2 Transition-State Theory and Free-Energy Relationships

The following sections deal with a much more difficult problem, that of reconstructing the part of the reaction profile between minima, and, in particular, that of determining transition-state structure, i.e. the arrangement of atoms corresponding to the energy maximum along the reaction profile. The difficulty arises because molecules traveling from minimum to minimum spend only a tiny fraction of their time in the region of the transition state. Except for some special systems [30], it is not possible to observe molecules when they are close to the transition state. One way around the problem is to resort to theory. With *ab initio* or force-field methods, transition states involving a few small molecules in the gas phase and composed mainly of light atoms may sometimes be calculated with some claim to credibility. This is hardly a situation typical of mainstream chemistry.

It is thus of interest to try to combine experimental data with theoretical concepts in order to obtain a description of events along the reaction profile. The general approach involves the choice of a model relating structure and energy, the determination of its parameters with the help of structural data from diffraction experiments and of energetic data from thermodynamic measurements (ΔG, etc.), from kinetic measurements ($\Delta G^{\pm}$, etc.) and from electronic or vibrational spectroscopy. The usefulness of the model (including its numerical parameters) is to be judged on the basis of its ability to summarize and classify a broad range of experimental data and to make predictions on reactivity from structural data or vice versa. Two concepts are especially important in this approach. The first is transition-state theory formulated by Eyring [31], which relates rate constants k of reactant consumption or product formation to the energy of the transition state:

$$k = (k_B T/h) \exp(-\Delta G^{\ddagger}/RT) \tag{5.30}$$

(k_B is the Boltzmann constant, a transmission coefficient of unity is assumed). The second concept is the equilibrium-rate theory of Marcus [32], which may be considered as an interpretation of the empirical relationships between (thermodynamic) free energy of reaction and (kinetic) energy of activation [33], the so-called free-energy relationships (FER). This theory states that

$$\Delta E^{\ddagger} = \Delta E^{\ddagger}_{enc} + \Delta E^{\ddagger}_0 + \Delta E/2 + \Delta E^2/(16\,\Delta E^{\ddagger}_0) \ . \tag{5.31}$$

$\Delta E^{\ddagger}_{enc}$ is the energy necessary to bring the reactants together in the encounter complex. $\Delta E^{\ddagger}_0$ is the intrinsic activation energy when ΔE, the energy of the reaction, is zero. According to eq. 5.31 $\Delta E^{\ddagger}$ decreases and then increases again for increasingly negative ΔE. The literature on this topic is vast. A summary which proved useful in the context of this work is given in [34]. Note, that ΔE is used here in a generalized sense to denote the energy function appropriate for a given problem, i.e. ΔG, ΔH, ΔV, etc.

The rationalization of FER by Marcus is based on a simple model of the reaction profile, that of two intersecting parabolas (Figure 5.10) [35]. In most applications of the Marcus equilibrium-rate theory, the reaction coordinate is a normalized quantity between 0 and 1, measuring in a generalized way "the progress of reaction"; it is usually poorly defined from a geometrical, structural point of view. Indeed, when the word "structure" is used in works on FER, it refers mainly to the connectivity

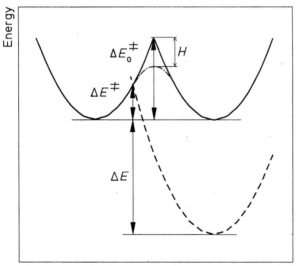

Fig. 5.10. Reaction profile, model of two intersecting parabolas. The activation energy is $\Delta E^{\ddagger}_0$ if the reaction energy ΔE is zero (solid parabola on the right). If the reaction energy ΔE is negative (dotted parabola on the right) $\Delta E^{\ddagger}$ is smaller than $\Delta E^{\ddagger}_0$ and the transition state, i.e. the intersection of parabolas, is closer to the reactants. A modification of the simple model in which the two potentials interact (resonance energy H) is shown as dash-dotted line

of the molecules under investigation rather than to their metrical aspects. The rate-equilibrium theory is consistent, however, with Leffler's postulate [36a] that the slope of a FER, e.g. the Brönsted coefficient in a proton transfer [36b], measures the position of the transition state along the reaction coordinate and with Hammond's postulate [37] that this position is displaced toward the reactants as the reaction becomes more exergonic. The rate-equilibrium theory has been shown to be valid for a broad range of barrier shapes [38] and has been extended to two-dimensional energy surfaces [34], thus providing a quantitative basis for More O'Ferrall diagrams [39].

The lack of structural definition of the reaction coordinate has several unfortunate consequences:

(1) Differences between relevant distances and angles of related reactants are difficult to introduce into interpretations of FER.

(2) There is no obvious way to make use of force constants which describe, essentially, the change in energy as a function of interatomic distance.

(3) Frequently, the reaction coordinate is vastly oversimplified and is mistakenly identified with a change in a single bond distance, bond angle or other structural parameter. The reaction path is not a line but a curve that winds its way through $(3N-6)$ dimensional configuration space; as shown in Section 5.4, it usually samples many dimensions rather than just one. In the following sections, some examples will be reviewed, which illustrate ways of combining structural and energetic information in the determination of reaction profiles.

5.5.3 Structural Reorganization in Degenerate Reactions

If reactants and products are isometric, the reaction is symmetric and the reaction energy is zero. In this simple case, the transition state is symmetric and so is the reaction profile [40]. A well known example is electron self-exchange [41], which can occur whenever a chemical species exists in two or more oxidation states. Hexacoordinate metal complexes in oxidation states $2+$ and $3+$ provide an example (Figure 5.11). Initially, the metal-ligand distance d_2 in ML_6^{2+} is long and the corresponding force constant f_2 is relatively low, whereas the distance d_3' in ML_6^{3+} is short with a high force constant f_3'. On the way to the transition state d_2 decreases and d_3' increases, until, at the transition state, $d = d' = d^{\ddagger}$. The electron is then transferred from one complex to the other, and both species relax to their energetically most favorable geometry at d_3 and d_2'. Assuming harmonic potentials, the transition-state reorganization energy $\Delta E_0^{\ddagger}(R^{\ddagger})$ can be expressed [41] entirely in terms of the ground-state properties d_2, d_3, f_2 and f_3

$$\Delta E_0^{\ddagger}(R^{\ddagger}) = 3(d_2-d_3)^2 f_2 f_3/(f_2+f_3) \tag{5.32}$$

Since the reaction usually occurs in a polar medium, the reorganization energy $\Delta E_0^{\ddagger}(S^{\ddagger})$ of the solvent shell has to be taken into account as well.

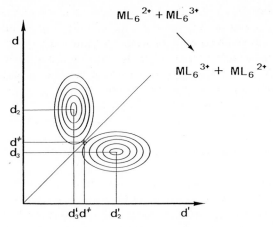

Fig. 5.11. Two-dimensional energy surface for the inner-sphere reorganization in an electron self-exchange reaction. Reorganization of reactants and products is assumed to follow harmonic potentials symbolized by contour lines. The two surfaces intersect along the diagonal mirror line. The point of lowest energy on the intersection parabola is the transition state; it is marked with a cross

$$\Delta E_0^{\ddagger}(S^{\ddagger}) = (\Delta e)^2(1/D_{op} - 1/D_s) \cdot (1/(2r_2) + 1/(2r_3) - 1/r) \qquad (5.33)$$

Apart from the amount of charge transferred (Δe) and the static and optical dielectric constants, (D_s, D_{op}), only structural quantities are involved here: the radii r_2 and r_3 of the two complex ions in the two oxidation states and the center-to-center distance r in the encounter complex, usually taken as ($r_2 + r_3$) [41].

For very large ions or vanishing solvent polarity ($D_{op} \approx D_s$) the solvent reorganisation energy is negligible. If the difference ($d_2 - d_3$) between the ground-state structures of the reactants is large, the molecular part $\Delta E_0^{\ddagger}(R^{\ddagger})$ of the activation energy is also large, because going to the transition state requires substantial reorganization of the metal-ligand complexes. The latter conclusion also holds for organic S_N2 reactions [42], as will be shown after the following digression.

Implicit in Figure 5.11 is a reaction profile that connects one minimum across the transition state with an equivalent one and looks somewhat like the intersecting parabola model of Figure 5.10 (solid lines). In so far as the harmonic approximation holds, the activation energy can be estimated from spectroscopic measurements of vertical transition energies [41] between the two surfaces. For equal, intersecting parabolas with force constant f and a separation Δd between the minima it may be shown that the vertical excitation energy from the minimum of one parabola to the other parabola is $h\nu = f\Delta d^2/2$. The corresponding thermal activation energy for going from one minimum to the other is given by

$$\Delta E_0^{\ddagger} = h\nu/4 . \qquad (5.34)$$

This example shows how electronic excitation energies ($h\nu$) and vibrational force constants combined with ground-state structural data (Δd) can lead to an estimate of activation energy and transition-state structure for thermal reactions.

A prerequisite for such estimates is that the encounter complex is sufficiently long lived for $h\nu$ to be measured. Another condition is that the resonance energy H at the point of intersection of the two surfaces is small (Figure 5.10). Both conditions are met in certain oligonuclear mixed-valence metal complexes.

Excitation energies have also been incorporated, with appropriate modification, into a similar model of the reaction profile for organic S_N2 reactions [42]:

$$X^{\cdot-} + R-X \rightarrow X-R + :X^- .$$

The excited state to be considered is now $X^{\cdot} + (R-X)^-$, and $h\nu$ is in the far UV rather than in the near IR (as in mixed-valence metal complexes). The excitation energy $h\nu$ is usually expressed as the sum of the ionization energy of $X^{\cdot-}$ and the electron affinity of $R-X$. In contradistinction to electron transfer and other weakly coupled mixed-valence systems, the resonance energy H at the intersection of the two energy surfaces (Figure 5.10) can no longer be considered small compared with $\Delta E_0^{\ddagger}$. However, it has been shown that H is almost constant for a series of related reactions and that differences in activation energy depend mainly on the energy required to deform the $R-X$ species from its geometry in the ground state to that in the transition state, i.e. activation energy is essentially controlled by structural reorganization. Further refinements of the argument are found in [42].

In the last example of this section, symmetry considerations are combined with structure-structure correlation and with activation energies to obtain a structure-energy correlation within a family of reaction profiles pertaining to the same conformational interconversion process in a series of similar molecules. NMR-measurements of (s-cis-η^4-butadiene) metallocene complexes (Figure 5.12) show coalescence in the methylene and cyclopentadienyl proton signals at different temperatures depending on the hydrocarbon substituents R [43]. The observations have been interpreted in terms of a degenerate inversion of the metallacyclopentene ring. Structure correlation by principal component analysis (see Chapter 4) on twelve such compounds (15 independent molecules) shows two main changes in molecular structure

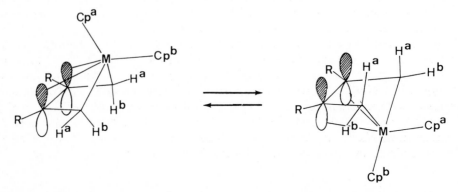

Fig. 5.12. Ring inversion in (s-cis-η^4-butadiene) metallocene complexes (M = Zr, Hf)

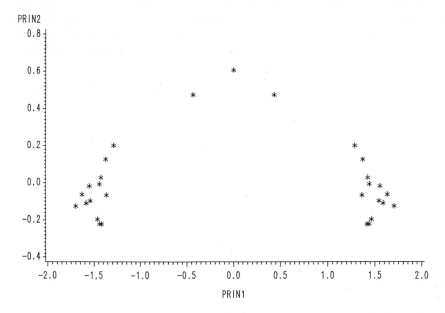

Fig. 5.13. Result of principal component analysis on the metallacyclopentene fragment of (*s-cis-η⁴*-butadiene) metallocene complexes (Figure 5.12). The coordinate PRIN1 measures the dihedral angle between the CCCC and the CMC planes, PRIN2 measures the bite distance CH₂ ... CH₂. The point at PRIN1 = 0 corresponds to a planar fragment. The distribution of data points is roughly semicircular and symmetrical with respect to a vertical mirror line. The length of the arc from this line to any point is defined as the "distance" x_0 between ground-state and transition state

(Figure 5.13) [44]. One coordinate, PRIN 1, describes the change in the dihedral angle between the 2-butene and the C−M−C plane in terms of the deviations of these atoms from their best plane. The other, PRIN 2, measures the change in the CH₂...CH₂ bite distance. Both coordinates are given in Å-units. In this coordinate system the observed structures are found to span, approximately, a semicircle, whose two halves are related by a mirror line. The points on the left- and right-hand sides, respectively, represent the two degenerate conformations of Figure 5.12; the point in the center represents a molecule with a planar, C_{2v}-symmetric metallacyclopentene fragment. (In this particular molecule CH₂ is replaced by O, C_5H_5 by $C_5(CH_3)_5$ and R = *t*-but.) The semicircle is a model of the reaction coordinate. Initially, the main change is in PRIN2, a decrease in the CH₂ ... CH₂ bite distance. As the reaction proceeds, the metallacyclopentene flattens until it becomes planar at PRIN1 = 0.

The distance x_0 in Å-units of each point along the semicircle from the unique point in the middle is correlated with the measured activation energy in Figure 5.14. The dependence of activation energy on the distance x_0 can be expressed as

$$\Delta E^{\ddagger}(x_0) = f x_0^4 \ . \tag{5.35}$$

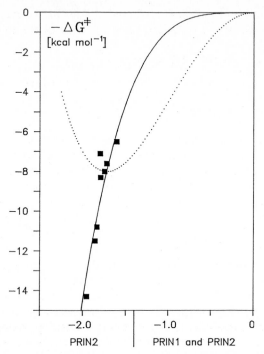

Fig. 5.14. Correlation of the "distance" x_0 between ground state and transition state (abscissa, in Å) with activation energy. The solid line is $\Delta G^{\ddagger} = f x_0^4$, which follows if the reaction profiles of individual molecules are assumed to be $\Delta G = f x^4 - f x_0^2 x^2$ (dotted line). Only one half of the symmetric reaction profile is shown

Note that in the region of the observed structures the activation energy shows a sharp decrease ($>50\%$) associated with only a relatively modest shift in molecular structure towards the transition state. The fourth-order dependence between structure and energy follows if the symmetric, double-well reaction profiles for the molecules involved are assumed to have the following simple algebraic form

$$\Delta E(x) = f x^4 - a x^2 \ . \tag{5.36}$$

The first term, important only at large x, is repulsive; in the example, it can be identified, approximately, with the repulsion between Cp and R in the highly puckered rings. The second term is attractive; it can be interpreted in terms of the binding energy between the RC=CR double bond and M; it increases as the five-membered ring puckers and as the doubly bonded atoms come closer to M. The function has minima at $x_0 = \pm (a/2f)^{1/2}$, representing the position of equilibrium structures along the reaction coordinate. All experimental points in Figures 5.14 lie close to the same fourth-order curve, indicating that the repulsive contribution to the reaction profile ($f x^4$) is common to all molecules in the series whereas the attractive term ($a = -f x_0^2 x^2$) varies from molecule to molecule.

If we choose an arbitrary molecule of the set as reference, then the reaction profiles of all related molecules may be expressed in terms of the reference profile as

$$\Delta E(x) = fx^4 - 2f(x_0 + \Delta x_0)^2 x^2$$
$$= fx^4 - 2fx_0^2 x^2 - (4fx_0 \Delta x_0 - 2f\Delta x_0^2)x^2$$
$$\approx fx^4 - 2fx_0^2 x^2 - 4fx_0 \Delta x_0 x^2 \ . \tag{5.37}$$

Here x_0 represents the equilibrium structure of the reference molecule, Δx_0 the structural difference between the reference molecule and its relative. The terms fx^4 and $-2fx_0^2 x^2$ express the repulsive Cp...R and the attractive M...C=C interactions of the reference molecule. The term $-(4fx_0\Delta x_0 - 2f\Delta x_0^2)x^2$ may be considered as a perturbation on the reference profile. This term performs the fine tuning of the M...C=C attraction and the Cp...R repulsion. It accounts for small differences in the electronic and steric nature of the substituted double bond. Depending on the sign of Δx_0, it may be positive or negative.

In this section, structural aspects of symmetrical electron-transfer processes, symmetrical $S_N 2$-reactions and a symmetrical conformational interconversion have been correlated with activation energies. Three general conclusions emerge: First, differences in reactivity are accompanied by differences in structure: the larger the structural difference between ground state and transition state, the larger the reorganization energy between the two. Second, a small change in ground-state structure is associated with a large change in activation energy, and last, the reaction profiles of related molecules are themselves closely related and can be expressed simply as more or less perturbed versions of a reference profile.

5.5.4 Structural Reorganization in Nondegenerate Reactions. Determination of Transition-State Structure

Degenerate equilibrium reactions are the exception rather than the rule. For kinetically controlled reactions equilibrium data are often lacking, whereas structural and kinetic data may be available. This raises the question whether there are still correlations between structures and rate constants, analogous to those discussed in the preceding section and to those between equilibria and rate constants. As will be shown, very much the same conclusions may be reached as for degenerate reactions.

Consider the dissociatively activated substitution (I_d) of NH_3 in S-bonded $[(NH_3)_5 \, Co(III) \, SO_3]^+$ and $[(NH_3)_5 \, Co(III) \, SO_2 R]^{2+}$ complexes [45]. In the sulfito complex the $Co-N$ distance *trans* to $Co-S$ is 0.089(3) Å longer than the $Co-N$ distances *cis* to $Co-S$. In the sulfinato complex the difference is 0.054(12) Å. The rate constant for substitution of the *trans*-NH_3 in the sulfito complex is higher by a factor of $100-200$ compared with the sulfinato complex. The corresponding difference in $\Delta H^{\neq}$ has been estimated from kinetic measurements to be about 9 kcal mol^{-1}. The two observations have been connected with the help of the shifted parabola model (Figure 5.15): Two identical parabolas for the *trans* Co$-$N bonds with force constant 245 kcal mol^{-1} Å^{-2} are displaced sideways by ca. 0.035 Å ($= 0.089$ Å $- 0.054$ Å). The two reactions are assumed to have the *same transition-state structure*, leading to a difference in activation energies of $0.035 \times 245 \times \Delta x^{\neq}$ kcal mol^{-1} or again around 9 kcal mol^{-1} if the lengthening $\Delta x^{\neq}$ of the Co$-$N distance between ground and transition state is ca. 1 Å. The latter value is said to be in excellent agreement with expectations based on volumes of activation [45].

This is a remarkable study, for its seems to be the first to combine ground-state structure with a force constant and with activation energies to estimate the value of a structural parameter in a transition state of an unsymmetrical reaction. Although the experimental basis is limited, the study illustrates the kinds of results that are obtainable.

A similar comparison has been made for nickelamine-, nickelaqua- and mixed amineaqua-complexes [46], where $\Delta G^{\neq}$ decreases with increasing nickel-ligand distance at a rate of approximately 50 kcal mol^{-1} Å^{-1} (Table 5.1). In another series of studies, the dissociatively activated exchange of the same neutral nitrogen base B in different BRCo(III)(dimethylglyoxime)$_2$-complexes has been investigated, where R is a variable alkyl group in the second axial coordination site. It was found that the exchange rate constant depends sensitively on the electronic and steric properties of the alkyl group and correlates well with the length of the scissile Co$-$N bond [47].

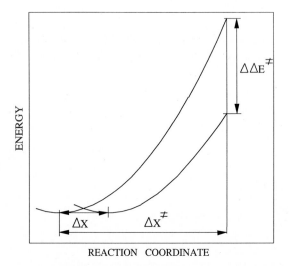

REACTION   COORDINATE

Fig. 5.15. Reaction profile: model of two shifted parabolas, representing related reactants undergoing the same reaction. If the same transition state geometry is assumed for both, the activation energies differ by $\Delta\Delta E^{\neq} \approx k \Delta x \Delta x^{\neq}$

Table 5.1.

Bond	$\Delta E^{\neq \text{ a)}}$ [kcal mol^{-1}]	$\Delta x_0{}^{\text{b)}}$ [Å]	$\Delta\Delta E^{\neq}/\Delta x_0{}^{\text{c)}}$ [kcal mol^{-1} Å^{-1}]	$f_2{}^{\text{d)}}$ [kcal mol^{-1} Å^{-2}]	$\Delta x^{\neq \text{ e)}}$ [Å]
Ni(II) $-$ OH$_2$	9 ($\Delta G^{\neq}$)	0.09	ca. 30 (70)	100	0.75
Ni(II) $-$ N(amine)	17 ($\Delta G^{\neq}$)	0.065	ca. 60 (120)	150	0.8
Co(III) $-$ N(pyridine)	20 ($\Delta G^{\neq}$)	0.08	ca. 110 (170)	250	0.7
Co(III) $-$ NH$_3$	30 ($\Delta G^{\neq}$)	0.04	ca. 175 (210)	250	0.85
R$_3$B $-$ NR$_3$	10 ($\Delta H^{\neq}$)	0.05	ca. 190 (190)	600	0.3
C $-$ OR	28 ($\Delta G^{\neq}$)	0.07	ca. 310 (370)	800	0.45

a) median of range of observed $\Delta E^{\neq}$s
b) distance between smallest and largest M$-$L distance observed
c) observed (calculated)
d) approximate values of harmonic M$-$L stretching force constants
e) estimated difference in M$-$L distance between ground and transition state

A lengthening of this distance by 0.08 Å is paralleled by a decrease in $\Delta G^{\neq}$ of ca. 9 kcal mol^{-1} (Table 5.1). Note that the rather modest increase in bond length is accompanied by a rate acceleration of more than six orders of magnitude. In the case of an associatively activated ligand exchange at boron (Scheme 5.1) the forward and backward reactions differ in $\Delta H^{\neq}$ by about 10 kcal mol^{-1}. The difference in B$-$N distances is $\sim$0.05 Å and the rate of change $\Delta\Delta H^{\neq}/\Delta x_0$ is ca. 200 kcal mol^{-1} Å^{-1} [48]. This ratio is even larger for the spontaneous cleavage of acetals [49], which will be discussed in detail later in this section. As may be seen from Table 5.1, $\Delta\Delta E^{\neq}/\Delta x_0$ correlates with the force constant f_2 of the bond to be broken and with the median of the range of $\Delta E^{\neq}$.

These trends can be portrayed by a simple model [49]. The reaction profile is written as the simplest polynomial function containing a minimum and a maximum, representing ground and transition state, respectively (Figure 5.16):

$$E = a(x-x_0)+f_2(x-x_0)^2/2+f_3(x-x_0)^3 \ . \qquad (5.38)$$

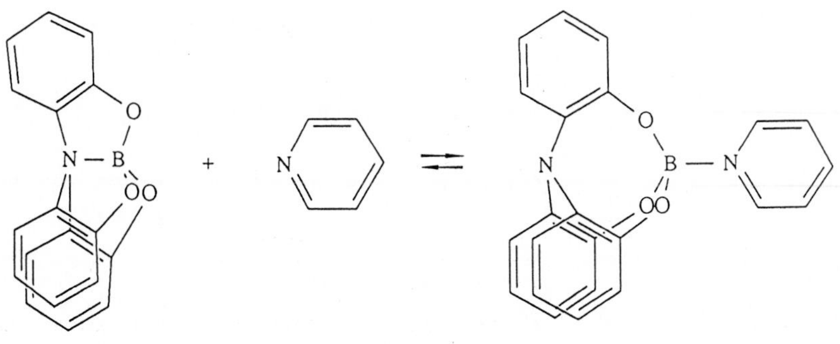

Scheme 5.1

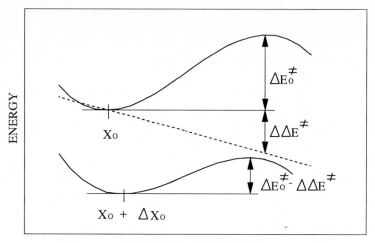

REACTION COORDINATE

Fig. 5.16. Reaction profile: model of cubic curve (third-order polynomial) to describe asymmetric reactions. The lower profile illustrates the influence of a linear perturbation (dashed line) on the upper profile

The model potential is a function of the deviation $(x-x_0)$ from the reference distance x_0, which is the equilibrium value when a is zero. The quadratic and cubic force constants f_2 (positive) and f_3 (negative) are taken as the same for all molecules in a related series, whereas a is a perturbation expressing steric and electronic differences in ligand-metal interactions between related molecules (dotted line in Figure 5.16). The perturbation a affects $\Delta E^{\ddagger}$ and shifts the equilibrium distance to $x_e = x_0 + \Delta x_0$. In linear approximation the relationship is

$$\Delta\Delta E^{\ddagger}/\Delta x_0 = -(6\,\Delta E^{\ddagger}(x_0)f_2)^{1/2} \tag{5.39}$$

where $\Delta E^{\ddagger}(x_0)$ is the activation energy for the reference equilibrium distance x_0 and may be chosen as the median of the range of observed activation energies. Comparison of observed and calculated values (Table 5.1) confirms the dependence of $\Delta\Delta E^{\ddagger}/\Delta x_0$ on the approximate magnitude of $\Delta E^{\ddagger}$, i.e. on $\Delta E^{\ddagger}(x_0)$, and on the bond stretching force constant f_2.

The observed values $\Delta\Delta G^{\ddagger}/\Delta x_0$ are consistently smaller than the ones estimated from the model. If the comparison is between $\Delta\Delta E^{\ddagger}/\Delta x_0$ and $\Delta\Delta H^{\ddagger}/\Delta x_0$, which is more directly related to an energy surface, the agreement is better. This is to be expected since, for decreasing $\Delta G^{\ddagger}$, $\Delta S^{\ddagger}$ is often observed to decrease as well [50], i.e. $\Delta\Delta G^{\ddagger}/\Delta x_0$ is usually an underestimate of $\Delta\Delta E^{\ddagger}/\Delta x_0$.

The model also predicts the lengthening $\Delta x^{\ddagger} = x^{\ddagger} - x_0$ of the scissile bond in going from the ground to the transition state.

$$\Delta x^{\ddagger} \approx (6\,\Delta E^{\ddagger}(x_0)/f_2)^{1/2} \ . \tag{5.40}$$

For a given f_2, i.e. for a given type of bond, reorganization $\Delta x^{\ddagger}$ is large if $\Delta E^{\ddagger}$ is large. For a given $\Delta E^{\ddagger}$, $\Delta x^{\ddagger}$ is large if the bond is soft, i.e. if f_2 is small. The relative magnitudes of $\Delta x^{\ddagger}$ seem reasonable, small for the S_N2-type or associative ligand substitution at B, larger for the dissociative reactions at Co(III) and Ni(II) (Table 5.1).

Similar arguments have been used to explain qualitatively the retardation (!) of spontaneous aquation on going from $[Cr(NH_3)_5Cl]^{2+}$ to the more crowded $[Cr(NH_2CH_3)_5Cl]^{2+}$ and the acceleration on going from $[Co(NH_3)_5Cl]^{2+}$ to $[Co(NH_2CH_3)_5Cl]^{2+}$ [51]. In the Co-complexes the Co–Cl distance is the same in both compounds and $\Delta H^{\ddagger}$ is the same to within 0.5 kcal mol^{-1}. In the more crowded methylamine Cr-complex the Cr–Cl distance is shorter (!) by 0.028 Å than in the less crowded one and $\Delta H^{\ddagger}$ is correspondingly higher, by 4 kcal mol^{-1}. For both pairs $\Delta S^{\ddagger}$ is somewhat larger in the more crowded molecule, thus accounting for the acceleration in the series of Co-complexes. Here the combination of structural with kinetic data has disposed of arguments invoking a difference in mechanism between the two Cr-complexes.

Finally, the entry Co(III)-NH$_3$ in Table 5.1 calls for comment. The ratio between rate constants of base-catalyzed and spontaneous hydrolysis of $[Co(NH_3)_6]^{3+}$ and many $[Co(NH_3)_5X^-]^{2+}$ complexes is consistently ca. 10^5 [52]. The distances in the mono-deprotonated cage complex $[Co(diNOsar)(-H^+)]^{2+}$ (diNOsar = 1,8-dinitro-3,6,10,13,16,19-hexazabicyclo[6,6,6]eicosane, Scheme 5.2) are Co–NR$_2^-$: 1.946(7) Å, Co–NHR$_2$ (*cis*): 1.974(7) Å, Co–NHR$_2$ (*trans*): 2.016(7) Å, i.e. the Co–N bond *trans* to the deprotonated site is 0.04 Å longer than the bonds in *cis*-position [53]. We consider this structure as a mimic of the reactant structure in base-catalyzed hydrolysis, $[Co(NH_3)_5NH_2^-]^{2+}$ or $[Co(NH_3)_4NH_2^-X^-]^+$, which have not been isolated. In terms of the simple model discussed above, the elongated *trans*-bond is estimated to react 10^6 times faster than the shorter *cis*-bond: the observed ratio is 10^5.

Local perturbations may affect other kinds of structural parameters as well. Deviations of trigonal atoms from the planes of their three bonded neighbors have been correlated with the direction of electrophilic addition to the trigonal centers involved;

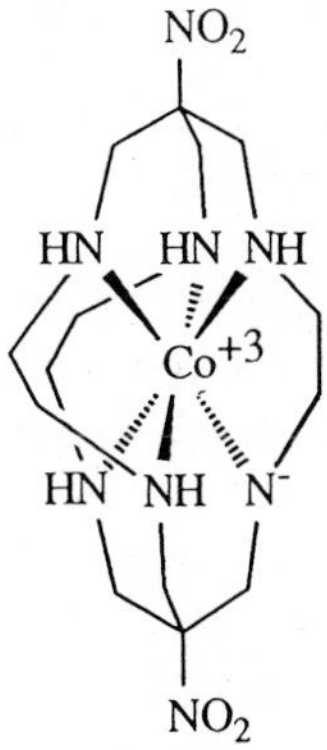

Scheme 5.2

even for very small ground-state pyramidalities, addition takes place preferentially on the diastereotopic face corresponding to development of the incipient lone pair orbital [54]. Once again, a small perturbation in the ground state can be extrapolated to a chemically significant stabilization of the transition state.

In the examples discussed in this section, the reaction coordinate has been identified with a single structural parameter, a bond distance. This is an oversimplification, as pointed out earlier, but not one which is inherent in the model, which may be generalized to any number of dimensions. As an example, we discuss a two-dimensional energy surface for the spontaneous hydrolysis of acetals.

The following observations have been made on the family of tetrahydropyranyl-acetals shown in Scheme 5.3. From top to bottom the pK_a of the exocyclic substi-

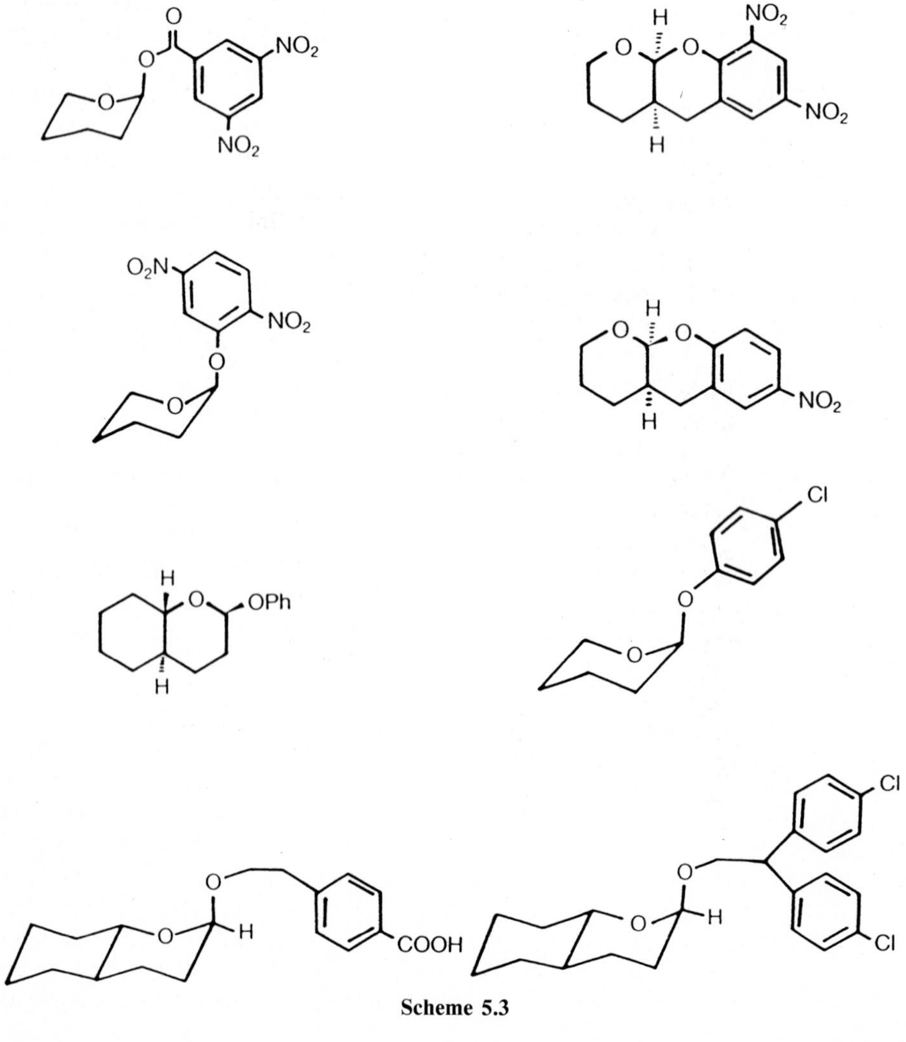

Scheme 5.3

tuents (benzoate, phenolates, alcoholates) increases, the rate for the spontaneous cleavage of the exocyclic $C-O$ bond decreases (by fifteen orders of magnitude!), the lengths of the $C-O$ bond exocyclic to the tetrahydropyranyl ring decreases ($1.48-1.41$ Å), and the endocyclic $C-O$ distance increases ($1.38-1.42$ Å) [55]. These trends may be understood qualitatively in terms of the electron-accepting abilities of the variable acetal substituent. A good acceptor withdraws electron density from the exocyclic oxygen substituent in the ground state, thereby lowering the energies of the $\sigma(C-O)$ and $\sigma^*(C-O)$ orbitals. This permits delocalization of the axial lone pair of the ring oxygen into $\sigma^*(C-O)$ of the axial exocyclic oxygen. As a consequence, the endocyclic ketal $C-O$-distance is shorter than the exocyclic one. On going to the transition state, a negative charge develops on the leaving oxygen. This charge is stabilized more by a good electron-acceptor than by a poor one, thus lowering the energy of the transition state and accelerating the reaction. In this argument, the same factor, namely electron-accepting ability, is made responsible for both the structure and reactivity trends. It should therefore be possible to find a model which relates structure and reaction rate directly and can account for the magnitude of the effects: small changes in $C-O$ distances and an enormous change in reaction rate.

We start by expressing the energy as a function of the two $C-O$ distances, r_1, r_2, or rather as the deviations, $\Delta r_1, \Delta r_2$, from some reference molecule [49]:

$$E(\Delta r_1, \Delta r_2) = f_{11}(\Delta r_1^2 + \Delta r_2^2)/2 + f_{12}\Delta r_1 \Delta r_2$$

$$+ f_{111}(\Delta r_1^3 + \Delta r_2^3) + f_{112}(\Delta r_1^2 \Delta r_2 + \Delta r_1 \Delta r_2^2)$$

$$+ a(\Delta r_1 - \Delta r_2) \ . \tag{5.41}$$

In the quadratic and cubic part of the potential it is assumed that the $O-C-O$ fragment shows approximate C_{2v}-symmetry, i.e. $f_{11} = f_{22}$, $f_{111} = f_{222}$ and $f_{112} = f_{122}$. In the analysis by Bürgi and Dubler-Steudle [49] the numerical values of f_{11} and f_{12} were taken from normal-coordinate analyses and from *ab initio* calculations; f_{111} and f_{112} were obtained from the following conditions: (1) together with f_{11} and f_{12} the cubic constants must reproduce the observed activation energy $\Delta E_0^{\ddagger}$ of the reference molecule; (2) the distances $\Delta r_1^{\ddagger}$, $\Delta r_2^{\ddagger}$ in the transition state must obey the rule of constant Pauling bond order, $n_1^{\ddagger} + n_2^{\ddagger} = 2$ with $\Delta r_i^{\ddagger} = -c \log n_i^{\ddagger}$ [56]; (3) the stretching force constant calculated with the above potential for the shortened endocyclic $C-O$ distance $r_2^{\ddagger}$ in the transition state has to follow Badger's rule or an analogous relationship between interatomic distance and stretching force constant [57]. Actually, two of the three conditions suffice to fix f_{111} and f_{112}, but the third one serves as a useful test for the quality of the numerical parametrization in areas of the energy surface far from the minimum. The analytical form of the perturbation, $a(\Delta r_1 - \Delta r_2)$, expresses the observation that lengthening of one $C-O$ bond is accompanied by shortening of the other. The quantity a reflects changes in the electron-accepting ability of the exocyclic substituent. For the reference molecule $a = 0$. The general appearance of the corresponding energy surface between equilibrium and transition state is shown in Figure 5.17. With increasingly negative perturbation a, the minimum is shifted to larger equilibrium values of Δr_{10}, smaller ones of Δr_{20}, and the activation energy decreases. The changes in equilibrium structure and activation energy have been calculated numerically as a function of a and the dependence of $\Delta E^{\neq}$ on $q_0 = (\Delta r_{10}^2 + \Delta r_{10}^2)^{1/2}$ compared with the experimental data (Figure 5.18).

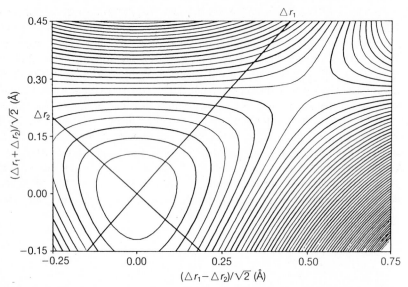

Fig. 5.17. Energy contour diagrams for the spontaneous cleavage of tetrahydropyranyl acetals as a function of the two acetal $C-O$ bond distances. Energy minimum and transition state are shown (note that Δr_1 and Δr_2 are oriented diagonally)

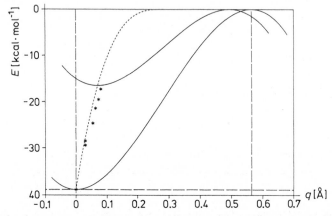

Fig. 5.18. Correlation between change in ground-state structure ($q_0 = (\Delta r_{10}^2 + \Delta r_{20}^2)^{1/2}$) and activation energy for spontaneous cleavage of tetrahydropyranyl acetals (dotted line). Reaction profiles for the slowest and fastest reaction are also shown

The quality of the agreement confirms that the model is able to reproduce the interdependence of structure and reactivity. The influence of the perturbation on Δr_{10} and Δr_{20} separately is shown in Figure 5.19. The shortening of Δr_{20} is initially equal to the lengthening of Δr_{10}, but with increasing perturbation the lengthening exceeds the shortening because of the anharmonic terms in the potential. The model reproduces the experimental observations to within their standard deviations (ca. 0.006 Å).

The changes in distance between ground and transition state of the reference molecule are calculated to be $\Delta r_1^{\ddagger} \approx 0.54$ Å and $\Delta r_2^{\ddagger} \approx -0.15$ Å, corresponding to $r_1^{\ddagger} \approx 1.95$ Å and $r_2^{\ddagger} \approx 1.27$ Å. These estimates compare well with the results of *ab initio* calculations on the transition states of the water-assisted decomposition of $H_2CO_3 \cdot H_2O$ to $CO_2 \cdot 2H_2O$ [59], $CH_2 = C(OH)_2 \cdot H_2O$ to $CH_2 = C = O \cdot 2H_2O$ [60] and $CH_2(OH)_2 \cdot H_2O$ to $CH_2O \cdot 2H_2O$ [61]. These reactions proceed through a six-membered transition state in which the water molecule acts both as a hydrogen-bond donor and acceptor. The water molecule, or rather an incipient H_3O^+-species, stabilizes the leaving OH^--group by incipient donation of a proton (Scheme 5.4). The calculated distance changes in going to the transition states are 0.17, 0.34, 0.38 Å for $\Delta r_1^{\ddagger}$ and -0.07, -0.11, -0.16 Å for $\Delta r_2^{\ddagger}$, somewhat smaller than for the acetals. An analogous argument for the latter must take into account that the solvent water molecule available for stabilization of the incipient alcoholate anion is

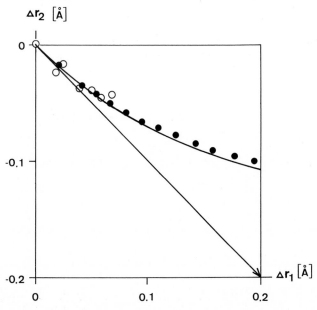

Fig. 5.19. Acetal C–O distances perturbed by the stereoelectronic effects of various leaving groups. Empty circles: observed; filled circles: calculated from two-dimensional model discussed in text. The solid line is a curve of constant bond order. The diagonal is at $\Delta r_1 = -\Delta r_2$. The Δr_i's are deviations from reference distances [49]

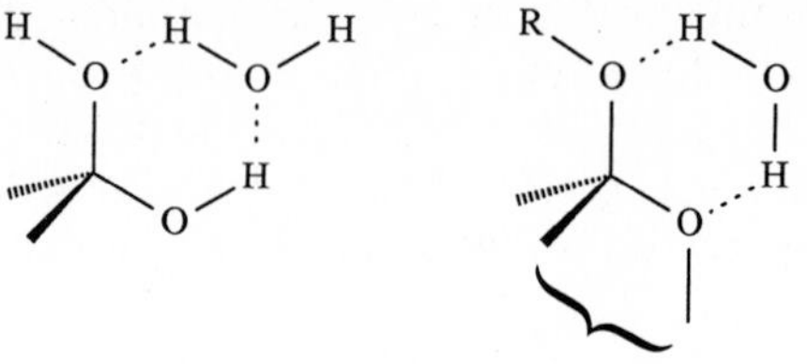

Scheme 5.4

presumably less acidic than the incipient H_3O^+ of the diol examples (Scheme 5.4). Stabilization becomes important further along the reaction coordinate, and, correspondingly, the transition state is later for acetal hydrolysis.

As mentioned earlier, the one- and two-dimensional models of reaction profiles discussed above can be generalized to any number of dimensions. The only prerequisite is that sufficient information on potential constants, especially anharmonic ones, is available.

What is perhaps the most remarkable feature of these examples is the sharp sensitivity of the activation energy of bond-breaking processes to quite small changes in ground-state bond distances. An analogous observation applies to equilibrium distances r_0 and dissociation energies D_0 of many classes of chemical bonds [62]. It can be accounted for by a simple modification of the Morse equation

$$V(r-r_0) = D_0 \{\exp\left[-2B(r-r_0)\right] - 2n^q \exp\left[-B(r-r_0)\right]\} \tag{5.42}$$

where B is the Morse constant (ca. 2 Å^{-1}), q is another constant (ca. 0.4–0.55) and n is the Pauling bond order (or fractional bond number) [1a]. For $n = 1$ ($n^q = 1$), the above relationship is the same as the unmodified Morse equation, for $n < 1$ the importance of the (negative) attractive term decreases and for $n > 1$ it increases relative to that of the (positive) repulsive term. The (non-linear) perturbation of the original Morse equation is thus

$$2(1 - n^q) \exp\left[-B(r-r_0)\right] \tag{5.43}$$

The perturbed equilibrium distance is

$$\Delta r_0(n) = -qB^{-1} \ln n \tag{5.44}$$

This equation is evidently the same as the Pauling relation connecting bond distance and bond order, mentioned earlier in this section. A change of bond order by a factor of 2 corresponds to $\Delta r_0 \approx 0.1 - 0.2$ Å. The new dissociation energy $D_0'(n)$ is

$$D_0'(n) = D_0 n^{2q}$$

$$= D_0 \cdot \exp\left[-2B \, \Delta r_0(n)\right] \; . \tag{5.45}$$

This expression is the same as that used by Johnston and Parr [63] in their bond-energy – bond-order (BEBO) method for describing potential energy surfaces of simple chemical reactions. It shows an exponential dependence of dissociation energy on change in ground-state distance. Stretching force constants $k(n)$ and equilibrium distance are related as follows

$$\Delta r_0(n) = [\ln (2B^2 D_0) - \ln k(n)]/2B .\qquad (5.46)$$

An empirical equation of the same analytical form has been shown to hold for a wide variety of diatomics, including covalent, polar and van der Waals molecules [57b].

Other types of interatomic potential functions may be modified in a similar way. For example, the general inverse power potential may be modified to

$$V(r-r_0) = [D_0/(l-m)][m(r_0/r)^l - s^q l(r_0/r)^m] \quad (l>m)\qquad (5.47)$$

leading to a new equilibrium distance $r_0'(s)$ such that

$$s = (r_0'(s)/r_0)^{(m-l)/q} .\qquad (5.48)$$

The modified equation defines bond valences s that are similar to Pauling's bond numbers n and are used throughout Chapter 10 to correlate experimental bond distances.

The above discussion shows that relationships between bond valences, bond lengths, dissociation energies and stretching force constants may also be connected, at least conceptually, to our ideas on energy surfaces and energy perturbations described in sections 5.5.3 and 5.5.4. There we have shown that for rather diverse reactions small structural differences of reactant molecules are associated with large changes in activation energies and that this correlation can be rationalized in terms of simple models of reaction profiles. As a spin-off from such models, structural parameters of transition-state structures have been estimated.

5.6 The Principle of Structure-Energy Correlation

Many molecular properties such as structure, flexibility and reactivity are "local", i.e. they can be associated with specific parts of a molecule or molecular fragment. This makes it possible to classify and group molecules into families of fragments exhibiting similar properties. In an attempt to analyze these similarities, we have used what we might call the principle of energy correlation (in analogy to the principle of structure correlation, Section 5.4): If related molecules undergo related reactions, then the respective reaction coordinates and the energy profiles along them are similar. Differences among them, both in energy and in the location of stationary points, may be described in terms of simple, continuous energy perturbations [64].

In practice, an arbitrary reference molecule is chosen from a family of related ones; its reaction profile is reconstructed, using as much experimental information as is available. The changes in structure and energy for varying degrees of perturbation may then be obtained and compared with experiment. Note that, although the rate of change in activation energy with respect to ground-state structure is derived from the properties of an arbitrarily chosen reference molecule (structure, force constants, energy), it applies to the entire family of molecules and is thus a collective property of this family. For our approach to be useful, the perturbation should have a simple algebraic form involving only a few parameters, while still accounting for all observed correlations among properties. The models described in the preceding section meet these conditions.

The perturbation models do not make explicit reference to electronic or steric pecularities of individual members of the group. These are absorbed, in parametric form, into the potential constants associated with the perturbation. A qualitative understanding of steric and electronic factors may nevertheless be helpful, not only for finding a suitable functional form of the perturbation, but also for assessing the possible meaning of the corresponding numerical constants in physical and chemical terms.

We conclude this section by discussing the relationship between the principles of structure correlation and of energy perturbation. In Section 5.4 the principle of structure correlation was paraphrased by saying that "observed structures tend to concentrate in low lying regions of the potential energy surface". This statement implies a general shape of the energy surface applicable to all fragments in a family. The discussion leading to the energy perturbation principle has shown that, although the energy surfaces vary somewhat between members of a family, they can nevertheless be represented by a basic reference surface, modified by appropriate perturbations. Thus the basic assumption underlying the principle of structure correlation is substantiated.

Another point concerns the critical discussion of structural scatterplots in Section 5.4. There it was argued that a quantitative measure of relative energy is almost impossible to obtain from scatterplots alone because of the difficulty of evaluating the perturbations and distinguishing them from experimental error for a sample of structures drawn from a wide variety of crystal and molecular environments. This seems in contradiction to the detailed interpretation of structures, force constants, and energies given here. There is a difference, however. While it is true that we have dealt with relatively small structural differences, their significance was judged not only in terms of experimental error, but also in terms of their correlation with activation energy. It is the generality of the trends observed for a variety of fragment families and their chemical reactions that lends credibility to the perturbation model, to its quantitative details, and to the energy perturbation principle in general.

5.7 Conclusions

This chapter has dealt with physical and chemical interpretations rather than with statistical analyses of structure correlation. The treatment is based on the observation that molecules and molecular fragments can be grouped into families based on chemical fragment composition and the three-dimensional arrangement of the constituent atoms. The energy surfaces associated with individual fragments are assumed to be broadly similar within a family and to be interrelated by energy perturbations. Energy perturbations affect structures most strongly where the energy surface is shallow and much less where it is steep. This is the basis for structure-structure correlation (Section 5.2). The assumption of energy perturbations has been substantiated by studying their effects on distant points on the energy surface, namely ground and transition states. Thus, the same perturbation approach is also the basis for structure-energy correlation (Section 5.4).

A further point that emerges from perturbation models is an interpretation of enzymatic reactions. If one says that "binding energy is utilized to lower the energy of the transition state", it is implied that the transition state binds to the enzyme better than the ground state does. In terms of our model, the enzyme may be considered as *perturbing* the structure of the substrate in the enzyme-substrate groundstate complex in the direction of the transition-state complex. Compared to the uncatalyzed or unperturbed reaction, the effects on reaction rate can be profound (see Chapter 13).

Both types of correlation analysis, structure-structure and structure-energy correlations, have much in common with FER's or energy-energy correlations [65]. The three types of correlation are analogous in the sense that they all uncover relationships between observables for which there is no *a priori* reason to be related. They are thus providing genuine, new information, which needs to be interpreted.

Structure correlations and FER's are well established by now. The same is not true for structure-energy correlations, possibly because in most cases either only kinetic or only structural data are available for a given compound; possibly also, because structure-energy correlations require expertise in different experimental techniques. These difficulties should not be serious obstacles, however, to further development of such correlations: X-ray facilities are available in many chemical laboratories, and the X-ray crystallographer will usually be more than willing to collaborate with the chemist.

With $\Delta\Delta G^{\ddagger}/\Delta x_0$ values of the order of a few hundred kcal mol^{-1} Å^{-1}, bond length differences of a few hundredths of an Å between corresponding bonds in related structures can have an enormous influence on the rates of corresponding bond-breaking reactions. This underlines the need for accurate bond-length measurements in structural studies bearing on chemical reactivity. Since careful X-ray analyses, especially at low temperatures, are perfectly capable of determining interatomic distances with a precision of the order of 0.001 to 0.002 Å, interesting results can be expected from the combination of structural and kinetic studies.

Many of the ideas on which this work is based could only be sketched in the available space; some of the interpretations are preliminary or incomplete, and some problems, for example, the role of solvent, have hardly been mentioned. In view of the title of this book, "Structure Correlation", this seems permissible, and, in any case, the loose ends are amenable to exploration combining reliable experimental data with sound theoretical ideas.

References

[1a] Pauling, L., *The Nature of the Chemical Bond*, Cornell University Press, Ithaca, NY, 1st edn. **1939**, 2nd edn. **1940**, 3rd edn. **1960**

[1b] Updates produced with the help of computerized data bases are to be found in appendix A to this book. See also: Allen, F. H., Kennard, O., Watson, D. G., Brammer, L., Orpen, G. A., Taylor, R., *J. Chem. Soc. Perkin* **1987**, *2*, S1–S19; Orpen, A. G., Brammer, L., Allen, F. H., Kennard, O., Watson, D. G., Taylor, R., *J. Chem. Soc. Dalton* **1989**, S1–S83

[2a] Schuster, P., Zundel, G., Sandorfy, C. (eds.) *The Hydrogen Bond*, North-Holland, Amsterdam, **1976**, Vol. I–III

[2b] Jeffrey, G. A., Saenger, W., *Hydrogen Bonding in Biological Structures*, Springer, Berlin, **1991**

[3] Etter, M. C., *Acc. Chem. Res.* **1990**, *23*, 120–126; Etter, M. C., Reutzel, S. M., *J. Am. Chem. Soc.* **1991**, *113*, 2586–2598; Etter, M. C., *J. Phys. Chem.* **1991**, *95*, 4601–4610; Desiraju, G. R., *Acc. Chem. Res.* **1991**, *24*, 290–296

[4] Bent, H. A., *Chem. Rev.* **1968**, *68*, 587–648

[5] Harris, P. M., Mack, E. Jr., Blake, F. C., *J. Am. Chem. Soc.* **1928**, *50*, 1583–1600; Kitaigorodskii, A. I., Khotsyanova, T. L., Struchkov, Y. T., *Zh. Fiz. Khim.* **1953**, *27*, 780

[6] Bolton, W., *Acta Cryst.* **1965**, *18*, 5–10

[7] Rosenfield, R. E., Parthasarathy, R., Dunitz, J. D., *J. Am. Chem. Soc.* **1977**, *99*, 4860–4862

[8] Alcock, N. W., *Bonding and Structure*, Horwood, Chichester, **1990**

[9a] Nyburg, S. C., *Acta Cryst.* **1979**, *A35*, 641–645

[9b] Nyburg, S. C., Faerman, C. H., *Acta Cryst.* **1985**, *B41*, 274–279

[10] Murray-Rust, P., Motherwell, W. D. S., *J. Am. Chem. Soc.* **1979**, *101*, 4374–4376

[11] Rosenfield, R. E. Jr., Swanson, S. M., Meyer, E. F. Jr., Carrell, H. L., Murray-Rust, P., *J. Mol. Graph.* **1984**, *2*, 43–46

[12a] Bürgi, H. B., Shefter, E., *Tetrahedron* **1975**, *31*, 2976–2981. For a similar example see: Nørskov-Lauritsen, L., Bürgi, H. B., Hofmann, P., Schmidt, H. R., *Helv. Chim. Acta* **1985**, *68*, 76–82

[12b] Bürgi, H. B., *Inorg. Chem.* **1973**, *12*, 2321–2325

[13] Murray-Rust, P., Bürgi, H. B., Dunitz, J. D., *Acta Cryst.* **1978**, *B34*, 1793–1803

[14] Mooney Slater, R. C. L., *Acta Cryst.* **1959**, *12*, 187–196; Slater, J. C., *Acta Cryst.* **1959**, *12*, 197–200

[15] Auf der Heyde, T. P. E., Bürgi, H. B., *Inorg. Chem.* **1989**, *28*, 3960–3969

[16a] Murray-Rust, P., Bürgi, H. B., Dunitz, J. D., *J. Am. Chem. Soc.* **1975**, *97*, 921–922

[16b] Bürgi, H. B., Dunitz, J. D., *Acc. Chem. Res.* **1983**, *16*, 153–161

[17] Bürgi, H. B., *Angew. Chem.* **1975**, *87*, 461–475; *Angew. Chem. Int. Ed. Engl.* **1975**, *14*, 460–473

[18] Bartenev, V. N., Kameneva, N. G., Lipanov, A. A., *Acta Cryst.* **1987**, *B43*, 275–280

[19] Murray-Rust, P., in: *Molecular Structure and Biological Activity*, Griffin, J. F., Duax, W. L. (eds.), Elsevier, New York, **1982**, pp. 117–133

[20] Bürgi, H. B., Dunitz, J. D., *Acta Cryst.* **1988**, *B44*, 445–448

[21] Bürgi, H.B., Blanc, E., Schwarzenbach, D., Liu, S., Lu, Y., Kappes, M.M., Ibers, J.A., *Angew. Chem.* **1992**, *104*, 667−669; *Angew. Chem. Int. Ed. Engl.* **1992**, *31*, 640−643

[22] Johnson, R.D., Yannoni, C.S., Dorn, H.C., Salem, J.R., Bethune, D.S., *Science* **1992**, *255*, 1235−1238

[23] Destro, R., *Chem. Phys. Lett.* **1991**, *181*, 232−236

[24] Simmons, C.J., Hathaway, B.J., Amornjarusiri, K., Santarsiero, B.D., Clearfield, A., *J. Am. Chem. Soc.* **1987**, *109*, 1947−1958; Simmons, C.J., *Struct. Chem.* **1992**, *3*, 37−52

[25] Sim, G.A., *J. Chem. Soc. Chem. Comm.* **1987**, 1118−1120; Sim, G.A., *Acta Cryst.* **1990**, *B46*, 676−682

[26] For reviews see: König, E., *Progr. Inorg. Chem.* **1987**, *35*, 527−622; Bloomquist, D.R., Willett, R.D., *Coord. Chem. Rev.* **1982**, *47*, 125−164

[27a] Ohashi, Y., *Acc. Chem. Res.* **1988**, *21*, 268−274 (Racemization in crystals of chiral cobaloxims)

[27b] Uchida, A., Dunitz, J.D., *Acta Cryst.* **1990**, *B46*, 45−54

[28] Gavezzotti, A., *Acta Cryst.* **1987**, *B43*, 559−562; Gavezzotti, A., Bianchi, R., *Chem. Phys. Lett.* **1986**, *128*, 295−299

[29] See, for example, the collection of papers by Schmidt, G.M.J., et al., in: *Solid State Photochemistry*, Ginsburg, D. (ed.) Verlag Chemie, Weinheim, New York, **1976**

[30] Zewail, A.H., *Science* **1988**, *242*, 1645−1653

[31] Eyring, H., *J. Chem. Phys.* **1935**, *3*, 407. For reviews on the development of transition state theory, see: Laidler, K.J., King, M.C., *J. Phys. Chem.* **1983**, *87*, 2657−2664; Truhlar, D.G., Hase, W.L., Hynes, J.T., *J. Phys. Chem.* **1983**, *87*, 2664−2682

[32] Marcus, R.A., *J. Phys. Chem.* **1968**, *72*, 891−899

[33a] Hammett, L.P., *Physical Organic Chemistry*, McGraw Hill, New York, 1st. edn. **1940**, 2nd edn. **1970**. For more recent discussions see: Exner, O., *Correlation Analysis of Chemical data*, Plenum, New York, **1988**; Hansch, C., Leo, A., Taft, R.W., *Chem. Rev.* **1991**, *91*, 165−195

[33b] For FER of inorganic reactions see, for example, Swaddle, T.W., *Coord. Chem. Rev.* **1974**, *14*, 217−268; Chapman, N.B., Shorter, J., *Correlation Analysis in Chemistry*, Plenum, London, **1978**; Chen, Y.T., *Coord. Chem. Rev.* **1987**, *79*, 257−278

[34] Grunwald, E., *J. Am. Chem. Soc.* **1985**, *107*, 125−133

[35] Earlier models based on two intersecting curves include notably those of Evans, M.G., Polanyi, M., *Trans. Farad. Soc.* **1936**, *32*, 1340 and Bell, R.P., *Proc. Roy. Soc. London*, **1936**, *A154*, 414. For a review, see Warhurst, E., *Quart. Rev. Chem. Soc.* **1951**, *5*, 44−59

[36a] Leffler, J.E., *Science* **1953**, *117*, 340; *J. Org. Chem.* **1955**, *20*, 1202; *J. Chem. Phys.* **1955**, *23*, 2199

[36b] For a review, see: Bell, R.P., *The Proton in Chemistry*, Chapman and Hall, London, 2nd edn. **1973**

[37] Hammond, G.S., *J. Am. Chem. Soc.* **1955**, *77*, 334−338

[38] Kurz, J.L., *Chem. Phys. Lett.* **1978**, *57*, 243−246; Murdoch, J.R., *J. Am. Chem. Soc.* **1983**, *105*, 2667−2672; Magnoli, D.E., Murdoch, J.R., *J. Am. Chem. Soc.* **1981**, *103*, 7465−7469

[39] More O'Ferrall, R.A., *J. Chem. Soc. B* **1970**, 274−277

[40] Alternatively, the transition state and reaction profile may not be symmetric, but then there are additional, symmetry related trajectories and transition states. Taken together, they conserve the symmetry of the energy surface as a whole. See: Salem, L., *Acc. Chem. Res.* **1971**, *4*, 322−328

[41] For a review on electron transfer reactions, see: Marcus, R.A., Sutin, N., *Biochim. Biophys. Acta* **1985**, *811*, 265−322

[42] Shaik, S.S., *Pure Appl. Chem.* **1991**, *63*, 195−204; *Acta Chem. Scand.* **1990**, *44*, 205−221

[43] Erker, G., Engel, K., Krüger, C., Chiang, A.-P., *Chem. Ber.* **1982**, *115*, 3311−3323; Krüger, C., Müller, G., Erker, G., Dorf, U., Engel, K., *Organometallics* **1985**, *4*, 215−223

[44] Bürgi, H.B., Dubler-Steudle, K.C., *J. Am. Chem. Soc.* **1988**, *110*, 4953−4957

[45] Elder, R.C., Heeg, M.J., Payne, M.D., Trkula, M., Deutsch, E., *Inorg. Chem.* **1978**, *17*, 431−440

[46] Schwarzenbach, G., Bürgi, H.B., Jensen, W.P., Lawrance, G.A., Moensted, L., Sargeson, A.M., *Inorg. Chem.* **1983**, *22*, 4029−4038

[47] Bresciani-Pahor, N., Forcolin, M., Marzilli, L. G., Randaccio, L., Summers, M. F., Toscano, P. J., *Coord. Chem. Rev.* **1985**, *63*, 1−125; Randaccio, L., Bresciani-Pahor, N., Zangrando, E., Marzilli, L. G., *Chem. Soc. Rev.* **1989**, *18*, 225−250

[48] Müller, E., Bürgi, H. B., *Helv. Chim. Acta* **1987**, *70*, 499−510

[49] Bürgi, H. B., Dubler-Steudle, C. K., *J. Am. Chem. Soc.* **1988**, *110*, 7291−7299

[50] Halpern, J., *Bull. Chem. Soc. Japan* **1988**, *61*, 13−15

[51] Lay, P. A., *Coord. Chem. Rev.* **1991**, *110*, 213−233; *Comments Inorg. Chem.* **1991**, *11*, 235−284

[52] Lawrance, G. A., *Adv. Inorg. Chem.* **1989**, *34*, 145−194

[53] Geue, R. J., Hambley, T. W., Harrowfield, J. M., Sargeson, A. M., Snow, M. R., *J. Am. Chem. Soc.* **1984**, *106*, 5478−5488

[54] Seebach, D., Zimmermann, J., Gysel, U., Ziegler, R., Ha, T.-K., *J. Am. Chem. Soc.* **1988**, *110*, 4763−4772

[55] Jones, P. G., Kirby, A. J., *J. Am. Chem. Soc.* **1984**, *106*, 6207−6212

[56] See [1 a]. The constant c has been chosen as 0.56 Å to reproduce the $C−O$ distance in the hydrolysis intermediate $−C=O^+−$ ($\sim$ 1.25 Å, shortening ca. 0.17 Å, see [49])

[57a] Badger, R. M., *J. Chem. Phys.* **1934**, *2*, 128−131

[57b] Herschbach, D. R., Laurie, V. W., *J. Chem. Phys.* **1961**, *35*, 458−463

[58] Ferretti, V., Dubler-Steudle, K. C., Bürgi, H. B., in: *Accurate Molecular Structures*, Domenicano, A., Hargittai, I. (eds.), Oxford University Press, Oxford, **1992**, 412−436

[59] Nguyen, M. T., Hegarty, A. F., Ha, T.-K., *J. Mol. Struct. (Theochem)* **1987**, *150*, 319−325

[60] Nguyen, M. T., Hegarty, A. F., *J. Am. Chem. Soc.* **1984**, *106*, 1552−1557

[61] Williams, I. H., Spangler, D., Femec, D. A., Maggiora, G. M., Schowen, R. L., *J. Am. Chem. Soc.* **1983**, *105*, 31−40

[62] Bürgi, H. B., Dunitz, J. D., *J. Am. Chem. Soc.* **1987**, *109*, 2924−2926

[63] Johnston, H. S., *Adv. Chem. Phys.* **1960**, *3*, 131; Johnston, H. S., Parr, C., *J. Am. Chem. Soc.* **1963**, *85*, 2544−2551

[64] Bürgi, H. B., in: *Perspectives in Coordination Chemistry*, Williams, A. F., Floriani, C., Merbach, A. E. (eds.), Verlag Helvetica Chimica Acta, Basel, VCH, Weinheim, **1992**, 1−29

[65] This nomenclature, which expresses the analogy between the three types of correlation, was first suggested in [55]

Part II
Molecular Structure
and Reactivity

6 Organic Addition and Elimination Reactions; Transformation Paths of Carbonyl Derivatives

Andrzej Stanislaw Cieplak

6.1 Introduction

Sometimes even the most sophisticated ideas of the theory of bonding have humble and sweaty origins, born from a squalid experience of an organic chemist toiling at a cluttered laboratory bench. And so, it is in the early work on the constitution of two minor opium alkaloids where we have to trace back the concept of a molecular structure arrested in an intermediate stadium of a chemical reaction.

The two principles of the opium filtrate remaining after isolation of morphine and thebaine were cryptopine and protopine, described in the late 19th century [1] (Scheme 6.1, p. 206), and subsequently found, especially the latter, to be among the most ubiquitous alkaloids of *Papaveraceae* and several other plant families [2]. Shortly before World War I, in the course of degradation studies of cryptopine carried out in the University Chemical Laboratories at Oxford, W. H. Perkin Jr. defined the dilemma posed by the structure and reactivity of these compounds [3]. Trying to establish the nature of cryptopine's five oxygen functionalities, he discovered that while the alkaloid does not form the expected addition products with hydroxylamine, semicarbazide or amyl nitrite and ethoxide, it must nonetheless contain a keto group neighboring an unsubstituted methylene group. Similarly, in the 1916 article (extending over 213 pages of The Journal of the Chemical Society) Perkin reported that in order to quaternize the tertiary amino group of cryptopine, he had to treat the compound with methyl iodide in methanol in a sealed tube in the waterbath for two days.

These two unusually sluggish groups had another distinction: they were incorporated into and across a ten-membered ring, hitherto unknown in natural alkaloids. The fact that such a ring allows for an effective interaction of two functional groups had already been recognized by Gadamer, working in Marburg, during his investigation of the degradation products of tetrahydroberberine [4]. In the case of cryptopine and protopine this interaction was reflected in the facility, for instance, with which they would form bicyclic systems, yielding hydroxy ammonium salts upon attempts to obtain keto ammonium salts via normal protonation procedures. The studies of the so-called anhydrobases of berberine [4a] had prompted Gadamer to generalize the notion of the conjugation of residual valences in enamines [5] to the transannular interaction of an amine and an olefin; he thus suggested that the keto and

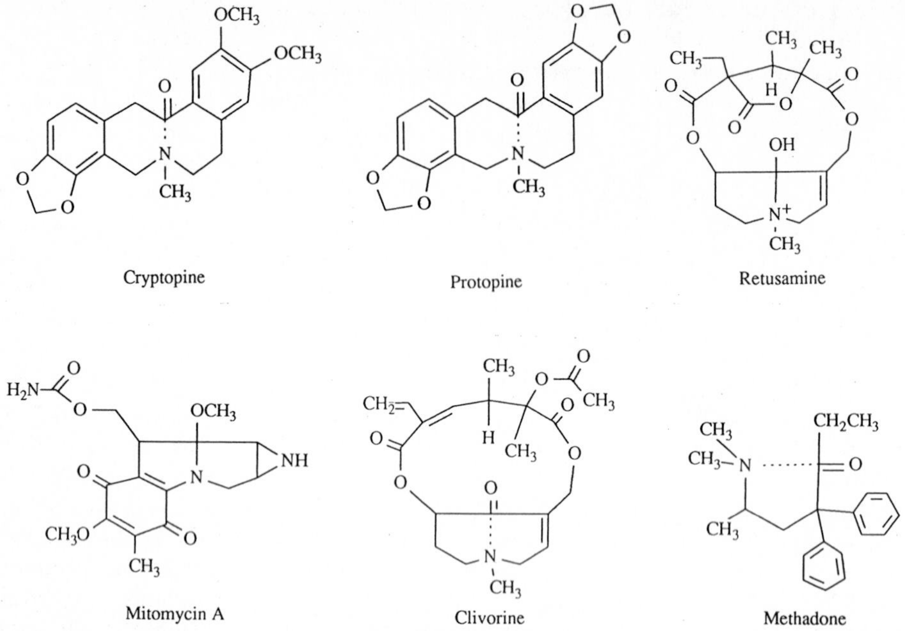

Cryptopine

Protopine

Retusamine

Mitomycin A

Clivorine

Methadone

Scheme 6.1

amino groups of protopine and cryptopine have a tendency to exist in the betaine form, which would explain their lowered activity [6]. The modern description of the phenomenon, however, was ultimately proposed by Sir Robert Robinson. Thinking along the same lines but in terms of the electronic theory of bonding, he formulated a novel concept of the through-space partial bond between the nitrogen and the carbonyl carbon [7].

Infrared and ultraviolet spectroscopies, introduced in the chemistry of natural products about three decades later, provided the first evidence to support this idea [8, 9]. The discovery of similar modification of reactivity of amino and keto groups across a nine-membered ring of vomicine [10], a strychnine congener [11], pointed to the generality of such transannular interactions in medium-size rings. This then became the topic of systematic and extensive investigations by Leonard and coworkers [12].

The continued interest in such interactions eventually led to X-ray crystallographic studies, and several structures of related natural products had been determined by 1971 (see Scheme 6.1) [13–17]. These determinations confirmed the hypothesis of N...C=O interaction, revealing short distances between the N and carbonyl C atoms and non-planarity of the carbonyl group. Indeed, in the 5-azacyclooctanone ring of clivorine, the N...C=O distance turned out to be nearly in the range of covalent bonding, 1.993(3) Å [17]; the C=O bond is markedly elongated, 1.258(3) Å, and the distance between carbonyl C and the plane of its three ligands is 0.213 Å, about half of the analogous distance in a tetrahedral carbon atom. On the other hand, in the crystal structure of retusamine [14], where an analogous 5-azabicyclooctanone ring is transformed by protonation into a quaternized aminoalcohol, the C−N bond at 1.64 Å is significantly longer than expected. Inevitably, both structures were discussed in terms of partial formation and cleavage of the C−N bond [14, 17b]. The conceptual framework for such a discussion was already available in crystallography. In 1959, Rose Mooney Slater, discussing structural diversity and asymmetry in triiodide ions, observed that pairs of corresponding I−I bond distances map out a hyperbolic curve [18]. Consequently, J. C. Slater proposed that the observed distribution reflects the shape of the potential energy surface for linear triatomic molecules or ions [19], as revealed by calculation of energy contours for linear configurations of H_3 [20]. Ten years later Bent, in his monumental survey of the structural aspects of donor-acceptor interactions, articulated the idea that crystal structures might represent early stages of chemical reactions [21].

However, it was not until the detailed examination of the crystal structure of methadone in its free base form, that the full appreciation of the data on geometry of N...C=O interactions emerged [22]. As a result, the idea of ascribing both static and dynamic properties of related molecules to one potential energy surface coalesced into a formalized concept [23]. Bürgi, Dunitz and Shefter proposed that the data for clivorine and the other five structures shown in Scheme 6.1 provide sample points on or close to a reaction coordinate, and thus the entire set should map out the reaction pathway for nucleophile addition to a carbonyl group. Indeed, structural parameters for intramolecular N...C=O interactions obtained from these six crystal-structure analyses readily fit a simple quantitative model invoking conservation of bond order defined as a logarithmic function of bond length [24, 25]. The model successfully related the C−N bond order and pyramidalization of the carbonyl carbon.

This result substantiated the notion that the distribution of experimental geometry parameters tends to be concentrated in low-lying regions of the potential energy surface and showed how surprisingly useful the simple relations invoking bond-order conservation might be [26]. As for the mechanism of nucleophilic addition to carbonyl, it suggested that even the earliest stages of the approach of the reagents are governed by stringent stereoelectronic requirements. Already at the van der Waals distance the trigonal center is detectably pyramidalized in response to the interaction while the nucleophilic centre and the C=O group form an angle roughly equal to the tetrahedral valence angle, which stays nearly constant along the reaction coordinate.

The ideas espoused in the Bürgi-Dunitz-Shefter paper became immensely influential. The notion that partial pyramidalization is related to incipient bond order lent support to proposals of Fukui [27] and Mock [28] and was seminal in the development of current views on non-planarity and intrinsic preferences in face selection (see Section 6.5.3.3). The conclusion that there is a strongly preferred orientation for nucleophilic attack inspired Baldwin's approach vector analysis [30], Liotta's and Burgess's trajectory analysis [31], and became a cornerstone of numerous models of asymmetric induction by Nguyen Trong Anh [32], Houk [33] and Heathcock [34], and of diastereoselection in aldol condensation by Seebach [35], Nguyen Trong Anh [36], and Eschenmoser [37, 38]. Generally considered as an indispensable argument in discussions of stereoselectivity, the hypothesis of obtuse-angle nucleophilic approach to carbonyl has now become a staple of undergraduate courses in organic chemistry [39].

The facility with which this status has been achieved is somewhat unusual considering the at times tortuous relations of crystallography and organic chemistry. It is still recalled that when the X-ray structure of penicillin, determined by Dorothy Hodgkin [40], was cited as a proof of the presence of the β-lactam ring in the molecule, Sir Robert Robinson, a long time opponent of this idea, thundered that it is not a proof of penicillin structure in solution [41]. Since then the claims of X-ray crystallography about connectivity in organic molecules were no longer disputed, but the notion of going beyond that point remained controversial.

A number of factors seem to have been essential for the reception of Bürgi and Dunitz's contribution. The most important probably was the support by *ab initio* calculations, which reproduced the dependence of partial pyramidalization on the length of the incipient bond and the approximate constancy of the angle of attack during the final stages of hydride addition to formaldehyde [42]. In general, the hypothesis of the non-perpendicular approach trajectory easily fitted in the qualitative sense into the modern models of reactivity and ground-state electronic structure of the carbonyl group. In fact, it is consistent both with the frontier molecular orbital model assuming σ-π separation and invoking the contour of the π^* MO [32], and with the bent-bond model due to the directionality of the displacement [43], as well as with models deriving reactivity indices from the total electron density due to the contours of the Laplacian of the charge density in Bader's approach and of the Fukui function in Parr's local density functional method [44, 45].

Furthermore, the postulate of a strongly preferred orientation for the nucleophilic attack corroborated the concepts invoked at the time to understand the high kinetic accelerations of enzymic or intramolecular reactions, such as Koshland's concept of orbital steering in enzyme catalysis [46]. Finally, the hypothesis offered grounds for attractive interpretations of a puzzling pattern of regioselectivity in the reactions of imides and cyclic anhydrides [47, 48].

Not surprisingly, the two decades since the initial publications and even the decade since the most recent review by Bürgi and Dunitz [29, 49] have witnessed an accumulation of new results. One of the important developments is the extension of the structure correlation analysis to the reactions of carbonyls which offer an alternative pathway to the acyl substitutions via tetrahedral intermediates, namely the pathway of substitution via an acylium intermediate [50]. This effort completed the description of the cycle of carbonyl addition-elimination reactions (see Scheme 6.2) in terms of the geometric coordinates of the reaction pathways and is claimed to demonstrate generality of the stereoelectronic effects governing these pathways. A major contribution towards understanding the nature of such effects was the discovery of conformational influences on the hydrolysis of orthoesters and the subsequent formulation of Deslongchamps' theory [51]. A new direction in structure correlation studies developed when Jones and Kirby and their coworkers were able to demonstrate quantitative correlations of ground-state geometries and activation energies for the breakdown of tetrahydropyranyl acetals [52]. Efforts to develop functional models accounting for these results [53] led to the formulation of the first empirical potential energy surface for the hydrolysis of acetals by combining experimental ground-state

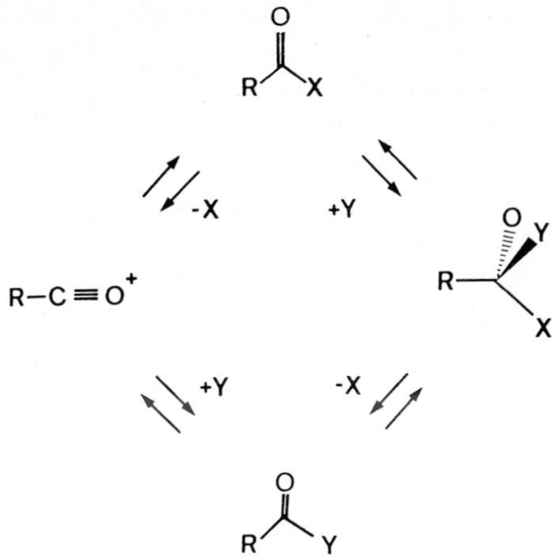

$$
\begin{array}{c}
\text{O} \\
\parallel \\
R \diagdown C \diagup X
\end{array}
$$

$-X$ $+Y$

$R-C\equiv O^{+}$

$$
\begin{array}{c}
\text{O} \quad Y \\
R-C \\
X
\end{array}
$$

$+Y$ $-X$

$$
\begin{array}{c}
\text{O} \\
\parallel \\
R \diagdown C \diagup Y
\end{array}
$$

Scheme 6.2

structural data, valence force constants and activation energy estimates (see Chapter 5) [54].

Apart from the progress in specific areas of structure correlation studies, methodological and technological advances have literally transformed experimental X-ray crystallography and theoretical chemistry during the last decade. Improvement in X-ray equipment for automatized collection of data for single-crystal structure determination and in standardized computer software for structure solving have led to a great increase in the quality and number of reported structures. Structure correlation investigations in the organic and organometallic areas are now backed by the Cambridge Structural Database (CSD), which has quadrupled in size since the early 80's to about 100 000 entries with average residual factor of about 6% (January 1992 issue). This database offers software facilities developed specifically for studies of intermolecular contacts [55] (see Chapters 3 and 4). It is indeed possible now to depart, if desired or necessary, from the tenets of the early structure correlation studies whereby influences of molecular and crystal environments on a fragment of interest were indistinguishable, fragment definitions tended to be very general, and the searches were limited to those where effects were large enough to overcome diversity of molecular embedding and limited quality of the X-ray determinations.

In the meantime, computational exploration of potential energy surfaces for organic reactions has progressed from exhaustive grid searches to location of stationary points by gradient-based techniques, and from simple energy minimization to characterization of saddle points by analytical evaluation of the first and second derivatives of the SCF energy [56–58]. These developments, along with wider availability of more powerful computers, have led to a vast increase of computational studies of transition-state structures, reaction pathways, catalysis, and medium and solvation effects. More recently, increasingly elaborate basis sets and faster MP2 geometry optimizations have made it possible to study even relatively small stereoelectronic effects at the required level of theory.

This chapter will examine structure correlations pertinent to carbonyl substitution-elimination reactions as well as results of computational studies of potential energy surfaces and transition-state structures for such reactions. We shall then assess the current standing of the original postulates as well as that of the derivative models employed in the discussion of stereoselection and regioselectivity.

6.2 Reaction Pathway for $sp^2 \rightleftharpoons sp^3$ Transformations of Carbonyls

6.2.1 Initial Stages of Nucleophilic Addition to a Carbonyl

6.2.1.1 Correlation of Partial Pyramidalization and the Incipient Bond Distance

The initial publication reported data for six structures [23]. In each the N, C and O atoms lie in an approximate local mirror plane and the angles α and β (see Figure 6.1 for the definition of symbols) do not vary by more than a few degrees from their mean values. On the other hand, the significant pyramidalization of the carbonyl group (Δ), measured by the distance of the C atom from the plane defined by R, R' and O, increases as the C...N distance (d_1) decreases. The correlation plot of Δ vs. d_1 is shown in Figure 6.2a. The points lie close to the smooth curve

$$d_1 = -1.701 \log \Delta + 0.867 \text{ Å} \tag{6.1}$$

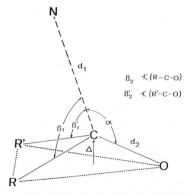

Fig. 6.1. Definition of geometrical parameters describing the disposition of nucleophile N relative to RR'C=O

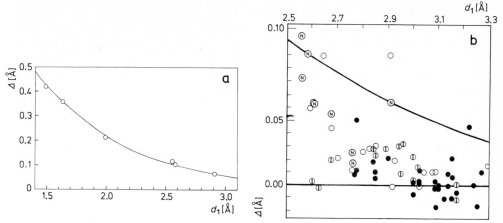

Fig. 6.2. a: Initial plot of pyramidalization Δ vs. N...CO distance d_1
b: revised plot of partial pyramidalization Δ vs. N...CO and O...C=O at distances $d_1 > 2.5$ Å
(N:N...CO interactions; remaining symbols refer to O...C=O interactions, ○: intramolecular,
⌀: dicarboxylic acids; ●: intermolecular; data from [29] and [63])

obtained by least-square analysis. Using standard bond lengths as constants and Pauling's coefficient relating C–C bond length and bond number, this equation can be rewritten and a similar one for C–O distances can be given

$$d_1 = -1.701 \log n + 1.479 \text{ Å} \tag{6.2a}$$

$$d_2 = -0.71 \log (2-n) + 1.426 \text{ Å} \tag{6.2b}$$

where $n = \Delta/\Delta_{max}$, $\Delta_{max} = 0.437$ Å. The success of this model in reproducing the observed distances d_1 and d_2 implies that the sum of the C–O and C–N bond numbers equals two for all the structures under discussion.

Subsequent investigations have provided two kinds of additional data. First, crystal structures of nine more compounds containing *tert*-amine and carbonyl fragments have become available since 1973 [59–62]. The basic types of molecular embedding employed to enforce the contacts are shown in Scheme 6.3. Second, examination of N...C=O interactions was soon followed by an analogous survey of O...C=O interactions manifested by short intermolecular or intramolecular contacts [63]; subsequently, many more examples of crystal structures containing such contacts were reported [61, 64–68]. Finally, attempts to capture initial stages of the hydride transfer in hydroxyketones were reported. Although results for hydroxybicyclo[3.3.1]nonanone and its methyl ether were inconclusive because of disorder [69], the study of four polycyclopentane-based hydroxyketones showing rapid intramolecular hydride transfer in solution (see Scheme 6.4) revealed significant pyramidalization of the carbonyl groups towards the hydrogen atom undergoing the exchange [70].

The revised plot of partial pyramidalization Δ vs. N...C=O and O...C=O distance d_1 for the now available data is shown in Figure 6.2b. Unquestionably, there

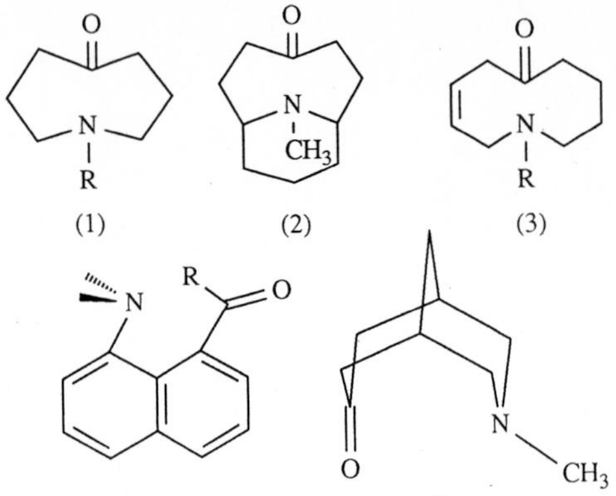

(1) (2) (3)

(4) (5)

Scheme 6.3

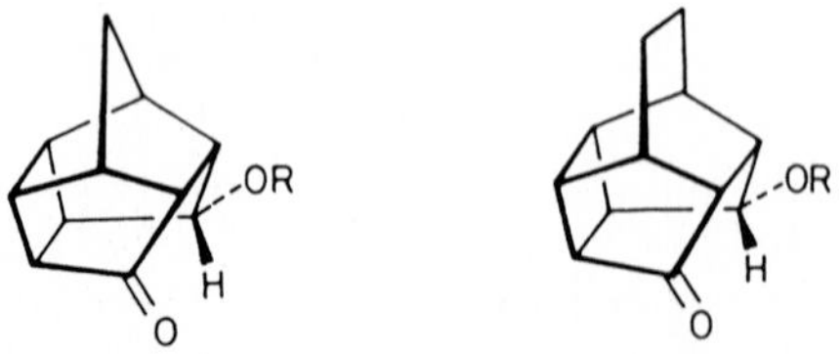

(1) (2)

k_{rel}(R = Na) 1 10^2

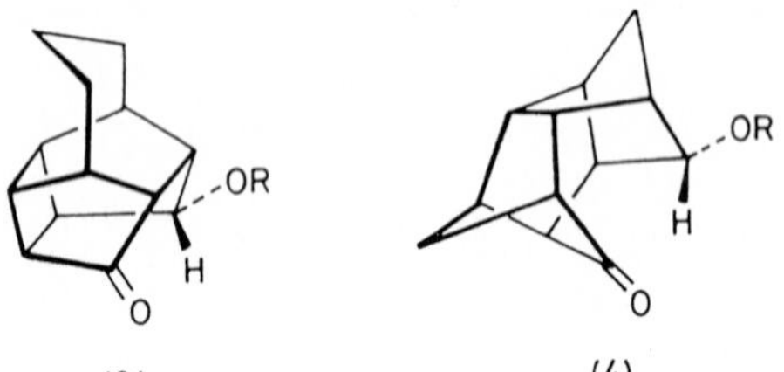

(3) (4)

10^3 10^6

(R = H, Na, p-NB)

Scheme 6.4

is a tendency for Δ to increase with decreasing distance d_1, but it seems clear that the original logarithmic curve overestimates non-planarity of the carbonyl group at large distances. Moreover, apart from the methadone point, the N...C=O points with $d_1 > 2.5$ Å show a quite similar distribution to the O...C=O interactions. Thus, in contrast to the earlier conclusion [63], there is no essential difference between the two. Originally, non-planarity of carbonyls in contact with O and N was believed to reflect the difference in nucleophilicity between the two atoms.

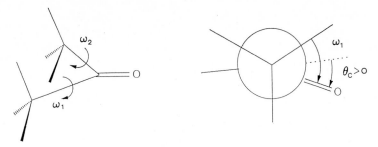

Scheme 6.5

In view of the importance of the methadone point in the original set of six crystal structures, the notion that it is an outlier point is quite intriguing. The high accuracy of the X-ray determination seems beyond question as it is confirmed by an independent study [71]. A possible explanation has been suggested by Cieplak and Bürgi [72] who found that aliphatic ketones show substantial pyramidalization depending on the conformation of the $C(sp^3)$ ligands. Loss of local C_s or C_2 symmetry of such a fragment is always accompanied by a slight deviation from planarity of the carbonyl group. The basic, general pattern of distribution of the distortion parameter θ_c as a function of the torsion angles ω_1 and ω_2 (see Scheme 6.5) is shown in the scattergram plotted for hexa-substituted acyclic and cyclic acetone derivatives $((CX_3)_2C=O$, $X =$ any atom from main groups IV–VII, Figure 6.3). For each fragment all 9 combinations of ω_1 and ω_2 were taken into account. In addition to each triple $\theta_c, \omega_1, \omega_2$, the triples representing isometric structures were added: $-\theta_c, \omega_2, \omega_1$; $-\theta_c, -\omega_1, -\omega_2$; and $\theta_c, -\omega_2, -\omega_1$ (see Section 2.6.3, Chapter 2). The scattergram reveals a distribution of data points reminiscent of the expected "egg-

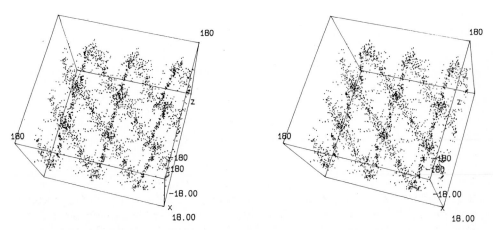

Fig. 6.3. Scatterplot of $\theta_c(=x)$, measuring the pyramidalization of $RR'C=O$, vs. torsion angles $\omega_1(=y)$ and $\omega_2(=z)$ about the bonds $R-C(=O)$ and $R'-C(=O)$. The pattern has the shape of an egg-carton [72a]

carton" surface in the three-dimensional configuration space. By analogy with similar observations for peptides [73], a simple model was proposed in which the pyramidalization of the carbonyl is a function of the conformation of its two $C(sp^3)$ ligands. For the acyclic pentan-3-ones where the α,α' substitution is limited to H and C, pyramidalization is found to depend on the torsion angles ω_1 and ω_2 about the $C_\alpha - C(=O)$ and $C_{\alpha'} - C(=O)$ bonds according to

$$\theta_c = 8.1 \sin [3(\omega_1 + \omega_2)/2] \cos [3(\omega_1 - \omega_2)/2] = 8.1 \, FF$$

where the measure of pyramidalization, θ_c, is defined as

$$\theta_c = \omega_1 (C_\beta - C_\alpha - C - O) - \omega_1' (C_\beta - C_\alpha - C - C_{\alpha'}) + 180°$$

and FF is an abbreviation for the product of trigonometric functions.

The definition implies out-of-plane displacements of O in the direction of a staggered conformation about $C_\alpha - C(=O)$ for positive θ_c (see Scheme 6.5). To check whether this effect could be responsible for the non-planarity of the carbonyl group in methadone, a linear regression was made on the data available for methadone salts [72b, 74], leading to $\theta_c = 10.8(2.8) \, FF$, $r = 0.85$ ($N = 8$). The θ_c value predicted for methadone from this equation is about 7.5°, whereas the observed values are 8.0° and 9.1°, the differences being insignificant (the regression line for the total sample is $\theta_c = 11.4 (1.5) \, FF$, $r = 0.93$ ($N = 10$)). In other words, all or almost all the pyramidalization in methadone can be explained by the local asymmetry of the keto group environment. The scatterplot for the salts and the methadone points (two independent determinations) is shown in Figure 6.4.

Similarly, the effect of a locally asymmetric environment might be a source of out-of-plane distortion of the carbonyl group in the crystal structures of polycyclopentane hydroxyketones (see Scheme 6.4). Distances between the carbinol and carbonyl C atoms in these molecules range from 2.48 to about 2.67 Å, and the H...C=O distances are correspondingly short, between 2.18 and 2.49 Å. Except for the least accurate structure, the carbonyl C is displaced out of the plane of its ligands towards the hydrogen atom (by 0.014 (4), 0.019 (5) and 0.074 Å). It is tempting to interpret these distortions as initial stages of the hydride transfer reaction; however, the carbinol hydrogen is expected to have little nucleophilic character in the *p*-nitrobenzoyl esters used for X-ray determination. The apparent discrepancy is resolved if one considers the environment of the carbonyl groups in these structures. Indeed, the largest distortion, in compound **4** (Scheme 6.4), is not unusual either in its sense or magnitude in the context of the general behavior of the bicyclo[2.2.1]heptan-2-ones, which were analyzed as a separate class in the study by Cieplak and Bürgi. In this skeleton, the carbonyl group is flanked by a tertiary carbon and a methylene group (Scheme 6.6, p. 216). The assumption of equivalency of the H and C ligands no longer holds here and the simple model described above fails. It appears that pyramidalization depends here almost exclusively on the orientation of the methylene group:

$$\theta_c = -7.1(0.8) \sin 3\omega_2 \, , \quad r = 0.872 \, , \quad N = 28 \, ,$$

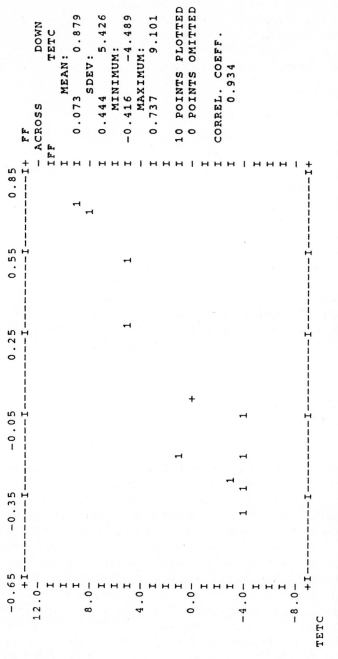

Fig. 6.4. Regression of (TETC) = θ_c vs. (FF) = $\sin[3(\omega_1+\omega_2)/2]\cos[3(\omega_1-\omega_2)/2]$ (see Scheme 6.5) for methadone (+) and its derivatives

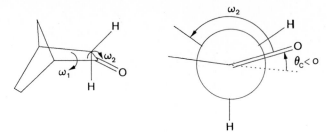

Scheme 6.6

where negative amplitude signifies that the carbonyl group prefers pseudo-eclipsing with respect to the closest C−H bond. The observed θ_c value for the polycyclic ketone in question is −10.2°, about half of which is explained as a bicyclo[2.2.1]-heptan-2-one conformational effect ($\omega_2 = -143.2°$). In **4**, however, the vicinal bond collinear with the carbonyl C p_z-axis is not a C−H bond, but a significantly elongated C−C bond (1.597(6) Å); thus, possibly even the entire distortion can be attributed to the local asymmetry. It appears that for nucleophile-carbonyl group contacts at large distances, other interactions become of comparable magnitude and have to be taken into account to properly quantify the description of pyramidalization of the carbonyl group.

6.2.1.2 Distribution of the Nu...C=O Angle in Intra- and Intermolecular Contacts

Figure 6.5 shows relative positions of N, C and O atoms and of the RCR′ plane in 15 molecules. In their original report, Bürgi, Dunitz and Shefter observed that the line of approach of the nucleophile is not perpendicular to the C−O bond but forms an angle of about 107° with it. This was considered to be a strongly preferred orientation for nucleophilic attack, in accord with Koshland's proposal of orbital steering. The mean N...C−O angle α is not appreciably different in the additional nine structures; in all cases, N lies approximately in the bisector plane (local mirror plane) and α is between 98.6 and 114.5°. A closer scrutiny of the larger sample suggests, however, that to some extent the approach angle depends on the molecular context. Thus, nine of the fifteen examples involve a 5-azacyclooctanone ring and in these α is nearly constant at around 112°; in one such structure where no N...C=O interaction is expected or manifested, α deviates by only a few degrees from this value (107.6°). For the remaining six structures, α is significantly smaller: two are 6-azacyclodecanones ($\langle\alpha\rangle = 102°$), three are 1,8-disubstituted naphthalenes ($\langle\alpha\rangle = 102°$) and one is the acyclic 4-aminoketone methadone ($\alpha = 105°$).

A similar picture emerges from a reassessment of the data on the O...C=O interactions [63]. Of particular interest are the structures containing intermolecular nucleophile-carbonyl group contacts. For a rather heterogeneous collection of structures, $\langle\alpha\rangle$ was found to be about 96°; for distances shorter than 3.1 Å $\langle\alpha\rangle$ is slightly

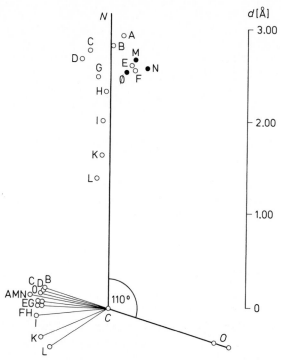

Fig. 6.5. Relative positions of N, C, and O atoms and of the RCR' plane in molecules of the types shown in Schemes 6.1 and 6.3 (data from [29])

smaller (93°). For intramolecular O...C=O interactions ⟨α⟩ was again found to depend on the molecular context [63]. For 1,2-dicarboxylic acids it was quite small (93°). For a sample that included diverse carbonyl derivatives such as nitro, hydroxy and alkoxy ketones, esters, etc. the range of α for the intramolecular contact is correspondingly wide, from 68 to 134°, with mean value 104°; if two structures of low accuracy are excluded, ⟨α⟩ for this sample drops to 101°. For a sample of 1,8-disubstituted naphthalenes ⟨α⟩ is 100° and for a more recent sample of bicyclo[1.1.1]pentane diesters it is 101° (intramolecular contacts in both cases) [61, 68].

Further evidence on the geometry of intermolecular nucleophile-carbonyl interactions comes from crystal structures of protonated ketones [75]. A number of these structures were found to contain contacts of weakly nucleophilic halogens with the keto groups; the authors point out that such interactions do not display the obtuse angle of approach. Recently, to address the question of geometry of intermolecular N...C=O interactions, we have examined a large number of suitable structures retrieved from the CSD [76]. Our survey reveals that, regardless of the type of nitrogen nucleophile, ⟨α⟩ is always close to 90°; *tert*-aliphatic amines, ⟨α⟩ = 95°; *tert*-anilines, ⟨α⟩ = 91°, trigonal *tert*-amines, ⟨α⟩ = 84°; oximes and hydrazones, ⟨α⟩ = 89°, triazoles, ⟨α⟩ = 90°; cyanides, ⟨α⟩ = 95°; azides, ⟨α⟩ = 83°. Finally, in order to test a uniform sample of relatively effective O nucleophiles, we also examined intermolecular contacts of nitro group oxygens with carbonyls, obtaining ⟨α⟩ = 92°.

Thus, it is an inescapable conclusion at this point that the large obtuse angle of nucleophile approach to a keto group ($\alpha = 112°$), found in the original studies to be constant over a wide range of N...C=O distances, is unique to the transannular interaction in the 5-azacyclooctanone ring. No other fragments reported to contain intramolecular N...C=O or O...C=O interactions show such high α angles, the available $\langle\alpha\rangle$ values ranging from 93 to 102°. Furthermore, the intermolecular contacts of neutral N and O nucleophiles with carbonyls appear to occur along the local C_{3v} axis of the trigonal C atom; at least at distances of 2.8–3.2 Å, $\langle\alpha\rangle$ is invariably close to 90°. It seems now that the questions if, when, and to what extent this local threefold symmetry is lost along the reaction pathway, require further study.

The proposition that the bisector plane offers a minimum energy pathway is supported by the data on intermolecular O...C=O interactions. The β_1 and $\beta_{1'}$ angles (see Figure 6.1 for definitions) differ mostly by no more than 10°. This is no longer true for intramolecular interactions, where the skeletal constraints can make it impossible to attain the optimal geometry.

The symmetry of approach should also depend on the nature of the ligands of the carbonyl group [30, 31]. This seems to be the case [68], but again, the question needs more detailed examination.

The last aspect of the structural expression of the N...C=O interaction is the direction of the lone-pair orbital on the nucleophile. Although it cannot be observed, for the tertiary amino group it may be assumed to lie close to the local threefold axis. Indeed, in the initial study, this direction was found to coincide with the N...C direction within a few degrees, except in the case of methadone, where the angle is about 21°. This question does not seem to have been pursued since then.

6.2.2 Initial Stages of Spontaneous Hydrolysis of Acetals

6.2.2.1 Distortions from C_{2v} Symmetry in Acetals:
Correlation of the Antisymmetric Stretching
and Bending Displacement Coordinates

The acetal fragment, consisting of a tetrahedral C carrying two oxygen substituents, is the product of completed O...C=O addition. Thus, according to the principle of microscopic reversibility, distortions of its geometry toward spontaneous cleavage can be expected to map out the final stages of the reaction pathway for nucleophilic addition. Since the O...C=O contacts did not cover the region intermediate between the van der Waals distance and the covalent bonding distance, such distortions complete the picture of the reaction path at the product end.

The sample originally examined was a heterogeneous collection: diols, hemiacetals and acetals of different degree of substitution at carbon (see Table 5 of reference [63]). The C–O bond length ranged from 1.33 to 1.48 Å, with the more extreme values in the less accurate determinations. Similarly, the C–C–O valence angles varied between 100 and 124°. The structural heterogeneity of the sample may seem too high

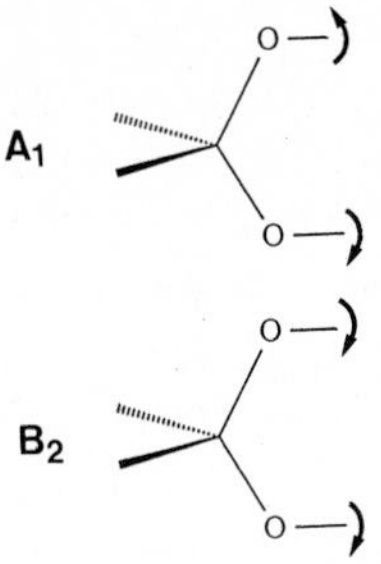

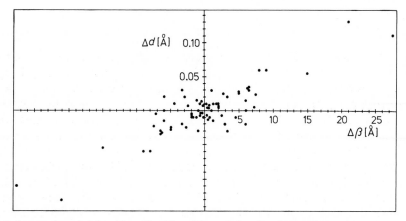

$$S_1 = (d_1 + d_2) / \sqrt{2}$$

$$S_{10} = (\beta_1 + \beta'_1 + \beta_2 + \beta'_2) / 2$$

$$S_2 = (d_1 - d_2) / \sqrt{2}$$

$$S_7 = (\beta_1 + \beta'_1 - \beta_2 - \beta'_2) / 2$$

Scheme 6.7

to expect correlations involving the bond lengths. Nevertheless, the authors detected a correlation between two parameters proportional to symmetry displacement coordinates (see Chapter 2, Section 2.3) for a tetrahedron of C_{2v} symmetry: the difference of the averaged $C-C-O$ valence angles of the two $C-O$ bonds, $\Delta\beta = (\beta_1 + \beta_{1'}) - (\beta_2 + \beta_{2'})$ and the difference of the two $C-O$ bond lengths, $\Delta d = d_1 - d_2$. The interest in this set of structural parameters stems from the analysis of covariance in terms of symmetry coordinates. A di-oxygen-substituted tetrahedral C fragment is described by four bond distances and six bond angles. These internal coordinates can be transformed into a system of linear combinations that are unchanged (apart from possible sign reversal) by the C_{2v} symmetry operations. The symmetry coordinates (Scheme 6.7) can couple with one another at the harmonic approximation level only if they belong to the same irreducible representation of the local symmetry. Thus, examination of the irreducible representations of the C_{2v} point group reveals what couplings of the structural parameters are possible. Only the B_2 representation is of interest here, since it corresponds to the anti-symmetric stretch-bend mode and the changes it describes could occur in the initial stages of a cleavage reaction. The scatterplot of $\Delta\beta$ vs. Δd (Figure 6.6) shows a tendency for

Fig. 6.6. Scatterplot of $\Delta\beta = (\beta_1 + \beta'_1) - (\beta_2 + \beta'_2)$ vs. $\Delta d = d_1 - d_2$ (see Figure 6.1)

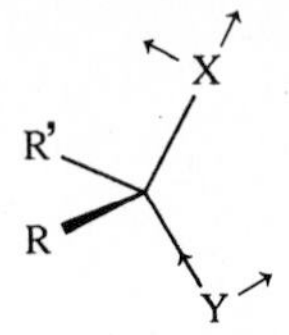

Scheme 6.8

large positive values of Δd to be associated with large positive values of $\Delta\beta$. Thus, the longer bond tends to be associated with smaller $C-C-O$ angles, the shorter bond with larger ones, i.e. as the $C-O$ bond becomes shorter it tends to lie closer to the $C-C-C$ plane. In contrast, no trend was apparent in the variation of the $O-C-O$ angle, which was then taken as more or less constant.

This picture draws on the analogy to vibrational coupling of the antisymmetric stretching and bending modes of a tetrahedral molecule with C_{2v} symmetry (see Chapter 5, Section 5.3.2). From the scatterplot of $\Delta\beta$ vs. Δd, the quadratic coupling constant for the two B_2 symmetry displacement coordinates should have a negative sign, the same as determined spectroscopically for methylene dihalides [77]. This analogy implies that along the minimum energy path for decomposition of a tetrahedral intermediate, the leaving group X does not simply depart along the line of the $C-X$ bond. Instead, as the bond $C-Y$ shortens and the bond $C-X$ weakens, the coupling between the asymmetric bend and stretch coordinates comes into operation. The shortening $C-Y$ bond moves towards the RCR′ plane and the direction of departure of the leaving group X is continuously adjusted to retain an XCY angle of about 110° (see Scheme 6.8).

Since then, extensive studies of structural variance in the acetal fragment based on crystallographic data and on *ab initio* calculations have been made by several authors [78–82]. None of these, however, was concerned with the correlation described here between the antisymmetric stretching and bending coordinates. Recently, Irwin and Dunitz re-examined the X-ray data on acetal geometry, focusing on the effects of ring constraints and substitution at the acetal carbon [83]. Stringent criteria of quality and accuracy of structure determinations were applied (*AS = 1 or *RFAC <0.05 if *AS = 0, 2; all relevant H-atoms located and reported; no atom heavier than Cl; see Chapter 3). Scatter plots of the antisymmetric stretch/bend (B_2) coordinates confirm the earlier conclusion, i.e. they show a positive correlation between $\Delta\beta$ and Δd for the fragments carrying two carbon substituents at the acetal carbon (referred to as ketal fragments). The regression slope, however, depends strongly on the presence or absence of a ring constraint. It is relatively large (comparable to the slope reported in the earlier study) only for the acyclic ketals and much smaller for cyclic ketals such as, for instance, 5-membered ring ketals (2,2-dialkyldioxolanes).

6.2.2.2 Structural Expression of the Anomeric Effect in Acetals [84]; Correlation of the $O-C-O$ Angle and Δd

Two lines of development have had an especially significant impact on the study and interpretation of structural variance in acetals. One was the discovery by Altona et al. that the pattern of bond distances and bond angles about the anomeric center in 2-halotetrahydropyrans and related molecules depends on the configuration at this center [85]. The other was Deslongchamps' discovery that the rate and direction of the hydrolytic breakdown of orthoesters and hemiorthoesters strongly depend on the conformation about the $C-O$ bonds [51]. Both kinds of evidence were interpreted in terms of the hypothesis that n,σ^* hyperconjugation plays a major role among the interactions at an anomeric center. This hypothesis has not gone unquestioned [86–91]. Attempts to demonstrate a definite structural expression of n,σ^* delocalization in glycosides and acetals have not been conclusive [80, 81], and kinetic studies have been claimed to contradict Deslongchamps' theory [87]. The task of confirming this hypothesis was undertaken by Kirby and Jones and their coworkers, who made crystal structure analyses and kinetic studies of spontaneous hydrolysis for a number of 2-aryloxy and 2-alkoxytetrahydropyrans [52]. Their results appear to provide convincing evidence for the validity of the n,σ^* hyperconjugation hypothesis.

Since these results are discussed in a later part of the chapter, it suffices here to say that the two acetal $C-O$ bond distances do display the expected hyperboloid relationship in the *axial* acetals but not in the *equatorial* ones, they do change in the expected way in response to increasing electron demand of the group attached to the exocyclic oxygen, and the largest bond length perturbation is accompanied by a flattening of the anomeric center and of the ring. It therefore seems that structural expression of the anomeric effect can indeed be found in crystal structures of acetals, provided there is an appropriate conformational and electronic bias in the fragment.

The question then arises anew as to what exactly is the structural variance associated with the initial stages of the spontaneous breakdown of an acetal. In particular, is the $O-C-O$ angle really constant along the progress coordinate? A detailed principal component analysis of the acetal fragment geometry for the Kirby-Jones tetrahydropyranyl acetals was carried out by Bürgi and Dubler-Steudle [54]. There is a considerable variation in the value of the $O-C-O$ angle in this sample. Extrapolation along the eigenvector of the first principal component, identified with the incipient hydrolysis of the acetal fragment, gives an $O-C-O$ angle of about $80-90°$ for the transition state instead of the expected $100-110°$.

This observation prompted our interest in testing the behavior of the $O-C-O$ angle in more narrowly defined subclasses of acetals. Since three types of compound account for a majority of the examples (2,2-dialkyldioxolanes, 2-oxy-2-alkyltetrahydrofuranes and 2-oxytetrahydropyranes), the survey focused on derivatives of these. As noted before, the enormous increase in the number of structures containing such fragments makes it possible to reexamine their experimental geometries at a fairly high level of accuracy. The search of the January 1992 issue of the CSD yielded $N = 98$ fragments of saturated 2,2-dimethyldioxolanes ($\langle\sigma\rangle < 0.005$ Å and *RFAC < 0.06), $N = 26$ furanose fragments with alkyl substitution at the exocyclic glycosidic oxygen and at the anomeric carbon (but no quaternary groups; $\langle\sigma\rangle < 0.01$ Å and *RFAC < 0.07), $N = 44$ axial and $N = 29$ equatorial pyranosyl glycoside fragments with phenyl or similar substitution at the glycosidic oxygen ($\langle\sigma\rangle < 0.01$ Å and *RFAC < 0.06). Structures containing metals or third row elements have been excluded [92]. Linear regression analysis reveals significant correlation: $\alpha(°) =$

$111.0(0.1) - 29.6(3.5)\,\Delta d$, $r = -0.86$ for the alkyl furanosides; $\alpha(°) = 112.6(0.3) + 59.8(7.2)\,\Delta d$, $r = 0.79$ for the axial aryl(acyl)pyranosides, where $\Delta d = r(CO, endo) - r(CO, exo)$ in Å. The scatterplots of Δd vs. $\alpha(O{-}C{-}O)$ are shown in Figure 6.7. On the other hand, in dioxolanes and equatorial pyranosyl glycosides, α is not correlated with the change in the bond lengths and is more or less constant.

The major difference between these two types of structures is in the conformation of the acetal fragment $C{-}O{-}C{-}O{-}C$. In dioxolanes and equatorial pyranosides

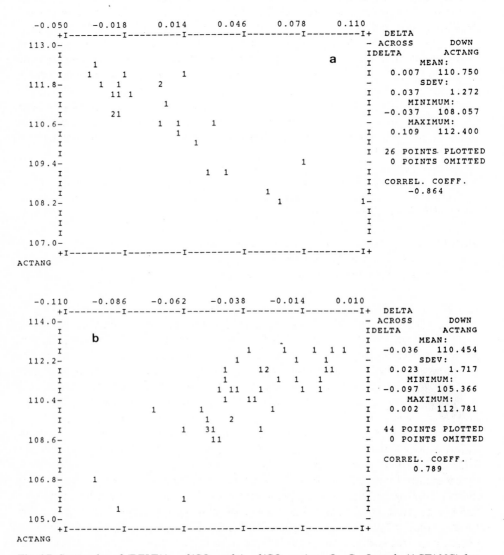

Fig. 6.7. Scatterplot of (DELTA) = $d(CO, endo) - d(CO, exo)$ vs. $O{-}C{-}O$ angle (ACTANG) for: a) alkyl furanosides, b) axial aryl (acyl) pyranosides

it is nearly planar, so that no anomeric-type interaction is possible. In furanosides and axial pyranosides the planes of the two ether moieties form an angle which is optimal for such an interaction, and thereby a stage for a spontaneous breakdown is set. Thus, the observed structural variance most likely reflects the distortions along the pathway of hydrolysis. Reversal of sign of the correlation can be explained in a simple way within the framework of the n, σ^* interaction hypothesis. Namely, it is expected as a consequence of the reversal of the donor-acceptor roles between the endocyclic and exocyclic oxygen atoms in alkyl furanosides and aryl(acyl) pyranosides: in the latter the exocyclic group (the aryloxy group) is clearly a better acceptor and poorer donor than the endocyclic ether group. In the former, the endocyclic group is a better acceptor since the $\sigma^*(C-O)$ energy level of a bond in a five-membered ring is lower than that of a bond in an open chain. The change of direction of the predominant n, σ^* delocalization reverses, of course, the direction of the incipient hydrolysis. Thus, these findings would appear to confirm the earlier observation that α decreases along the progress coordinate for the acetal breakdown. They also suggest that even the apparently intractable structural variance in the case of acetals such as alkyl pyranosides might be revealed by factor analysis to result, so to speak, from a superposition of two different incipient hydrolysis processes.

6.2.2.3 Inorganic Models

6.2.2.3.1 Ligand Addition to the BO₃ Group: Correlation of Partial Pyramidalization and the Incipient Bond Distance

In 1982, Zobetz examined the symmetry and frequency distributions of main geometrical parameters of BO_3 groups in accurate structures of minerals and synthetic inorganic borates [93]. He noted small but significant deviations from planarity in 20 out of 75 BO_3 groups included in the survey. The distortion is such that the boron atom is displaced from the plane of the BO_3 group towards the fourth oxygen atom, which lies vertically above it, at a distance shorter than 3.0 Å. Extreme distortions of this kind were found in cubic and orthorhombic boracites (O...BO_3 distance d_1 between 1.675 Å and 2.414 Å, Δ between 0.411 and 0.076 Å, respectively). However, a dependence of the magnitude of the deviation on the distance was also clearly recognizable in tourmalines, with longer O...BO_3 distances (d_1 between 2.733 and 2.835 Å, Δ between 0.018 and 0.006 Å).

A reexamination of these distortions with a much larger sample has now been reported [94]. The plot of boron pyramidalization Δ vs. the O...BO_3 distance d_1 in 256 borate groups is shown in Figure 6.8. It includes, in analogy to Figure 6.2, the bonding region for the BO_4 groups ($d_1 < 1.6$ Å). The curves shown in the plot are obtained by calculations based on Zobetz's repulsion model of the BO_4 geometry. The distribution bears a striking resemblance to the one found for the nucleophile-carbonyl group interactions.

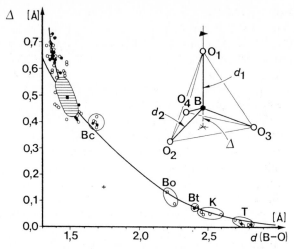

Fig. 6.8. Distance Δ of boron atom from the plane defined by O(2), O(3), O(4) vs. distance $d(B-O)$ between boron atom and O(1) (T: Tourmalines; K: Kotoit and related compounds; Bt, Bo, Bc: Borazites. The hatched area summarizes 896 data points)

6.2.2.3.2 Ligand Elimination from the Tetrahedral XY$_4$ Species: Bond Length and Valence Angle Correlations

Spontaneous breakdown of acetals is an example of the S_N1 type of reaction, unimolecular elimination from a tetrahedral center. Reaction pathways for this general type of reaction have been derived by analysis of deviations from the expected T_d symmetry of tetrahedral molecules of the YMX$_3$ and MX$_4$ formulation [95].

Only distortions maintaining approximate C_{3v} symmetry were considered since it is this type of distortion that occurs along the elimination coordinate. The YMX$_3$ species (YSO$_3$, YPO$_3$, YSnCl$_3$) automatically belong to this category, while the distorted MX$_4$ molecules (PO$_4^{3-}$, SO$_4^{2-}$, AlCl$_4^-$) have to be classified as approximately C_{3v}- or approximately C_{2v}-symmetrical according to some arbitrary criteria. The experimental sample points for the phosphate ions PO$_4^{3-}$ were taken from an exhaustive compilation by Baur published in 1974 [96]. From this list, 68 structures were defined as having approximate C_{3v} symmetry: in each case the distortion vector D_3 lies within 20° of the symmetry coordinate S'_{3a} or $-S'_{3a}$ (see Chapter 2, Section 2.4.1, for further explanation). The data for 26 SO$_4^{2-}$ and 17 AlCl$_4^-$ fragments of sufficient accuracy ($\sigma(d) < 0.02$ Å) were selected from the BIDICS 1969–72 bibliography of inorganic crystal structures [97]. For structures with approximate C_{3v} symmetry the appropriate bond lengths and angles were averaged to produce structures with exact C_{3v} symmetry, as indicated in Figure 6.9.

Scatterplots d_1 vs. α and d_2 vs. α for sulfate, phosphate and tetrachloroaluminate ions are shown in Figure 6.10a, b, c. In every case, as the axial bond distance d_1 increases, the central atom comes closer to the plane of the other three bonds, which become shorter.

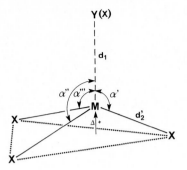

Fig. 6.9. Definition of geometrical parameters for YMX$_3$ fragments with approximate C_{3v} symmetry

The d_i vs. α correlations are also very similar for the YMX$_3$ species. Indeed, data sets for all the available structures of the YMX$_3$ formula, including YAlCl$_3$, YSO$_3$, YPO$_3$, OPX$_3$, YSnCl$_3$, YGeCl$_3$, YSiCl$_3$, YSnBr$_3$, YSnPh$_3$, YPF$_3$, YPCl$_3$ and YPPh$_3$, can be referred to a common origin after replacing d_1 by $\Delta d_1 = d(MY) - d_t(MY_4)$ and d_2 by $\Delta d_2 = d(MX) - d_t(MX_4)$ where d_t refers to the corresponding T_d species. As the plot in Figure 6.10d shows, all the Δd_1, α points lie close to one curve and all the Δd_2, α points close to another. It thus seems that, regardless of the nature of the central atom and the diversity of the crystal environment, all these tetrahedral molecules deform along the same path in the subspace maintaining C_{3v} symmetry.

Another remarkable feature of the presented plots is that the smooth curves are not just correlation curves of any arbitrary form drawn to fit the data as well as possible, but are derived from a simple model of bonding. The model is based on three assumptions. The first is Pauling's relation between bond length and bond number [24]:

$$\Delta d_i = -c \log n_i . \tag{6.3}$$

The second is the postulate of bond order conservation: it is assumed that for all points along the correlation curves, i.e. for all tetrahedral molecules in question, $n_2 + 3n_1 = 4$ [25]. Finally, the displacement Δ of the central atom from the plane of the three basal ligands is also taken as a measure of n; when $\Delta = \Delta_t$ (regular tetrahedron), $n_2 = 1$, when $\Delta = 0$ (trigonal planar molecule), $n_2 = 0$. The necessary relation between bond number and bond angle is then chosen as follows: $n_2 = (\Delta/\Delta_t)^2 = 9 \cos^2 \alpha$ and $n_1 = 4/3 - 3 \cos^2 \alpha$. This satisfies the desired conditions that $n_1 = 1$ for $\alpha = 109.47°$, $n_1 = 0$ for $\alpha = 90°$ and $\partial(\Delta d_1)/\partial(\Delta \alpha) = 0$ for $\alpha = 90°$. The same pair of functions is used in all the plots in Figure 6.10. The results of least-squares fitting give numerical expression to the qualitative similarity. The constants in the Pauling relation for PO$_4^{3-}$, SO$_4^{2-}$ and AlCl$_4^-$ are equal within experimental error.

In the previously mentioned study, Zobetz also examined distortion pathways for the approximately C_{3v} symmetrical O...BO$_3$ and BO$_4$ groups [94]. The correspond-

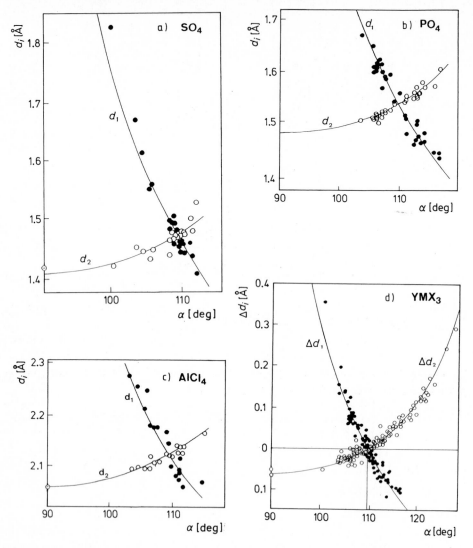

Fig. 6.10. Scatterplot of d_1 (●) and d_2 (○) vs. α (for definitions, see Fig. 6.9).
a: SO_4^{2-}, b: PO_4^{3-}, c: $AlCl_4^-$ ions and d: analogous correlation for a large set of YMX_3-species (see text)

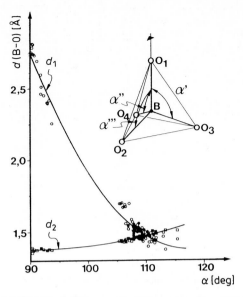

Fig. 6.11. Scatterplot of d_1 ($\bigcirc$) and d_2 ($\square$) vs. α for BO_4 groups (for definitions, see Figure 6.9). The solid lines represent quadratic regression curves

ing plot is shown in Figure 6.11. Clearly, the early conclusions [95] are valid for the boron-centered tetrahedral coordination as well.

6.3 Reaction Pathway for $sp^2 \rightleftharpoons sp$ Transformations of Carbonyls

6.3.1 Initial Stages of Nucleophilic Addition to sp Centers

Studies of the geometry of intramolecular contacts between nucleophilic O and N atoms and the sp C and N electrophilic centers were undertaken in a direct and straightforward extension of the earlier investigations of aminoketones and amino-acid derivatives. Four structures, namely 8-methoxy- and 8-nitronaphtho-1-nitrile (Scheme 6.9, structures **2** and **3**) [98], quinoline-8-diazonium-1-oxide tetrafluoroborate (**1**) [99], and 2,2'-bipyridine-3,3'-dicarbonitrile (**4**) [100], were analyzed to obtain information about preferred approach directions of the nucleophilic O and N atoms and the electrophilic sp centers in the $C \equiv N$ and $N \equiv N^+$ groups. For 8-methoxynaphtho-1-nitrile and quinoline-8-diazonium-1-ox-

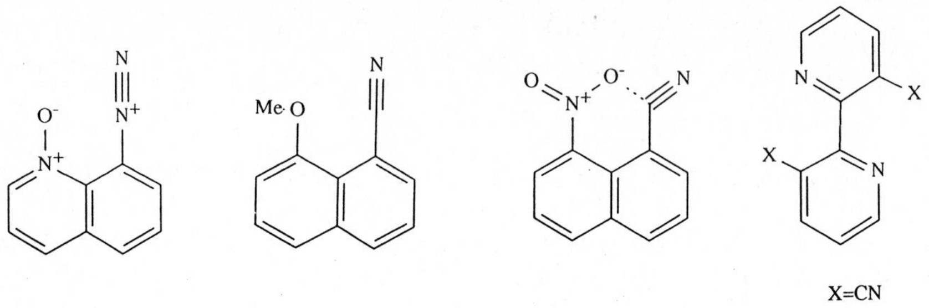

(1)   (2)   (3)   (4)

X=CN

Scheme 6.9

ide, the characteristic pattern of bond angle distortions for the *peri* substituents of naphthalene is observed: the exocyclic bond to the electrophilic center is splayed outward, the one to the nucleophilic center is splayed inward. A similar pattern is recognizable in the bipyridine fragment, but in 8-nitronaphtho-1-nitrile, where the nucleophilic center is separated by an additional bond from the nucleus, both groups are splayed outwards. In each case the contact distance is considerably shorter than the sum of corresponding van der Waals radii: 2.443(2) Å in the diazonium salt (**1**), 2.594(4) Å in the methoxy (**2**), 2.685(4) and 2.788(4) Å in the nitro (**3**), 2.695(2) and 2.740(2) Å in the bipyridine derivatives (**4**). In every case the central sp atom deviates from the internuclear axis of its two ligands towards the nucleophile while remaining in or very close to the plane defined by these three atoms. Thus the diazonium group deviates by 10.4° from linearity, and the cyano groups by 6.2, 5.8, 8.5 and 8.5°. The pattern of close contacts and deviations of the bond angles can be interpreted as indicative of attractive interactions between the neighboring functional groups. To the extent that such interactions represent the initial stages of nucleophilic addition to sp centers, these structures provide information on the stereoelectronic control of the approach angle. In two cases, combination of splaying and bend yields α angles of 103.5° and 104.4°. In the bipyridine derivative, the contact angles α are 107.6 and 108.0°, and the authors point out that a smaller angle of 99° could be obtained by internal rotation to a co-planar conformation without any change in the N...C≡N distance. On the other hand, the contact angles in two independent 8-nitro-naphtho-1-nitrile molecules are about 95°, and the differences in the N−O bond lengths suggest considerable involvement of the nitro groups in partial bonding. The authors warn, however, that the disorder present may have affected the quality of the structural data.

6.3.2 Initial Stages of Spontaneous Cleavage
of Ketene Acetal-Like Fragments

6.3.2.1 Bond Length and Valence Angle Correlations in Sydnones and Enamines

By analogy to the acetal group (see Section 6.2.2.1), an olefin carrying 1,1-hetero-atom substitution can be considered as a product of completed addition of a nucleophile to a ketene or ketene derivative. Consequently, distortions of the local C_{2v} symmetry of such a fragment would be expected to occur along the pathway of elimination driven by the bond−no-bond resonance: $C=CX_2 \leftrightarrow C=C=X^+ + X^-$.

The idea that the ground-state molecular structure may reflect contribution of a canonical resonance form which incorporates a ketene moiety, was originally advanced by J.C. Earl to rationalize the structure and properties of sydnones [101, 102]. It was swept away, in a triumph of convention over substance, by the concept of meso-ionic compounds [103], only to be vindicated twenty years later by X-ray determination of the crystal structure of 4,4'-dichloro-3,3'-ethylenebissydnone (Scheme 6.10) [104]. Thiessen and Hope pointed out that the pattern of bond lengths in the sydnone ring and the striking deformation of the bond angles about the carbonyl group are best accounted for by invoking the open-chain ketene form **IId** (Scheme 6.11) as a significant contributor to the resonance hybrid **II**.

The same point of view was taken by Gieren and Lamm in their interpretation of the crystal structure of 3-phenylthio-8-oxa-1-aza-2,5-diazoniaspiro[4.5]deca-1,3-diene-2,4-diolate (Figure 6.12) [105], a photoisomer of a sydnone. In spite of the disrupted conjugation, the molecule shows the same pattern of bond lengths and the unusual deformation of the bond angles about the carbonyl group $C(2)-O(2)$ as the sydnone structure. The authors describe the structure, adopting Bürgi's and Dunitz's parlance, as a result of addition of a tertiary amine to ketene, which is arrested shortly before completion.

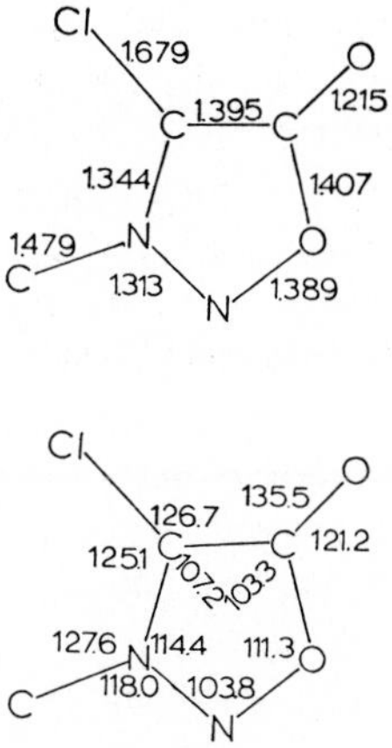

Scheme 6.10

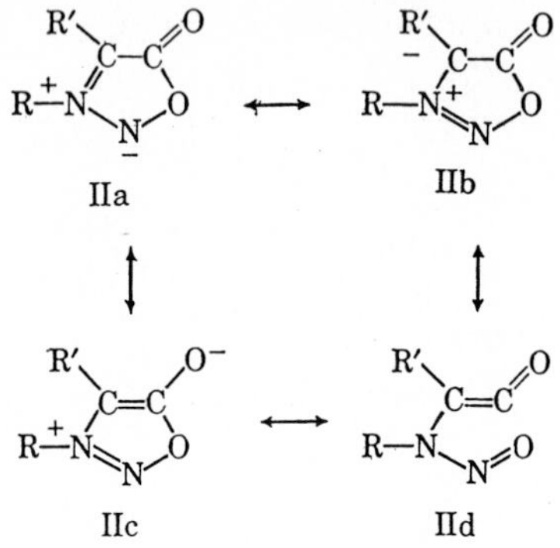

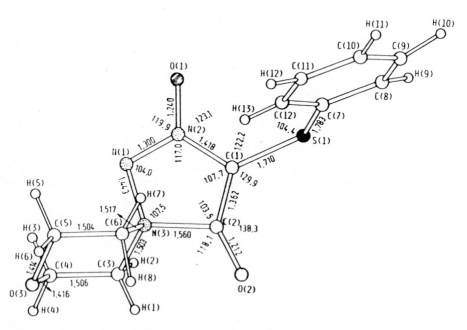

Scheme 6.11

Fig. 6.12. Bond distances and angles in 3-phenylthio-8-oxa-1-aza-2,5-diazoniaspiro[4,5]deca-1,3-diene-2,4-diolate

The question of structure correlations associated with the ring opening of syd-nones does not seem to have been pursued any further since Gieren and Lamm's 1978 article [106]. However, interesting evidence concerning stereoelectronic requirements that control the pathway of such an elimination was provided by the study of crystal-line α-substituted enamines, published a year later by van Meersche and Declerq and their coworkers [107]. The geometry of $C=CX-NMe_2$ fragments in five crystal structures of piperazine-bisenamines (Scheme 6.12; $X = CH_3$, CN, I, Cl, F), was

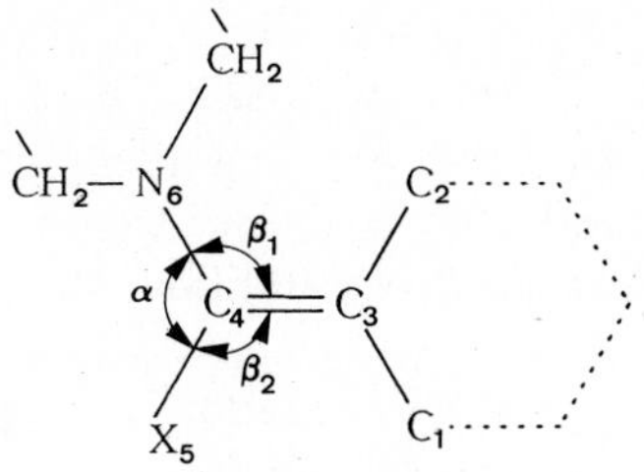

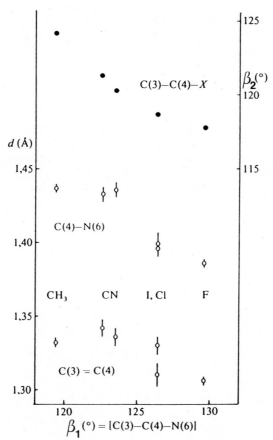

Scheme 6.12

Fig. 6.13. Scatterplot of β_2, $d(C-N)$ and $d(C=C)$ vs. β_1 in α-substituted enamines (for definition of symbols see Scheme 6.12)

analyzed in terms of the increasing keteniminium character. Figure 6.13 shows the pertinent bond lengths and valence angles as a function of the $C=C-N$ angle. The electronegativity of the group X is clearly a major factor in the structural variation of the fragment. The $C=C-N$ angle (β_1, see Scheme 6.12) increases by about 10°

on going from $X = CH_3$ to $X = F$, from $119.4°$ to $129.6°$. Thus, as the $C-X$ bond becomes more ionic, the $C=C-N$ moiety tends toward linearity. Taking the change as a manifestation of an incipient heterolytic elimination, one would expect concomitant changes in the bond lengths, and indeed, both the $C=C$ and the $C-N$ bonds shorten in this process (Figure 6.13). In addition, the N lone pair becomes more involved in the in-plane interactions, as revealed by the torsion angles. This lone pair is inclined by about $34°$ to the CCNX plane for $X = CH_3$, but by only $17°$ and $7°$ for $X = F$ and Cl. For $X = I$, it is almost in the plane ($5°$). All the $C-X$ distances are longer than in the corresponding vinyl derivatives $CH_2=CHX$. The elongation increases, however, when the N lone pair is closer to the CCNX plane: by 0.02, 0.05, 0.05 and 0.11 Å, on going from CH_3 to I. This is consistent with the hypothesis that n,σ^* interaction facilitates the cleavage of the $C-X$ bond. In regard to the question of the constant departure angle, the evidence is rather inconclusive. In fact, both remaining bond angles decrease, albeit the $C=C-X$ angle β_2 might be decreasing slightly faster than the $X-C-N$ angle α (compare $124.2-118.9°$ with $116.4-112.6°$).

6.3.2.2 Bond Length and Valence Angle Correlations in Ester Enolates

The third type of $C=CX_2$ olefin examined from the point of view of structure correlation was the ester enolate. Three crystal structures of lithium ester enolates, obtained in the course of synthetic studies by Seebach et al., provided a total of four fragments depicted in Scheme 6.13 [108]. The variation in the $C-O$ bond lengths and bond angles provides a picture of the incipient stages of elimination of an alkoxy ion (alcoholate ^-OR), yielding a ketene. The $C-OR$ bond length varies between 1.379(3) and 1.412(5) Å, and the $C=C-O^-$ angle increases from 125.4 to $128.2°$ (d_1 and β_1 in Scheme 6.14), i.e. the picture is similar to the enamine one. The other

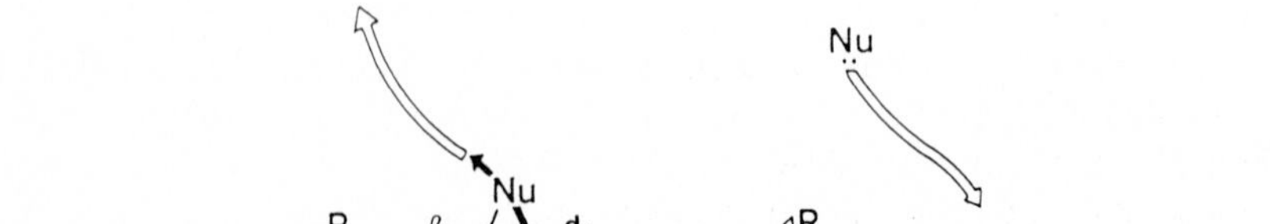

Scheme 6.13

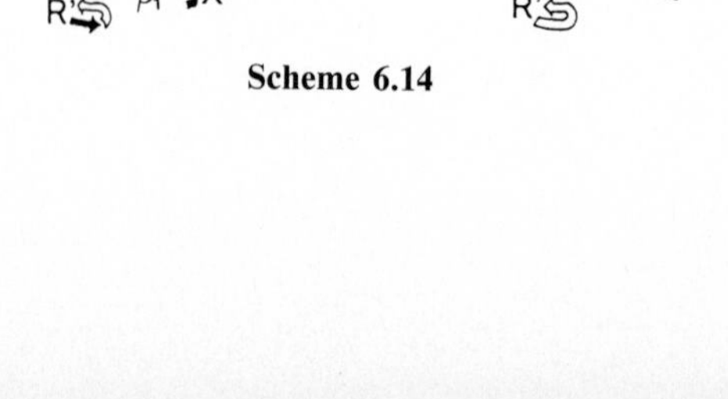

Scheme 6.14

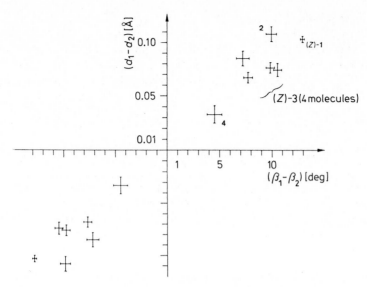

Fig. 6.14. Scatterplot of $(d_1 - d_2)$ vs. $(\beta_1 - \beta_2)$ in lithium ester enolates (for definition of symbols, see Scheme 6.14)

bond angles here change from 115.3 to 118.2° (CCO, β_2 in Scheme 6.14) and from 115.1 to 116.9° (OCO, α in Scheme 6.14). The concomitant changes in the C=C and C−O bond lengths are analogous to those for enamines although the correlations are poorer, perhaps due to the heterogeneity of the distal substitution in the enolates. The fragment geometry was analyzed in a manner analogous to that of the tetrahedral acetal unit. The difference in the C−O bond lengths vs. the difference in the C=C−O bond angles is plotted in Figure 6.14 (after symmetry expansion and addition of one data set for an ylide acetal). The correlation was interpreted in terms of the breakdown of a trigonal center (ester enolate) into a digonal center (ketene) and a leaving group (alcoholate), in analogy to the breakdown of a tetrahedral center (acetal) into a trigonal center (ketone) and a leaving group (alcohol or alcoholate). The shortening bond makes a larger angle with the C=C bond and so moves toward the direction characteristic of the ketene. Similarly, as the other bond begins to lengthen, the angle it forms with the C=C double bond decreases. The authors conclude that the leaving group does not depart along the initial direction of the C−OR bond, but rather the angle of departure is continually adjusted to maintain an approximately equal angle with the forming carbonyl bond, as indicated in Scheme 6.14. From the limited data available it would appear that for a given angular deviation the difference between the two bond lengths is larger for a trigonal center than for a tetrahedral one (compare Figures 6.6 and 6.14).

6.3.3 Initial Stages of Spontaneous Cleavage of Carbonyl Derivatives to Acylium Ion

6.3.3.1 Valence Angle Correlations in Lactones and Lactams

Analysis of structural characteristics of the carboxylic ester and carboxylic amide groups reveals an interesting trend relating the difference of the β_1 and α bond angles to ring size and thus to the β_2 angle (Figure 6.15) [109]. The carbonyl O atom usually deviates from the $C-C-X$ bisector and the deviation depends on the lactone or lactam ring size. A scatterplot of $\langle \beta_1 \rangle - \langle \alpha \rangle$ vs. $\langle \beta_2 \rangle$ shows positive differences for all lactones, the magnitude increasing with decreasing ring size (Figure 6.15). For lactams, the effect is smaller. There is a difference in behavior between acyclic esters and amides: the former deviate significantly from the trend observed for the lactones while the latter do not show a significant deviation. The limited data on thioesters and thiolactones show an even weaker or possibly reversed dependence on α. These changes may portray the early stages of a reaction which starts at the esters RCOOR′ and leads to linear oxocarbonium ions $RC{\equiv}O^+$ by expelling $^-OR'$. This proposal has led to a more detailed investigation of $RC(=O)X$ derivatives.

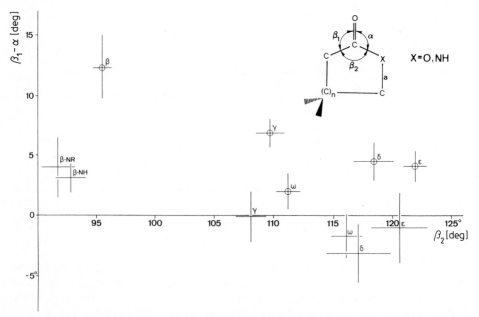

Fig. 6.15. Difference $\langle \beta_1 \rangle - \langle \alpha \rangle$ vs. $\langle \beta_2 \rangle$ (averaged experimental values, error bars indicate standard deviations of the populations). Esters and lactones (X=O) are marked by a circle; the remaining points refer to amides and lactams (X=NH, NR); ω indicates acyclic functionality

6.3.3.2 Bond Length and Valence Angle Correlations in RC(=O)X Derivatives

The following examination of the effect of the substituent X on the structure of RC(=O)X molecules is based on data from microwave spectroscopy, gas-phase electron diffraction and *ab initio* MO calculations. Ferretti et al. [50] have retrieved results published until 1984 for formyl and acetyl derivatives, R = H or CH_3 (see Tables 6.1 and 6.2). The valence angle distributions α, β_1, β_2 were analyzed by prin-

Table 6.1. Selected structural parameters for HCOX molecules [a,b]

X	β_2	β_1	α	d_2	d_1	Technique[c]
F	107.7	130	122.3	1.188	1.346	ED + MW
OCOH	108.8 112.5	125.3 126.7	125.9 120.8	1.197 1.183	1.360 1.388	MW
Cl	110.0	126.5	123.5	1.188	1.760	MW
OH (mon.)	111.9	123.3	124.8	1.201	1.340	MW
OH·H_2O	110.8	123.8	125.4	1.212	1.340	MO (4-21 G*)
OCH_3	109.3	124.9	125.8	1.200	1.334	MW
OC_2H_5	109.91	124.95	125.14	1.204	1.347	MO (4-21 G)
OC_3H_7	110.00	124.75	125.25	1.204	1.345	MO (4-21 G)
NH_2	112.7	122.5	124.7	1.220	1.350	MW
H	116.2	121.9	121.9	1.203	1.100	ED + MW
C_2H_5	115.01	120.76	124.23	1.211	1.513	MO (4-21 G)
C_3H_7	115.09	120.70	124.21	1.211	1.512	MO (4-21 G)
CH_3	115.3	120.5	124.2	1.207	1.512	ED + MW

[a] Bond distances are given in Å, angles in degrees. For references, see [50].
[b] Definition of symbols: see Figure 6.16 or 6.17.
[c] ED: gas-phase electron diffraction; MW: microwave spectroscopy; MO: *ab initio* MO calculations.

Table 6.2. Selected structural parameters for CH_3COX molecules [a,b]

X	β_2	β_1	α	d_2	d_1	Technique[c]
$OCOCH_3$	108.3	130.0	121.7	1.183	1.405	ED
F	110.5	128.8	120.7	1.185	1.362	ED + MW
Cl	111.6	127.2	121.2	1.188	1.798	ED + MW
Br	111.0	126.7	122.3	1.184	1.977	ED + MW
OH (mon.)	110.6	126.6	122.8	1.214	1.364	ED
OH (dim.)	113.0	123.6	123.4	1.231	1.334	ED
CN	114.2	124.6	121.2	1.205	1.474	ED + MW
H	115.3	124.2	120.5	1.207	1.114	ED + MW
$NHCH_3$	114.1	124.1	121.8	1.224	1.386	ED
NH_2	115.1	122.9	122.0	1.220	1.380	ED
CH_3	116.0	122.0	122.0	1.210	1.517	ED + MW
$OSiH_3$	115.9	122.4	121.7	1.214	1.358	ED

[a,b,c] See corresponding footnotes to Table 6.1.

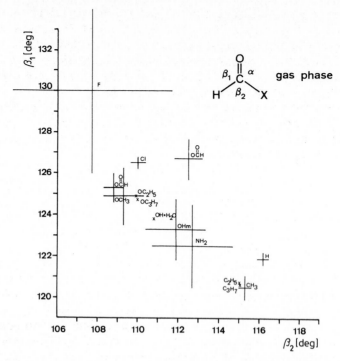

Fig. 6.16. Scatterplot of bond angles β_1 vs. β_2 for HCOX molecules. The nature of X is given next to each datapoint

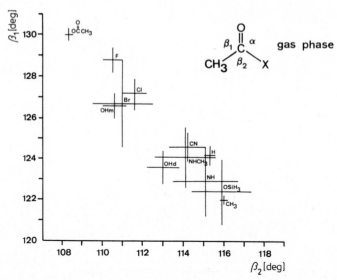

Fig. 6.17. Scatterplot of bond angles β_1 vs. β_2 for CH_3COX molecules. The nature of X is given next to each datapoint

cipal component analysis. The first principal component accounts for most of the total variance in both cases, 79% when R = H and 93% when R = CH$_3$. The eigenvectors (β_1, β_2, α) are (−0.715, 0.699, 0.015) and (−0.699, 0.715, −0.016), respectively. The negative correlation between β_1 and β_2 with slope approximately −1 is evident in Figures 6.16 and 6.17. The individual β_2 and β_1 values depend on the electronegativity of the group X; β_2 is largest for halides and anhydrides, smaller for esters, acids and amides, and smallest for ketones and aldehydes. This is analogous to the trend found for enamines (Section 6.3.2.1) and can be interpreted in the same way. The α angle does not seem to vary much with X and remains nearly constant for a given substituent R.

6.3.4 Reaction Path for Elimination

The data described above corroborate the qualitative picture of the incipient stages of elimination from a trigonal center resulting in the formation of a digonal center. Further details of the reaction path for RC(=O)OR′ can be derived by analyzing the distance and angle changes found in crystal structures. If α is assumed to be nearly constant along the reaction path, four parameters are relevant: the two bond distances d_2(C=O) and d_1(C−O), and the two bond angles β_1 and β_2. While these two bond angles are linearly related in the RC(=O)X molecules discussed in Section 6.3.3.2,

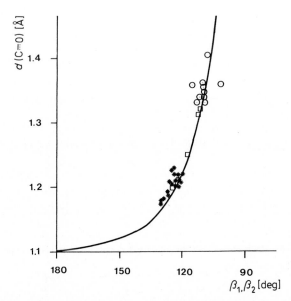

Fig. 6.18. Scatterplot of R−C−O angles vs. adjacent C−O distances. *: β_1 (angle R−C=O) vs. d(C=O); ○: β_2 (angle R−C−OR′) vs. d(C−O); □: average β_1, β_2, and d values for RCOOCH$_3$, RCOOH, and RCOO⁻ from [110]

extrapolation to the limit of complete dissociation is, of course, impermissible. One proposal is that the two angles are related by a hyperbolic function $\beta_1 \cdot \beta_2 = [(360° - \alpha)/2]^2$, which also accounts for the observed bond angles at the digonal centers of the compounds **1** and **2** in Scheme 6.9. The relationships between distances and angles can be obtained from a scatterplot (Figure 6.18), which includes gas-phase data for $CH_3C(=O)X$ and $HC(=O)X$ and average values from crystal structure determinations of $RCOO^-$, $RCOOH$ and $RCOOCH_3$ compounds [110]. The data follow an empirical relationship of the form $d(C-O) = 1.10 + 0.0031 \exp [0.0625(180 - \beta_1, \beta_2)]$ (Å). The constant 1.10 Å corresponds to the distance in $RC\equiv O^+$. The equation yields $O \ldots C$ and $O \ldots N$ distances in the structures **1** and **2** (Scheme 6.9) that are close to observed values, so the extrapolation by ca. 1 Å is still successful. The two $C-O$ distances are related, at least approximately, by a Pauling type relationship $(d_i = 1.334 - 0.2164 \ln n_i)$ and the rule of constant bond order $(n_1 + n_2 = 3)$.

6.4 Computational Investigations of Reaction Pathways for Carbonyl Additions and Eliminations and Related Reactions

The following five sections review two decades of development in the MO theory of potential hypersurfaces for nucleophilic addition to carbonyls. The early results of such studies, which shaped the concepts and imagery of present-day organic chemistry, suffer from limitations that become evident as the methodological evolution and accumulation of higher level *ab initio* calculations continue. Our review aims to enable the reader to judge how the selection of the model system, i.e. the nature of the nucleophile, carbonyl substrate, and the medium, as well as the inherent properties of different basis sets [111] affect the outcome of the calculations. The information given about the technical aspects of the calculations is fairly detailed, for otherwise the account would fail to convey the vivid sense of the depth and scope of the transformation of this field that has occurred over the last ten or so years.

The survey is limited to the studies of carbonyl additions. A number of studies of nucleophilic additions to digonal centers, pertinent to this chapter's topic, have also been reported in the literature. The examples include additions of hydride ion, lithium hydride, and methyllithium to acetylene [112], of water and lithium enolate to ketene [113], of water, lithium hydride, and methyllithium to carbon dioxide [114], and of hydride ion to hydrogen isocyanide [115]. The general conclusions of our review will apply to this work as well.

6.4.1 Formaldehyde

6.4.1.1 Anionic Nucleophiles

The first *ab initio* study of an addition to formaldehyde was carried out by Bürgi et al. using an SCF-LCGO-MO sp basis set [42a], comparable to an STO-3G minimal basis set, with hydride ion as a model nucleophile. The calculated minimum energy path is shown in Figure 6.19. There is no barrier along this path, the binding energy increasing continuously from 19.4 kcal mol^{-1} at 3 Å to 48.4 kcal mol^{-1} at the bonding distance of 1.12 Å in methanolate ion. The hydride ion initially approaches formaldehyde along the C_2 axis of the molecule, and the interaction is believed to be essentially coulombic. It is then forced out of the molecular plane by the repulsive interaction with the two H atoms of the formaldehyde molecule. At the distance of 2.5 Å the H...C=O angle reaches ca. 127°, at which point the calculated pyramidalization at the carbonyl C is still zero; consequently, H...C and the H–C–H plane form an angle of ca. 53°.

At around 2.0 Å, the calculated approach angle H...C=O is ca. 120° and the pyramidalization becomes appreciable, although still rather small; the angle between H...C and the H–C–H plane is ca. 65°. An angular displacement of 10° from the minimum energy path either in the H...C = O plane or in the lateral sense seems to require 1 to 1.5 kcal mol^{-1}. Below 2.0 Å this "funnel" narrows considerably and

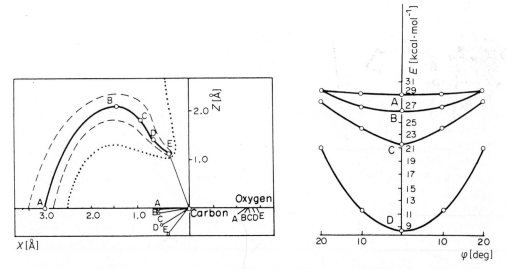

Fig. 6.19. Left: Minimum energy path for addition of hydride to formaldehyde. The points A, B, C, D, E correspond to H$^-$....C distances of 3.0, 2.5, 2.0, 1.5 and 1.12 Å for which the calculated binding energies (relative to infinitely separated H$^-$+CH$_2$O) are 19.4, 21.2, 26.7, 39.9 and 48.4 kcal mol^{-1}. The dashed and dotted curves show paths that are 0.6 and 6.0 kcal mol^{-1} higher than the minimum energy path. Right: Energy profiles for lateral angular displacements out of the XZ plane

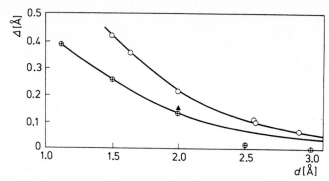

Fig. 6.20. Comparison of amine and hydride addition. ○: experimental points for amine addition; ⊕: theoretical points for hydride addition; ▲: theoretical point for amine addition. The smooth curves are those obtained from Equations (6.1) and (6.4)

the out-of-plane displacement increases faster according to a logarithmic relationship:

$$d = -1.805 \log \Delta + 0.415 \; (\text{Å}) \; . \tag{6.4}$$

The corresponding curve is shown along with the experimental one in Figure 6.20.

A reexamination of this reaction at a higher computational level was undertaken recently by Bayly and Grein [116]. The authors tested several basis sets: the minimal basis set STO-3G; the medium-sized split-valence basis sets 4-31G and 4-31G* (also used to test the effect of adding polarization functions on hydride and diffuse functions on oxygen and hydride); and finally the well extended 6-31G** and 6-31++G** sets. Reaction profiles obtained at the restricted Hartree-Fock level with the STO-3G and 4-31G basis sets are similar to the one described by Bürgi et al., i.e. no energy barrier is encountered along the addition path. However, introduction of the diffuse functions on oxygen and hydride changes this picture dramatically, see Figure 6.21. The profiles now display a maximum at about 2.0 Å and a secondary minimum at about 3.2 Å. The maximum was shown in every case to be a true saddle point on the potential energy hypersurface, since the corresponding structures had a single negative eigenvalue in the Hessian matrix. The geometry changes along the addition path are shown in Figure 6.22, using the structures optimized with the 6-31++G** basis set. Initially, a planar charge-dipole complex is formed at a distance of 3.23 Å. This complex is separated from the product, methanolate ion, by a small energy barrier whose maximum lies below the energy of the separated reactants. In the corresponding transition-state structure, the H...C distance is 2.007 Å, the H...C=O angle is 118°, the carbonyl group is slightly pyramidalized, and the angle between C...H and the HCH plane is 77°.

The authors also checked the effect of electron correlation on the SCF results. The energy profiles of the structures optimized with second order Möller-Plesset calculations using a 6-31G* basis set (MP2/6-31G*) augmented by the diffuse functions on oxygen and hydride, are clearly changed. Inclusion of electron correlation produces a flattening of the secondary minimum, the barrier being reduced to less than

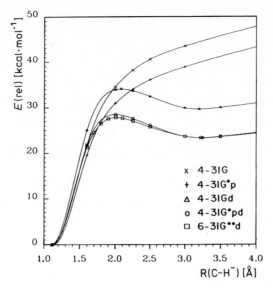

Fig. 6.21. Energy profiles of the $H^- + H_2CO$ addition calculated using various basis sets

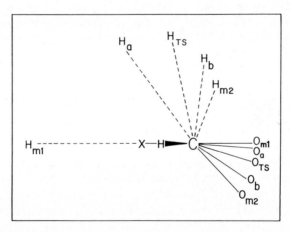

Fig. 6.22. Several structures of the $H^- + H_2CO$ addition, superimposed to show geometry changes over the course of the reaction. Geometries ($6\text{-}31 + + G^{**}//6\text{-}31 + + G^{**}$ calculations) are labeled as follows (with the $C-H^-$ distance given). **ml:** charge dipole complex (C_{2v}), 3.2268 Å; **a:** 2.5 Å; **TS:** transition state, 2.0071 Å; **b:** 1.6 Å; **m2:** global minimum, methanolate (C_{3v}), 1.1226 Å

1 kcal mol^{-1}, and the local maximum is shifted from 2.0 Å (RHF) to 2.7 Å at the MP2 level, and to 2.4 Å at the MP3 and MP4 levels. The authors also confirm earlier reports, *vide infra*, that zero-point energy corrections tend to offset correlation energy corrections.

A similar picture emerges from the studies of the addition of hydroxide ion to formaldehyde. This reaction was first examined by Stone and Erskine [117], using intermolecular SCF perturbation theory as an alternative to the SCF supermolecule approach, and by Williams, Maggiora and Schowen [118] using the approach of Bürgi et al. [42]. In the first study, the attack angle of the hydroxide ion, held at the constant distance of 2.0 Å, was varied (in the symmetry plane) without optimization of the substrate geometry. The energy minimum was found in the range 100 to 110°, the energy of the system being raised by 5 to 7 kcal mol^{-1} on reduction of this angle to 90°. Since the intermolecular interaction energy is computed in this scheme as a sum of electrostatic, exchange, exchange-repulsion, charge transfer, and polarization energies, the sources of the observed directionality could be identified. The authors conclude that the effect is due to the closed-shell repulsion, the electrostatic interaction, and the charge transfer; a frontier-orbital description of the charge-transfer term is found inadequate in general.

Schowen et al. [118] calculated interaction energies as differences between energies of the supermolecular complexes and the combined energies of isolated molecules after geometry optimization. The dependence of interaction energy and the O...C–O angle on the intermolecular distance closely resembles that found for the hydride addition. At an O...C distance of 4.0 Å, there is a slight preference for the anion to approach formaldehyde along the C_2 axis. At 3.0 Å, hydroxide ion is already forced out of the plane and the O...C–O angle is 130° if the O–H bond is synplanar with C=O and 126° if it is antiplanar. The energy difference between the two conformers is only 0.2 kcal mol^{-1}, and the potential energy barrier to rotation about the O...C bond is negligibly small. Essentially, the two moieties are still free to adopt any orientation with respect to each other. Further calculations were carried out with the constraint of C_s symmetry. At 2.0 Å the O...C–O angle has decreased to 112° and the O...C–H angles are 94°, so the carbonyl is appreciably pyramidalized. At this stage, the interaction energy has increased from ca. 9 kcal mol^{-1} to ca. 60 kcal mol^{-1}, with no detectable energy barrier along the reaction path.

The addition was later re-investigated with a larger basis set by Madura and Jorgensen (6-31+G*) [119], Bayly and Grein (MP2 (frozen core), MP2 and MP4//4-31G basis set augmented with polarization and diffuse functions) [120], and by Bernardi, Robb et al. (multi-configurational SCF theory, CAS-1 and CAS-2 MC SCF) [121]. As previously, Madura and Jorgensen determined the reaction path in C_s symmetry by carrying out a complete optimization at several values of the reaction coordinate d, the O...C distance. Evaluation of several alternative basis sets revealed that the interaction energy is particularly sensitive to the inclusion of diffuse functions. Since the zero-point and correlation energy corrections were found to nearly cancel, the reaction path was determined using the 6-31+G* basis set. Again, at large separation the approach is collinear (C_2 axis) up to an apparent minimum at $d = 2.74$ Å (structure **E** in Figure 6.23), which turned out to be a saddle point between two equivalent structures of C_1 symmetry ("hydrogen bonded" structure **F**) with, however, negligible energy difference (0.1 kcal mol^{-1}). From this minimum, the O atom rises out of the plane to the transition state at $d = 2.39$ Å (structure **C**) and finally forms the tetrahedral complex at $d = 1.47$ Å (structure **A**). The interaction energy is -19.14 kcal mol^{-1} at the 2.74 Å minimum and -35.15 kcal mol^{-1} in the tetrahedral complex. While the reaction profile is similar to the ones described previously, the higher-level calculations do reveal an intermediate and a saddle point

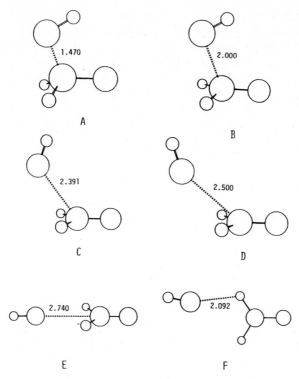

Fig. 6.23. Selected optimized 6-31+G* structures along the reaction path of hydroxide addition to formaldehyde

along the reaction path. The stationary points were characterized by frequency calculations at the 3-21+G level (the 3-21+G and 6-31+G* reaction profiles and geometries are very similar). The authors note that their findings are consistent with results of calculations by Wu and Houk for addition of methoxide to formaldehyde, which yield a barrier of 4 kcal mol^{-1} and a reactant-product energy difference of ca. 46 kcal mol^{-1} at the 3-21G level [122]. As for the approach geometry, the O...C=O angle is about 140° at $d = 2.5$ Å and decreases to 127° in the transition state. Pyramidalization at the carbonyl carbon is indirectly indicated but not analyzed.

The effect of including a correlation-energy correction on the reaction profile for this addition was studied by Bayly and Grein using the 4-31G basis set augmented with polarization d functions on the O atoms and the carbonyl C, and diffuse s and p functions on O; electron correlation calculations were made at the MP2 (frozen core), MP2 and MP4 levels [120]. For each fixed value of $d(\text{O}...\text{C})$, all other geometry parameters were optimized with enforced C_s symmetry. The structures corresponding to RHF energy maxima and minima included $d(\text{O}...\text{C})$ in the geometry optimization. The transition-state structure had a single negative eigenvalue in the Hessian matrix, showing it to be a true saddle point on the potential energy hypersurface. The minimum-energy reaction pathway found is very close to the one described previously. A small local energy maximum is found at $d(O...C) = 2.35$ Å, with the $O...C=O$ angle 124.4°. Introduction of electron correlation into the calculations changes $d(O...C)$ in the transition-state structure to 2.50 Å, while the position of the secondary minimum corresponding to the ion-dipole complex does not move appreciably. However, upon addition of electron correlation the barrier nearly disappears.

In another more recent study of this reaction, by Bernardi et al. [121], the OH lone pair was represented by two different spatial orbitals so that the two lone-pair electrons can either be spin-coupled together as a conventional lone pair, or one of the lone-pair electrons can be spin-coupled to one of the π electrons of the double bond. This removes the constraint implicit in an SCF computation that the reaction must involve a heterolytic charge transfer. The multi-configuration results, at the $3-21+G$ level and in C_s symmetry, are in qualitative agreement with those based on single configuration but indicate that the exothermicity of the reactions is overestimated by ca. 5 kcal mol^{-1} and that the minimum associated with the ion-dipole complex appears to be underestimated. The $O...C=O$ angle in the transition-state structure is smaller, 119.5 or 122.0° vs. 130.0°, and there is a slight pyramidalization at the carbonyl carbon, in contrast to virtual planarity at the $3-21+G$ level. Both changes might be related primarily to the fact that the transition state is more advanced, $d(O...C)$ being 2.124 and 2.200 Å (MC SCF) vs. 2.331 Å (SCF), and $d(C=O)$ being 1.263 and 1.261 Å vs. 1.241 Å, respectively.

A close second-row analog of the hydroxide ion is the hydrosulfide ion. A comparison of the energy profiles for the reactions of these two ions with formaldehyde was undertaken by Howard and Kollman [123]. The result is shown in Figure 6.24.

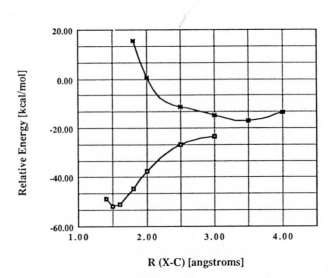

Fig. 6.24. Gas-phase reaction coordinates for addition of hydroxide ion to formaldehyde (lower curve) and of hydrosulfide ion to formaldehyde (upper curve). The curves were constructed by using a cubic spline interpolation of the calculated *ab initio* points. Both reaction coordinates were calculated at the RHF/4-31G//RHF/4-31G *ab initio* level

The only stationary point at the 4-31G level is the ion-molecule complex, 17.2 kcal mol^{-1} below the reactant level, with an S...C=O distance of 3.38 Å. Subsequently, the energy steadily increases, and no stable tetrahedral intermediate can be located; all attempts to find a corresponding stationary point ended in convergence to the ion-molecule complex minimum.

Since the mid 80's, the growing interest and capability to model more complex reactions has brought theoretical studies of addition of stabilized carbanions to formaldehyde. It turns out that even if the reaction is exothermic, the products and reactants are in this case usually separated by a significant activation-energy barrier. Consequently, relatively small basis sets are sufficient to locate the saddle points along the reaction path, or, in other words, it is not necessary to use basis sets augmented with diffuse functions. The first such study was reported by Volatron and Eisenstein, who examined alternative pathways (Wittig vs. Corey-Chaykovsky) for addition of phosphonium methylide and sulfonium methylide to formaldehyde [124]. For the latter pathway, acyclic transition-state structures were located and optimized with the 4-31G* basis set, 24.0 and 10.6 kcal mol^{-1} above the level of the reactants, respectively. In the phosphonium methylide addition, the transition-state structure corresponds to departure of phosphine in the intramolecular S_N2 reaction of the adduct. In the sulfonium methylide addition, the transition state is typical of a nucleophile adding to a carbonyl group, with a C...C=O angle of 106.6°, d(C...C) = 1.874 Å, and d(C=O) = 1.232 Å. The degree of pyramidalization at the carbonyl carbon was not reported.

An even smaller basis set is sufficient to locate the transition state for an aldol condensation. Houk et al. describe energetics and geometry of transition-state structures for the reaction of acetaldehyde enolate with formaldehyde at the 3-21G level, using the forming CC bond as reaction coordinate [125]. Optimizations were done with several constrained CC bond lengths, and the geometry corresponding to the maximum energy was selected as a trial geometry for transition-state optimization. Single-point energy evaluations were carried out with the 6-31G* or 6-31+G basis sets. The reaction is predicted to be exothermic by $10-12$ kcal mol^{-1}, depending on the basis set, and to have an activation energy of $12-15$ kcal mol^{-1} with respect to the hydrogen-bonded enolate-formaldehyde complex. Three distinct transition states (*anti* and *synclinal* rotamers) were located (see Figure 6.25, p. 246) with d(C...C) around 2.02 to 2.06 Å. The barriers to rotation about the incipient bond seem to be significant, but the energies of the three rotamers are, surprisingly, very similar. The C...C=O angles are $113-118°$, and both carbon atoms are pyramidalized. The values of the C...C=O angles in the transition states are consistent, it is emphasized, with the authors' earlier generalization that nucleophiles attack unsaturated bonds with an angle which is larger than tetrahedral [33].

6.4.1.2 Neutral Nucleophiles

Two neutral nucleophiles are of great importance in chemistry and biology: water and ammonia. Addition of the latter to formaldehyde is also a model for the reaction of tertiary amines with ketones. In an early study of this addition reaction, Yamabe

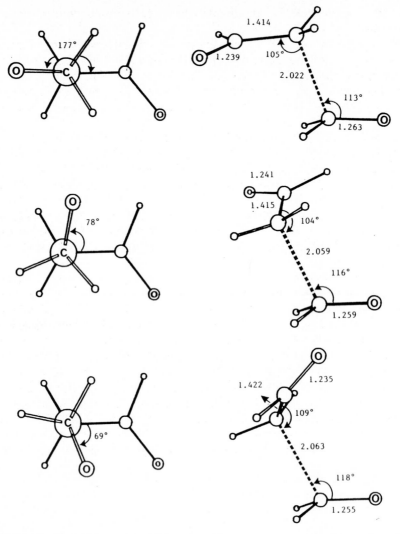

Fig. 6.25. 3-21G optimized geometries of three transition-state structures for the reaction of acetaldehyde enolate with formaldehyde

et al. employed extended Hückel and semiempirical ASMO SCF calculations [126]. The interaction energy for planar formaldehyde and ammonia approaching along the carbonyl C p_z-axis was calculated as a function of the N...C distance and the angle between this axis and the direction of the sp^2 lone pair of the N atom. The minimum corresponding to the tetrahedral addition product cannot be obtained by either method; in other words, the addition is strongly endothermic, and the energy monotonically increases as the distance $d(N...C)$ is decreased. However, at a given N...C distance, the interaction between the two molecules, in terms of the total energy, atomic populations, and bond orders, is largest when the lone pair is directed to the carbonyl C atom.

Fukui et al. studied the bond interchange in the course of ammonia addition to formaldehyde by means of a localized molecular orbital method, based on the analysis of the semiempirical INDO molecular orbitals of the isolated molecules and of the reacting system [127]. Here, the N...C=O angle was fixed at 107°, since preliminary calculations gave results in fairly good agreement with the experimental reaction path, i.e. the Bürgi-Dunitz trajectory.

Lipscomb et al. carried out MO calculations with the minimum basis set using the PRDDO procedure [128]. The N...C=O angle and the C=O bond length were optimized at several values of the reaction coordinate, $d(N...C)$. A rigid ammonia molecule was placed on the axis normal to the trigonal plane in a staggered conformation relative to formaldehyde. Only a shallow minimum at $d = 2.62$ Å was found. This result was reproduced in higher level calculations, and it was concluded that the bond formed between ammonia and formaldehyde is very weak. In contrast, MINDO/3 calculations predict the zwitterionic addition product to be a minimum with an activation barrier of 9.6 kcal mol^{-1} to its formation [129].

A measure of clarity has been achieved in this area, thanks to more extensive *ab initio* studies of the reactions of water, methanol, and ammonia with formaldehyde, carried out by Schowen et al. [130] and by Williams [131]. The STO-3G minimal basis was used for geometry optimizations, and the 4-31G split-valence basis was used for energy calculations for the optimized structures (4-31G//STO-3G) [130]. If a C_s symmetry constraint is imposed on the approach of water to formaldehyde, the interaction as a function of $d(O...C)$ is repulsive all along the formation path of the zwitterionic product. Thus, C–O bond formation and proton transfer must occur in a concerted manner, i.e. in a single transition state. The corresponding activated complex has a four-membered ring structure, as shown in Figure 6.26; the same is true for the addition of ammonia and amines. Geometries of all activated complexes are similar and suggest about 40–50% reaction progress, measured in terms of bond

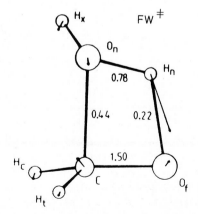

Fig. 6.26. Transition state for concerted addition of a water molecule to formaldehyde: numbers indicate bond orders

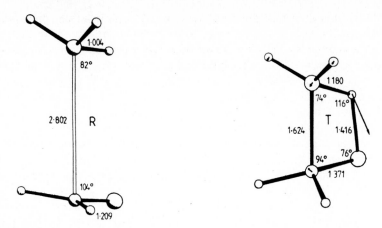

Fig. 6.27. Molecular complex of ammonia and formaldehyde (left) and transition state for concerted addition (right)

orders and of pyramidalization at the carbonyl C. The proton transfer is relatively less advanced. The transition-state geometry, in contrast to the energy, is not significantly dependent on the basis set. At the 4-31G level, the activation energy for water-formaldehyde addition is 44 kcal mol^{-1} and the reaction is exothermic by 16.8 kcal mol^{-1}.

Williams has subsequently reported 3-21G optimized geometries for reactant, activated and product complexes for ammonia addition to formaldehyde [131]. According to the *ab initio* results, the two molecules form a complex in which ammonia approaches the carbonyl C above the plane of its ligands at a distance of about 2.8 Å, Figure 6.27. The interaction of ammonia with formaldehyde in this complex was recently investigated by Kozmutza, Evleth and Kapuy [132]. The 6-31G* SCF-optimized structure is similar to the one reported by Williams. At the SCF level, the interaction energy is smaller than that of a hydrogen bond. The authors used the Boys-Bernardi counterpoise correction at both the SCF and various MP levels to eliminate the basis-set-superposition error; the dispersion energies with counterpoise correction were computed using 6-31G* and 6-311G** basis sets. The order of magnitude of the basis-set-superposition error is roughly equal to the interaction energy at the self-consistent-field level. Independently, Yamataka, Nagase et al. found addition of ammonia and amines to formaldehyde to occur in a single step through a four-centered transition state (6-31G*//3-21G) [133].

Vinylamine addition to formaldehyde was studied by Sevin, Maddaluno and Agami, using the minimal STO-3G basis, improved by a configuration interaction step involving all single and double excitations from the π and n occupied orbitals of the reactants to their two lowest vacant orbitals [134]. The authors conclude that formation of a zwitterionic intermediate is not likely in the absence of protic solvation. In analogy to the reactions of water and ammonia with formaldehyde, such a path is very endothermic, and no activation energy barrier is encountered except when the conformation about the incipient bond is eclipsed. For a concerted path, a region close to the actual saddle point was found by a grid search, involving variation of the C...C distance, C−C torsion and/or N−H and O...H distances. This region lies at an energy which is below that of the corresponding zwitterion; the transition-state structure was not fully optimized.

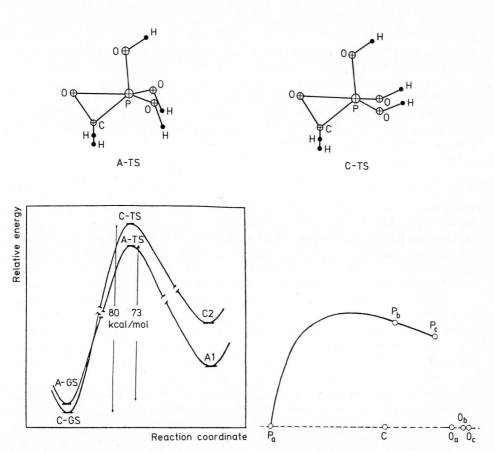

Fig. 6.28. Transition states (top), relative energy diagram (bottom left) and trajectory plot (bottom right) for the addition reaction of two different conformers of phosphite to formaldehyde

In contrast, Gorenstein et al. found that addition of phosphite to formaldehyde proceeds through zwitterionic transition states shown in Figure 6.28 [135]. The saddle points were located by studying the breakdown of the metastable pentacovalent phosphoranes, using a STO-3G basis set. The difference in activation energy between the **A** and **C** paths apparently reflects the difference in conformation about the P−O bonds [136]. The diagram and the trajectory plot show that the P...C=O angle is smaller than 90°.

6.4.1.3 Metal Hydrides and Organometallics

The first computational study of addition of a complex ion to formaldehyde was carried out by Dewar et al. who used semiempirical MNDO calculations to investigate the reaction of tetrahydroborate ion with formaldehyde [137]. Transition states were located approximately from plots of energy vs. a suitable reaction coordinate and refined by minimizing the scalar gradient. All stationary points were characterized by diagonalization of the Hessian matrix. According to this study, there is no concerted path for the reaction; the addition involves an endothermic transfer of hydride ion to give borane and methoxide ion. The geometry of the transition state was not reported.

Another study of the energy surface for the reaction in the absence of Li^+ was carried out by Eisenstein et al. with a 3-21G basis set [138]. It was found here that the borohydride line of approach is initially collinear with the C_2 axis, i.e. the direction of maximum electrostatic attraction. An energy minimum occurs at $d(BH...C) = 2.785\ \text{Å}$, where the two moieties display very little distortion from their monomer geometries. Three possible mechanisms of addition were evaluated; the energy diagram shown in Figure 6.29 suggests a single-step mechanism, the transition state for which is shown in Figure 6.30.

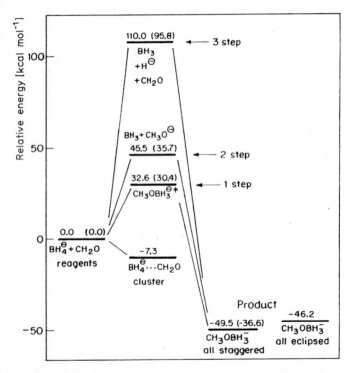

Fig. 6.29. Energy diagram for the possible reaction pathways of borohydride addition to formaldehyde. Activation energies for one-step, two-step, and three-step mechanisms are indicated for the 3-21G basis set. Values for the 6-31G* basis set are given in parentheses

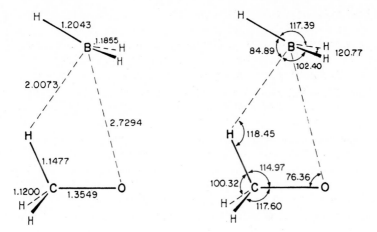

Fig. 6.30. Transition-state geometry for the one-step mechanism of borohydride addition to formaldehyde

A systematic *ab initio* investigation of the reaction of lithium borohydride with formaldehyde was carried out by Tomasi et al. [139], using primarily the minimal basis set STO-3G. In the first phase, the interaction between O and Li is dominant, resulting in formation of a linear, loose complex, $C=O \ldots LiBH_4$. The system was assumed to retain a symmetry plane bisecting the $H-C-H$ angle and preserving linearity of the $C-O-Li$ fragment during the rearrangement leading to the transition state. The saddle point corresponding to the transition state was located by two grid-search methods, both leading to the structure shown in Figure 6.31 which is consistent with the earlier conclusions of Dewar et al. [137]. The estimated activation barrier is 17.4 kcal mol^{-1}.

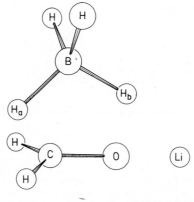

Fig. 6.31. Calculated transition-state geometry for the $H_2CO+LiBH_4$ reaction

The reaction of MgH_2 with formaldehyde as a model for the Grignard reaction was examined by Nagase et al., using the split-valence 3-21G basis set [140]. The stationary points on the potential surface were determined with the analytic gradient technique and identified as the equilibrium or the transition state (saddle point) by calculating normal vibrational frequencies. The reaction is calculated to be 63.4 kcal mol^{-1} exothermic. The reagents initially form a σ complex, which has to pass over an energy barrier of 12.7 kcal mol^{-1} to complete the addition.

The mechanism of organolithium addition to formaldehyde was studied by examination of model reactions with the monomers and dimers of lithium hydride and methyllithium [141]. Schleyer, Houk et al. used the 3-21G basis set and evaluated its performance for the monomeric LiH addition by comparing the results with the geometries and energies obtained at the (6-31G*//6-31G*) level and with the single-point (6-31+G*//6-31G) and (MP2/6-31+G*//6-31G*) values. All four reactions proceed in three stages: formation of the σ complexes, conversion to the transition-state structures for the addition (each characterized by a single imaginary frequency) with low activation energies, and the subsequent exothermic conversion to alkoxide complexes. The geometries of the transition-state structures are shown in Figure 6.32.

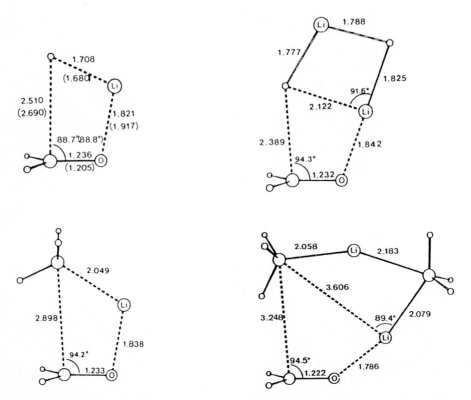

Fig. 6.32. Calculated transition-state geometries for addition of HLi, (HLi)$_2$, CH$_3$Li and (CH$_3$Li)$_2$ to H$_2$C=O

Similar results were reported by Streitwieser et al. [142] for the addition of LiH to formaldehyde, investigated at the 3-21G* level. The transition state for the monomer addition is very early, with about 10% reaction progress as measured by the change in the C=O bond length or by pyramidalization at the carbonyl carbon.

Houk et al. have made calculations for the reactions of lithium enolate and enol borinate of acetaldehyde with formaldehyde at the 3-21G level [125]. In the first case,

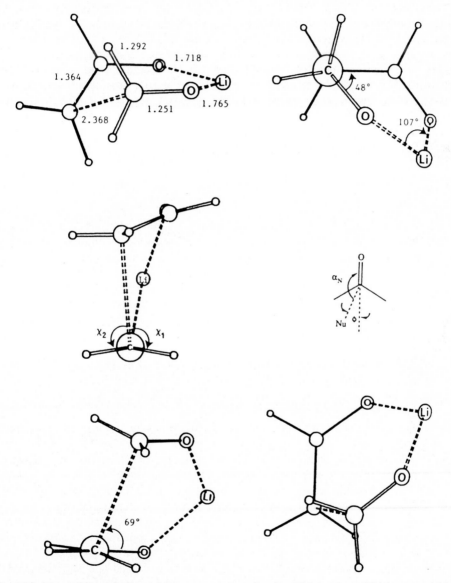

Fig. 6.33. Views of the 3-21G transition-state structure for the aldol reaction of lithium acetaldehyde enolate with formaldehyde

the activated complex is formed with a very low energy barrier (2 kcal mol^{-1}) from a coordinated complex of the reagents, which is stabilized by 28 kcal mol^{-1}. From the complex to the aldolate product, the reaction is exothermic by 12 kcal mol^{-1}. In spite of the low activation energy, there is an appreciable (20–30%) reaction progress in the transition state, which is shown in Figure 6.33. For the addition of enol borinate, two transition states were located, while the energetic profile of the reaction remains very similar to the one for the lithium enolate addition. The study of the enol borinate addition included calculations for acetone enol and for *E*- and *Z*-enols of propionaldehyde. The transition-state structures all have C...C=O angles about 101–108°, and the trigonal carbon atoms are appreciably pyramidalized. For these reactions, the asymmetry of the lithium and boron coordination might result in deviations of the approach trajectory from the symmetry plane normal to the plane of the formaldehyde atoms. An independent study of the reaction of formaldehyde with the lithium enolate of acetaldehyde by Tidwell et al. (at the 3-21G level) has corroborated these results [113b].

6.4.2 Acetaldehyde and Homologs

A major impetus behind the early studies of nucleophilic addition to homologs of formaldehyde was the interest in the mechanism of 1,2-asymmetric induction. Nguyen Trong Anh and Eisenstein modeled transition-state structures for hydride addition to 2-chloropropanal and 2-methylbutanal by "supermolecule" complexes in which hydride ion was placed at a fixed distance of 1.5 Å from the carbonyl C and the H...C=O angle was fixed at 90° [32]. Rotation about the C(1)–C(2) bond by 30° increments generates 12 conformers for each of the two alternative approaches of the hydride. Energy optimization was then carried out for the resulting 24 diastereomeric "transition states" using the minimal basis set STO-3G and standard geometries. For both aldehydes, the two most stable conformers of the "supermolecule" had the chloride or ethyl group *trans* to the hydride ion; in both, the methyl group slightly prefers the position *gauche* with respect to the nucleophile and carbonyl oxygen (Figure 6.34).

The calculations were then repeated for the two aldehydes complexed by a proton or by a lithium cation, using structure parameters obtained by partial geometry optimization of the corresponding complexes of formaldehyde. No significant change in the relative stabilities of the diastereomeric "transition states" was found.

Since the approach angle was not optimized in these calculations, a separate assessment of the factors governing deviation from the direction normal to the carbonyl plane was made. Energy optimization for the "model supermolecule" HCHO+H$^-$ (fixed distance 1.5 Å) and overlap calculations for H$^-$ and π^*(C=O) were carried out for approach angles from 90° to 115° by increments of 5°. The optimal angle was found to be 110°. The authors concluded that the deviation is due almost entirely to the interaction with the π^*(C=O) orbital, since maximal overlap is obtained for 105°.

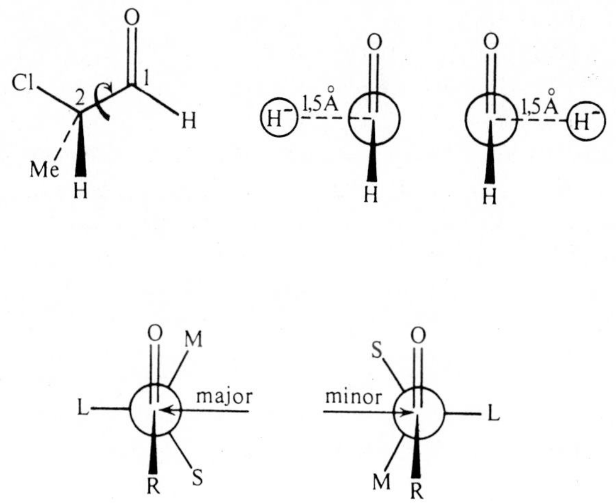

Fig. 6.34. Schematic view of the transition states for hydride addition to 2-chloro-propanal and 2-methyl-butanal (see text)

Subsequently, the optimal approach angles were also deduced for the diastereomeric hydride/2-chloropropanal "transition states", assuming quadratic interpolation of the energy on the approach angle (taken as 90°, 100° and 110°), and assuming a planar carbonyl group. The estimated angles range from 98 to 109°. For the complexes incorporating H^+ or Li^+, they lie between 95.5 and 103.5°. The results for 2-methylbutanal were similar.

The magnitude of the supermolecule stabilization relative to the isolated reactants was calculated for hydride addition to acetaldehyde, propanal, and chloroacetaldehyde, maintaining the H, CH_3 and Cl substituents, respectively, in the conformation *trans* to the hydride ion (Scheme 6.15). The increase in the stabilization energy, from 35.5 kcal mol^{-1} to 37.4 kcal mol^{-1} to 59 kcal mol^{-1}, was interpreted as a result of the increase in the interactions of the hydride, the $\pi^*(C=O)$ orbital, and the antiperiplanar σ^* orbital of the substituent.

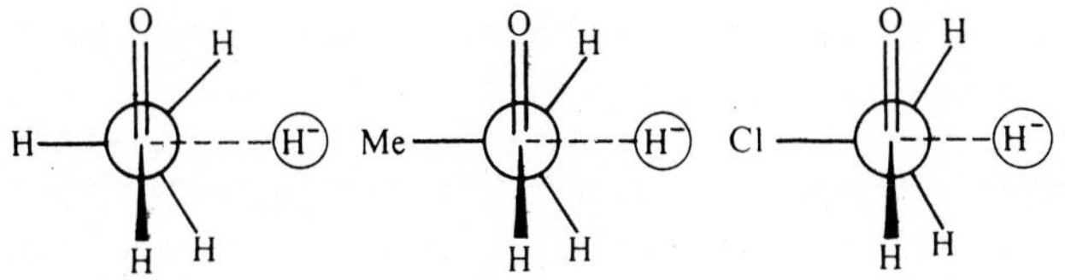

Scheme 6.15

The first systematic investigation of the reaction of an anionic nucleophile with substituted acetaldehydes, using the analytical gradient techniques, was reported recently by Wong and Paddon-Row [143]. The study focused on the addition of cyanide ion to aldehydes. This ion displays characteristics of a stabilized carbanion,

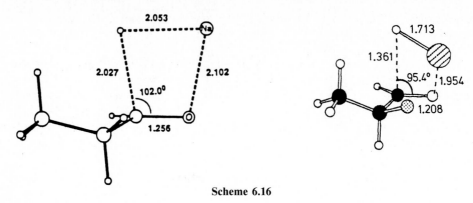

Scheme 6.16

insofar as its reactions with aldehydes appear to have intrinsic activation barriers; in any event, only a small split-valence basis set (instead of one augmented with the diffuse basis functions) was enough to obtain an activation energy barrier along the reaction path. Final geometry optimizations were carried out with the 6-31G* basis set, and single-point energies were obtained from MP2/6-31+G* (frozen core) calculations. The transition-state structures are characterized by large C...C=O angles and significant pyramidalization at the carbonyl C. Deviation from the local C_{3v} symmetry is gauged by the differences between the C...C=O ($\alpha = 112-114°$), C...C−C ($\beta_1 = 95-100°$) and C...C−H ($\beta_1' = 90-93°$) angles. In the series of the four substituted acetaldehydes X−CH$_2$−C(= O)H, these differences are slightly smaller for the more advanced transition states (X = CH$_3$, SiH$_3$) than for the earlier ones (X = F, CN). The high degree of rehybridization of the carbonyl C is reflected in the θ_c values, ranging from 37° to 43°. The order of relative stabilities of the +*sc*, −*sc* and *app* conformers found for this series is consistent with the Anh-Eisenstein model of 1,2-asymmetric induction.

Reactions of neutral nucleophiles such as water, ammonia and amines with acetaldehyde were examined in some of the previously reviewed studies of their reactions with formaldehyde [125, 130, 133]. In addition, Bestmann reported MNDO calcu-

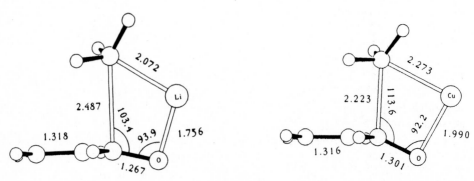

Scheme 6.17

lations of the reaction profiles for the addition of the ylide $H_3P^+-CH^--CH_3$ to acetaldehyde, which appears to proceed via a quasi-betain transition state [144].

Examples of metal hydride and organometallic additions include reactions of lithium hydride and sodium hydride with acetaldehyde and its homologs [145–147], reactions of methyllithium and methylcopper with acrolein [148], and aldol additions of lithium enolates [125]. The geometry of the four-center, cyclic transition states for such additions seems to depend on the size of the metal atom (see Schemes 6.16 and 6.17).

6.4.3 Acetone and Homologs

Sheldon reported that attempts to trace a trajectory similar to the one reported by Bürgi et al. [42] for the reaction of methoxide ion with acetone failed at the STO-3G level [149]. The investigation by Bayly and Grein [120] had included a study of the reaction profile of hydroxide ion addition to acetone with the explicit purpose of assessing the effect of alkyl substitution on the energy and geometry parameters along the reaction coordinate. Calculations were carried out using the 4-31G basis set augmented with polarization and diffuse functions on some atoms (see Section 6.4.1.1); electron-correlation calculations were made at the MP2 level. The secondary minimum was found to be significantly shifted, to 3.489 Å, and the difference between the barrier maximum and the secondary minimum was much higher for acetone than for formaldehyde. There are also significant changes in the transition-state structures, although the $O...C=O$ distances are very similar. Pyramidalization of the carbonyl group is twice as large in the acetone transition-state structure as in the formaldehyde one. The $O...C=O$ angles are 124.4 and 109.6° in the transition-state structures, and 180.0 and 145.7° in the reactant complexes, for formaldehyde and acetone, respectively.

A similar shift is noticeable if the transition-state structures for cyanide addition to acetaldehyde (see Section 6.4.2) and to cyclohexanone are compared [150]. The $C...C=O$ and $C...C-C$ angles for the axial and equatorial transition-state structures are $\alpha = 110.2°$, $\beta_1 = \beta_1' = 99.3°$ and $\alpha = 111.6$, $\beta_1 = \beta_1' = 98.7°$, respectively (6-31G*).

A few studies of metal hydride additions are also available. Houk et al. briefly reported location of the transition-state structure for LiH addition to acetone (3-21G) [145, 151]. Subsequently, alternative transition-state structures for addition of LiH to cyclohexanone, 3-fluorocyclohexanone, 1,3-dioxan-5-one and 1,3-dithian-5-one were reported [147, 152, 153]. The general features of these cyclic transition states are similar to those of the analogous structures reviewed earlier.

6.4.4 Carboxylic Acid Derivatives

The approach of OH$^-$ to formamide was studied by Tomasi et al. by SCF-LCAO-MO calculations with the STO-3G basis set [154]. For large separation distances, the approximation of using rigid charge distributions for both molecules was employed, so the interaction energy is only electrostatic. The results were in substantial agreement with SCF calculations with fixed geometries of the reactants, but allowing for mutual polarization, exchange contributions, and charge transfer. In both approximations, the most favored channels for the approach of OH$^-$ lie in the formamide plane and correspond to the directions of the three X$-$H bonds. The study did not find any path starting from the formamide plane and leading in a continuous fashion to the reaction intermediate. On the other hand, approach paths starting from positions placed above the formamide plane do not present any barriers. When the O...C distance is large or intermediate the interaction energy is less attractive than if OH$^-$ is in the molecular plane at equivalent distances. At the shorter distances the interaction energy monotonically increases and a smooth valley on the potential hypersurface leads eventually, without any local maximum, to the tetrahedral intermediate. This valley is quite wide for distances larger than 3.0 Å and becomes narrower as OH$^-$ approaches the C atom. At 3.0 Å the O...C=O angle is 94.0°, the O...C$-$N angle is 91.4°, and the O...C$-$H angle is 88.8°. As these values imply, the carbonyl C is distinctly pyramidalized; as for the N atom, it is nearly completely pyramidal. The bond angles then increase smoothly along the reaction coordinate, reaching 110.2, 103.8 and 97.4°, respectively, at the distance of 1.75 Å.

No energy barrier was found along the approach pathway for the addition of methoxide ion to formamide by the PRDDO study of Lipscomb et al. [128]. Partial optimization of the complex geometry at several values of the reaction coordinate d(O...C) shows that the O...C=O angle increases from 99.1° at 2.38 Å to 110.7° in the tetrahedral complex. In an attempt to describe orientational requirements of this reaction, the authors mapped out, for several values of the distance d, the surfaces in the spherical polar coordinate system where the energy relative to the separated molecules is zero. The force constant for misdirecting the methoxide at $d = 2.38$ Å away from the carbonyl oxygen in the C_s plane was estimated to be $k = 0.014$ kcal mol^{-1} deg^{-2}, compared with $k = 0.062$ kcal mol^{-1} deg^{-2} at $d = 1.55$ Å and with the experimental bending force constant for C$-$C$-$OR angles $k = 0.048$ kcal mol^{-1} deg^{-2}. The question of directionality of nucleophile addition to carboxylic acid derivatives was also addressed by Stone and Erskine [117] in the study which was described previously (see Section 6.3.1.1).

A CNDO/2 investigation of nucleophilic addition to *N*-methylacetamide and methyl acetate was reported by Aleksandrov and Antonov [155]. The direction of nucleophile approach was fixed along the normal to the amide or ester plane. No activation barrier obtains along the reaction path for any of the four nculeophiles studied: methoxide ion, hydrosulfide ion, methanol-methoxide ion, and water-formate ion (two models for the general base catalysis). The authors note varying degrees of pyramidalization at the carbonyl C, depending on the nature of the nucleophile (see Figure 6.35).

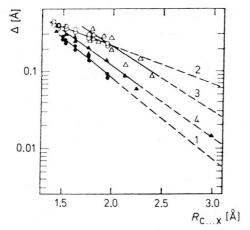

Fig. 6.35. Inductive influence of nucleophiles on the hybridization (pyramidalization Δ) of the carbonyl carbon atom of amide and ester substrates vs. distance $R_{C...X}$ between the carbonyl carbon atom of the substrate and the attacking atom of the nucleophile. Nucleophiles: 1) methanol; 2) methoxy anion; 3) methyl mercaptide anion; 4) system $H_2O-HCOO^-$

The reaction of hydroxide ion with formamide was reinvestigated by Kollman et al. by 4-31G geometry optimizations for both reactants held at distances of 6.0, 3.08, 2.08 and 1.48 Å (the latter corresponds to the tetrahedral complex) [156]. The lowest energy structure at 3.08 Å results from abstraction of one of the amide hydrogens by the hydroxide ion and corresponds to a hydrogen bond between water and formamide ion. The energy of the system monotonically decreases until the formation of the tetrahedral complex, which is 39 kcal mol^{-1} more stable than the reactants. The 6-31G* energies obtained by single-point calculations are very similar. Inclusion of the correlation energy, estimated at the MP2 level, results in several kcal mol^{-1} stabilization at the shorter distances. Subsequently, Kollman et al. compared reaction profiles for hydroxide ion and hydrosulfide ion additions to formamide [123], finding, as previously with formaldehyde (see Section 6.3.1.1), that no stable tetrahedral complex is formed by the latter. Instead, an ion-dipole complex is the energy minimum.

Maraver et al. report a MINDO/3 study of hydroxide ion addition to methyl formate and to methyl oxalate anion [157]. Geometry optimizations at selected values of the distance between the reactants were followed by the direct location of the stationary points. Hydroxide ion and methyl formate initially form a hydrogen-bonded complex, 10 kcal mol^{-1} more stable than the reagents, in which the $HO^-...H-C$ distance is 1.68 Å. This secondary minimum is separated from the global minimum, corresponding to the tetrahedral complex (ca. -50 kcal mol^{-1}), by a transition state at about 2.2 Å. The change in the $O...C=O$ angle along the reaction coordinate is shown in the diagram of the reduced potential surface in Figure 6.36.

Reactions of halide ions with acyl halides were investigated by Yamabe and Minato [158], and by Blake and Jorgensen [159]. The first study used the 4-31G basis set augmented with diffuse p functions on F and Cl. Potential energy profiles for the reactions of F and Cl with acetyl chloride were obtained by optimization of three

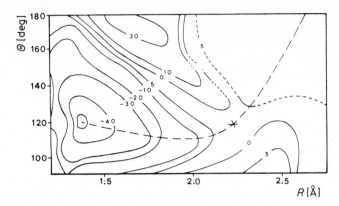

Fig. 6.36. Reduced potential surface of hydroxide ion addition to methyl formate (θ: O...C=O angle, R: O...C distance)

bond distances and the angle of displacement of the leaving group from the heavy-atom plane for different values of the angle between the incipient bond and the heavy-atom plane. No stable tetrahedral complex was found. The corresponding structure is a transition state along the double-well potential, where π-complexes at 2.55 Å (F, $\alpha = 102°$) and 3.53 Å (Cl, $\alpha = 110°$) are the minima. The barrier for addition of Cl$^-$ is 14.2 kcal mol^{-1}, whereas the one for F$^-$ is negligible. These results suggest concerted bond-breaking and bond-forming in the gas-phase reaction.

The second study by Blake and Jorgensen, using analytical energy gradients and the 3-21+G basis set, confirmed the double-well potential energy surface for chloride addition to formyl chloride [159]. In contrast to the earlier results, however, the minimum-energy structures were planar ion-dipole complexes (C...Cl, 3.26 Å along the dipole axis, i.e. approximately the C–H...Cl direction). Furthermore, at the 3-21+G level, the tetrahedral transition state has, surprisingly, slightly higher energy than the alternative completely planar transition-state structure. The relative stabilities are not altered with an extended split-valence basis set (6-31+G*) and are reversed only after inclusion of the correlation energy (MP2/6-31+G*).

As for the reactions of neutral nucleophiles with carboxylic acids and esters, Lipscomb et al. [128] examined the initial stages of esterification of formic acid by methanol, using the previously described approach. Only a very shallow well was found at $d = 2.28$ Å, with no energy barrier separating reactants from the highly endothermic zwitterionic product. The O...C=O angle at $d = 2.51$ Å was optimized to 92.6°; the other incipient bond angles are not reported. These authors assumed a staggered conformation about the incipient bond before proceeding with the variation of the O...C=O distance. In contrast, all degrees of freedom were used for geometry optimization in the study of the reaction of ammonia with formic acid by Loew et al. [160]. Exploration of a two-dimensional MNDO potential energy surface, followed by location of the minima and saddle points by analytical gradient methods at the STO-3G and 3-21G levels, established two pathways, stepwise and concerted, for the formation of the peptide bond. For both paths, transition states

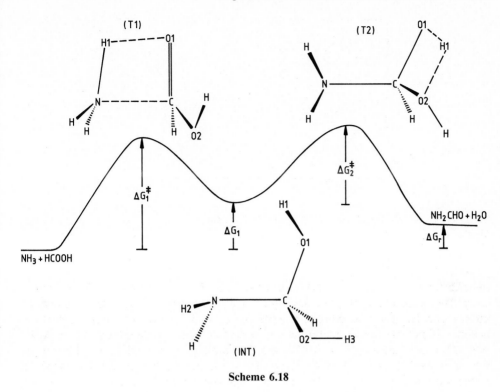

Scheme 6.18

involve simultaneous breaking and formation of several bonds in the planar four-membered cyclic structures, as shown in Schemes 6.18 and 6.19. The authors undertook an extensive examination of the basis set dependence (up to 6-31G**) and the effect of correlation (up to MP4), concluding that correlation energy is a significant factor in the stabilization of transition states.

6.4.5 Solvent Effects

Two phenomena that have received much attention in studies of solvent effects on carbonyl additions are catalysis of formaldehyde hydration and the effect of solvation on the energy profile for the addition of hydroxide ion to formaldehyde in aqueous solution.

   Williams et al. found that unless concerted proton transfer occurs during addition of water to formaldehyde in the gas phase, the interaction is repulsive all along the path to formation of the zwitterion (see Section 6.4.1.2) [130]. Subsequent examination of different modes of catalysis of this addition process by a single additional

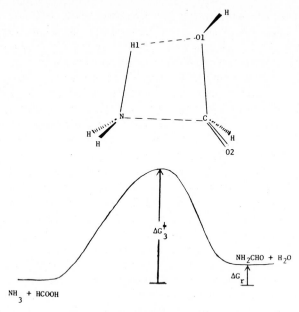

Scheme 6.19

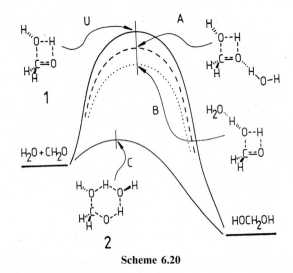

Scheme 6.20

water molecule led to the recognition that the greatest effect obtains upon incorporation of the ancillary molecule into the cyclic transition state (**2** in Scheme 6.20, see also Figure 6.37 b, c) [161]. At the STO-3G level, the potential energy barrier to concerted hydration is reduced from 42.2 kcal mol^{-1} to 0.8 kcal mol^{-1}. Both gas-phase and aqueous hydration are predicted to proceed via concerted addition of the water dimer with Gibbs free energies of activation 27 kcal mol^{-1} and 16 kcal mol^{-1}, respectively, in agreement with experiment in the latter case. It is also estimated that a mechanism involving three water molecules might be consistent with experimental results for dioxane solution. Similar findings were reported by Williams for catalysis of ammonia addition to formaldehyde by one or two molecules of water: the potential energy of the four-membered cyclic transition state is 37 kcal mol^{-1} higher than that of the reactants, while incorporation of a molecule of water lowers the energy barrier to 10 kcal mol^{-1}, and incorporation of two water molecules to less than 2 kcal mol^{-1} (full geometry optimization at the HF/3-21G level, see Figure 6.37e) [131]. In terms of relative Gibbs free energies of activation, the latter, two-water-catalyzed process is estimated to be preferred over the former by about 2.5 kcal mol^{-1} in the gaseous and aqueous phases and by about 6 kcal mol^{-1} in dioxane.

The effect of solvation on carbonyl additions involving hydroxide ion in water has been studied by Weiner, Chandra Singh and Kollman [156], Madura and Jorgenson [119], Howard and Kollman [123], and Hsiang-Ai Yu and Karplus [162]. Interaction energies along the gas-phase reaction path have been evaluated by means of molecular mechanics [123, 156], statistical mechanics (Monte Carlo simulations) [119], and a site interaction model [162]. In each study, a dramatic change in the reaction profile was found to take place in solution, resulting from the difference between solvation of the reactants and solvation of the transition state. The gas-phase profile for such reactions outlines a potential energy well; the additions proceed essentially without activation energy, except for a small barrier separating a hydrogen-bonded complex of the reactants from the tetrahedral complex. The shallow secondary minimum and the saddle point lie below the energy level of the isolated reactants, and formation of the tetrahedral intermediate is highly exothermic. In contrast, calculations of the free-energy profiles for the reaction in aqueous phase reveal a significant activation energy barrier; furthermore, the reaction is endothermic. These two situations are compared in Figure 6.38. Clearly, solvation of the separated reactants by water is more effective than solvation of either the transition state or the tetrahedral complex. Madura and Jorgensen argue that the total number of strong solute-water hydrogen bonds is nearly constant at 6 to 7 throughout the addition process, and that the stabilization effect arises because the hydrogen bonds for the hydroxide ion are stronger than those for all the other species along the reaction path [119]. However, Hsiang-Ai Yu and Karplus conclude that the reactant is solvated by one or two more water molecules than the transition state and the tetrahedral complex [162]. Both Monte Carlo simulation and the Karplus integral-equation study predict the same shift in the transition-state geometry, a decrease in the O...C=O distance from 2.39 Å in gas phase (6-31+G*) to about 2.0 Å in water, that is towards a relatively late transition state.

Besides the specific solvent effects such as catalysis and solvation by a protic solvent, the effects related to the properties of a solvent regarded as a continuous dielectric can be important along the reaction path. Until recently, the problem of quantification of such phenomena within the framework of the quantum-mechanical theory received relatively little attention. An early attempt to simulate solvent as a continuous medium in the *ab initio* geometry optimization was described by Tomasi et al. [139, 163]. They used an electrostatic interaction potential, added as a perturbation to the Hamiltonian of the reactive system, to account for the interaction between

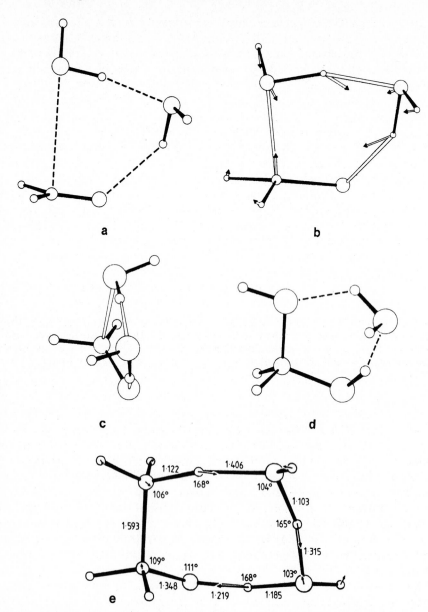

Fig. 6.37. ORTEP drawings for species involved in concerted nucleophilic additions to formaldehyde catalyzed by ancillary water molecules.

a: reactant complex $CH_2O \cdot (H_2O)_2$ with one ancillary water molecule

b: transition state with the reaction coordinate motion indicated by arrows – newly forming bonds are unshaded

c: transition state showing the coplanarity of the hydrogen-bonded fragment of the cyclic structure – newly forming bonds are unshaded

d: product complex $CH_2(OH)_2 \cdot H_2O$

e: transition state complex $CH_2O \cdot (NH_3) \cdot (H_2O)_2$ with two ancillary water molecules

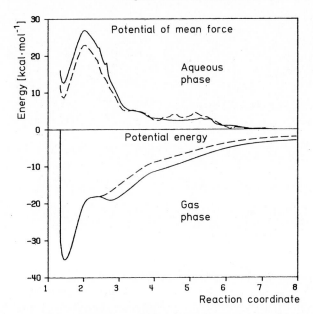

Fig. 6.38. Energy profiles in the gas phase and in aqueous solution for the hydroxide + formaldehyde reaction. Solid lines represent the colinear trajectory, while the dashed lines are for the perpendicular approach. The reaction coordinate is the distance in Å between hydroxy oxygen and carbonyl carbon. The quantity labeled *potential of mean force* is closely related to the free energy of the system

the solute and dielectric . The solvent is here considered as a bath whose internal energy is not affected by the solute. The cavity is described in terms of a suitable number of interlocking spheres of appropriate radii and centered, in general, at the positions of some nuclei. The transition state located with this model for the lithium borohydride reaction with formaldehyde has significantly lower energy relative to the initial coordination complex than the gas-phase transition state (42.8 kcal mol^{-1} vs. 64.3 kcal mol^{-1}, STO-3G minimal basis set). This is attributed mainly to weakening of the interaction between BH_4 and Li by the solvent, which presumably facilitates hydride transfer; the incipient bond distance and pyramidalization at the carbonyl C are not changed.

While this result confirmed the feasibility of the general approach, it did not precipitate wider exploration of dielectric medium effects. Recently, however, Wiberg et al. have incorporated the Onsager self-consistent reaction-field model into *ab initio* MO theory in an implementation which provides analytical gradients and second derivatives. The model considers just the dipole of the solute molecules and a spherical cavity whose radius is chosen for a given solute molecule from the molecular volume estimated at the 0.001 eB^{-3} electron-density contour (B is the Bohr radius), plus an empirical constant 0.5 Å to account for the nearest approach of solvent molecules [164]. Cieplak and Wiberg have used this model to probe solvent effects on the transition states for nucleophilic additions to substituted acetaldehydes [165]. For each

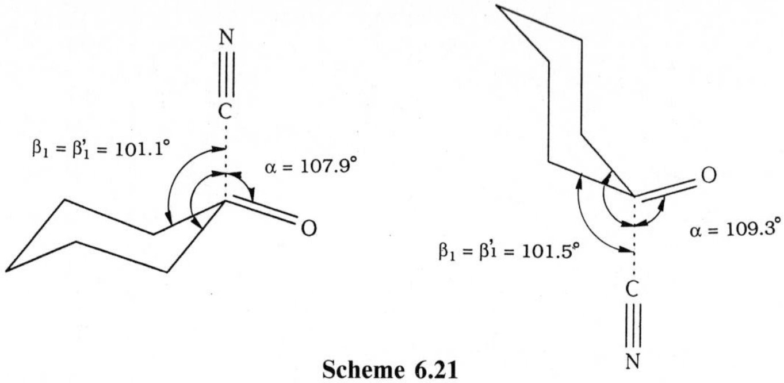

$\beta_1 = \beta_1' = 101.1°$

$\alpha = 107.9°$

$\beta_1 = \beta_1' = 101.5°$

$\alpha = 109.3°$

Scheme 6.21

of the four compounds examined earlier by Wong and Paddon-Row (Section 6.4.2) [143], a different conformer was predicted to be favored in solution. The effect of solvent is to increase the C=O bond length in the transition state, decrease the N≡ C...C=O distance, decrease the angle at which the CN attacks the carbonyl, and increase the dipole moment. Thus, the transition state in solution appears to be later than in the gas phase. The same is found for the transition states of axial and equatorial addition of cyanide ion to cyclohexanone (Scheme 6.21) optimized at the 6-31G* level (dielectric constant $\varepsilon = 78.5$, cavity radius $a_0 = 4.30$ Å). The gas-phase geometries were quoted earlier (see Section 6.4.3).

6.5 Reaction Pathways and Chemistry of Carbonyls

The early papers on structure correlation soon established the notion that nucleophilic attack on the carbonyl group proceeds with stringent stereochemical restraints [22, 23, 42]. This widely adopted hypothesis holds that the preferred approach path of a nucleophile is in the plane bisecting the carbonyl group and orthogonal to the molecular plane, and that the Nu...C=O angle is maintained constant along the later stages of the pathways at a value of $107 \pm 5°$. As the following sections adumbrate, this model, dubbed "rearside" attack model [47], has become very prominent in discussions of the mechanism of asymmetric induction and intramolecular reactivity.

6.5.1 Theoretical Elaborations of the "Rearside" Attack Model: Baldwin's and Liotta-Burgess' Trajectory Analyses

Among the first to explore implications of the "rearside" attack model was Baldwin [30]. In 1976, he proposed a qualitative method, based on the use of resonance struc-

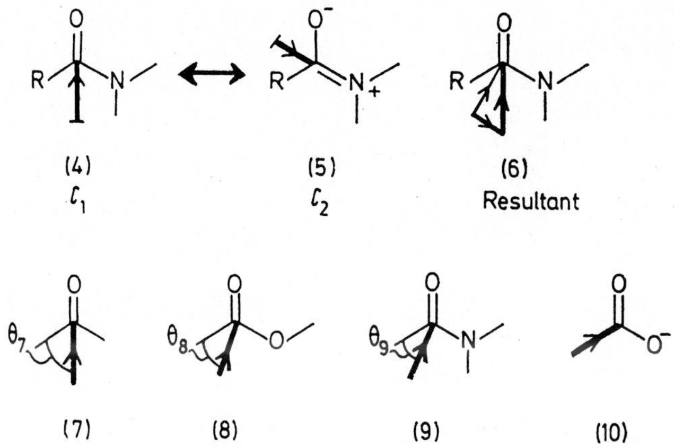

(4)   (5)   (6)

c_1 c_2 Resultant

(7)   (8)   (9)   (10)

$$[\theta_7 > \theta_8 > \theta_9;\ \theta_{10} = 0]$$

Scheme 6.22

tures, for assessing preferred geometries of nucleophile approach to unsaturated functions in general. The application of the method, known as "approach vector analysis", to the amide function is given in Scheme 6.22. The result of summing approach vectors is a shift of the nucleophile away from the bisecting plane and closer to the residue *R*. The magnitude of each vector is dependent on the relative contributions of the canonical Lewis structures; thus, the shift increases in the series of carboxylate ester, carboxamide, and carboxylate anion, as shown in Scheme 6.22. For conjugated cyclohexenones, Baldwin argued, such a shift leads to an increase in the effective bulk of H and CH_3 substituents in positions 5 and 6; the increase, however, is greater for the quasi-axial H at C(6) than for the quasi-axial H at C(5). As a consequence, nucleophile attack *trans* to the C(6)-H is preferred in the unsaturated ring by a greater factor than in the saturated ring. This assumption gave the correct configuration of the major hydride reduction product for 16 cases of cyclohexenone reduction.

Baldwin's model was subsequently restated in the language of molecular orbital theory by Stone and Erskine, and verified by perturbational and variational SCF calculations [117]. The results for hydroxide addition to formaldehyde, formic acid, formamide, and formate ion showed that the preferred direction of attack agrees with the predictions of the approach vector analysis. The frontier orbital term was found to dominate the stereoelectronic control, the n,σ* contribution being by far the largest among the terms depending on the approach geometry, although the electrostatic effect cannot be ignored.

Another formal treatment of the "rearside" attack model was proposed by Liotta, Burgess and Eberhardt [31]. These authors extended frontier orbital analysis of the interactions between a nucleophile and a π-electrophile by including the stabilizing two-electron interactions of the nucleophile HOMO with all the unoccupied π* and σ* MO's on the π-electrophile, as well as the net destabilizing four-electron interactions of the nucleophile HOMO with all the occupied π and σ MO's. The analysis yields then expressions defining separately the contributions of the charge-transfer and repulsive terms toward the nucleophile trajectory. The variables are orbital coefficients and energies, resonance integrals, and overlap integrals, all of which can be obtained from semi-empirical calculations, formal estimation procedures, and available tabulations. Thus, the model is intended to offer a semi-quantitative and simple way to predict the reactivity window of organic and biochemical reactions.

6.5.2 Consequences of Directionality of Nucleophilic Addition to Carbonyls

6.5.2.1 Directionality and Regioselection in Nucleophilic Addition to Dicarbonyl Compounds

It was known in the early 70's that unsymmetrically substituted anhydrides and imides can undergo highly regioselective transformations, e.g. to lactones and lactams, respectively (see Scheme 6.23) [166]. The growing importance of such products as synthetic intermediates raised the issue of control of regioselectivity, and several authors attempted to identify the factors at play [167–169]. The emphasis was on metal hydride reductions of succinic anhydrides to γ-lactones, and succinimides to hydroxypyrrolidinones. Suess [167], Rosenfield and Dunitz [47], and independently Kayser and Morand [169], proposed that the approach of a nucleophile to the less substituted carbonyl group can actually be more hindered if there is indeed a strong preference for "rearside" attack. The two pathways are compared in Scheme 6.24.

Although this explanation became widely accepted [170], continued examination of the phenomenon soon led Kayser and Morand [171], and Kayser and Wipff [172], to the conclusion that the steric effect is of minor importance in reactions of

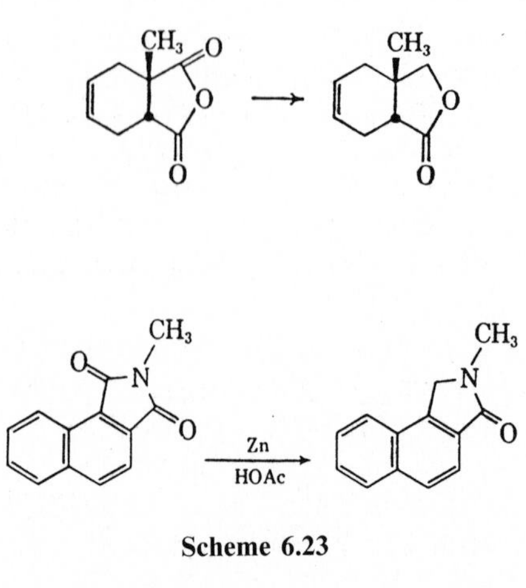

Scheme 6.23

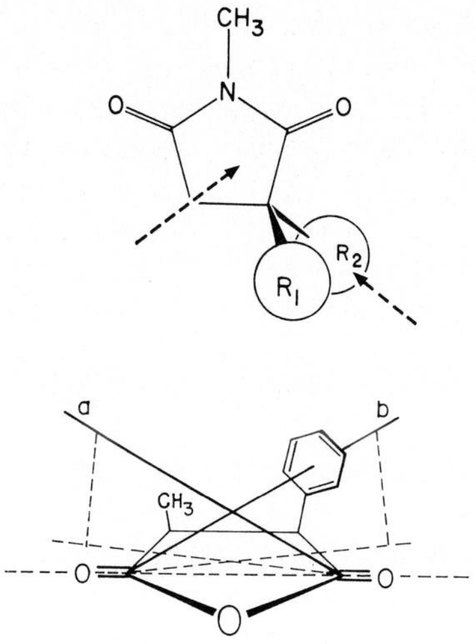

Scheme 6.24

NaBH$_4$ and LiAlH$_4$ with imides and cyclic anhydrides. Several observations are inconsistent with the "rearside" attack model. First of all, regioselectivity in the reduction of anhydrides by Li$^+$ or K$^+$ selectrides, which epitomize steric bulk, is reversed instead of enhanced [171a]. (Selectrides are tri-*sec*-butyl substituted borohydrides.) A pronounced temperature-dependence of the product ratios in these reductions confirms that the steric effect of large alkyl groups is involved [170f], since such an effect would encompass a significant entropic term. Thus, the more substituted sites are indeed more hindered. Second, regioselectivity is reversed in some rigid polycyclic derivatives where one of the anhydride faces is inaccessible. This suggests that the usual preference might involve face selection, related, for instance, to antiperiplanarity [171b, c]. Third, regioselectivity is *not* reversed in reductions of glutaric anhydrides [170f] or in reductions and organometallic additions of cyclohexen-1,4-diones [173], that is, in non-planar six-membered rings, where the model of steric hindrance to "rearside" attack is not applicable.

Furthermore, high regioselectivity is observed in unsaturated cyclic anhydrides, such as maleic anhydrides and phthalic anhydrides (Scheme 6.25) [166a, b, e, 174, 175] where substitution cannot introduce steric hindrance to the nucleophile attack. Kayser and Morand concluded that in such cases the most important factor is the intrinsic reactivity of the carbonyl group, which can be modeled by *ab initio* calculations of the LUMO coefficients for the ground-state geometries of the anhydrides

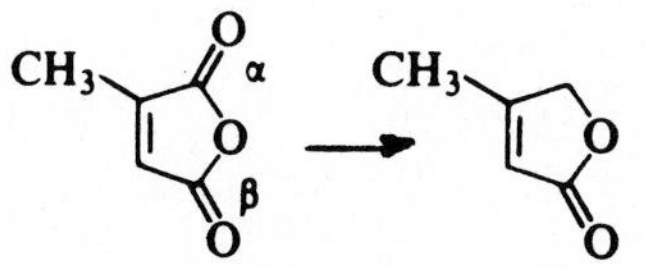

Scheme 6.25

[174c, 175g]. It has also been pointed out that chelation may be a factor [175g]. A similar conclusion was reached by Braun et al., who emphasized, however, the nature of the nucleophilic reagent [175f]. Their semi-empirical calculations of LUMO coefficients did not show significant differences between the two carbonyl groups of phthalic anhydrides.

Finally, the regioselection in metal-hydride reduction of unsymmetrically substituted acyclic diesters is not understood [176]; here, as well as in the reactions of cyclic anhydrides and imides, the effect of steric hindrance appears to play a minor role.

6.5.2.2 Directionality and Diastereofacial Selection in Nucleophilic Addition to Aldehydes and Ketones

Implications of the "rearside" attack model for asymmetric induction were first considered by Nguyen Trong Anh [32]. Scheme 6.26 shows two possible transition states for nucleophilic addition to an aldehyde (R=H), as postulated by Felkin et al. [177], where the symbols L, M and S stand for large, medium and small ligands at the stereogenic center. If a non-perpendicular attack is assumed, the steric hindrance encountered by a nucleophile is much more serious, according to Anh, in **B** than in **A**. Furthermore, when the aldehyde hydrogen is replaced by an alkyl group, the nucleophile is shifted towards the asymmetric carbon atom and can "'feel' better the difference between M (in **B**) and S (in **A**), which should lead to an increased stereoselectivity." The results of Anh's STO-3G calculations confirm this analysis. Anh also pointed out that optimizing the angles of attack favors the Cram stereochemistry, i.e. increases the energy difference between the two transition states in favor of **A**. Complexation of the carbonyl group by Li$^+$ reduces the values of the optimal angles of attack and thereby the stereoselectivity; in fact, maintaining the perpendicular attack along with electrophilic assistance predicts the anti-Cram stereochemistry. Anh therefore concluded that, the stronger the electrophilic assistance, the smaller the asymmetric induction.

Subsequently, the difference in the attack angle on a double bond between a nucleophile and an electrophile was postulated by Houk et al. [33] to be a major factor responsible for the different sense of 1,2-asymmetric induction. Since transition state **B** is less hindered than **A** in the case of hydroboration of alkenes or borane reduction of ketones (Scheme 6.27, right), it was proposed that such reactions should have a stereochemistry opposite to that obtained in metal hydride reductions (Scheme 6.27, left). The experimental results, indeed, confirm the reversal of the product ratios [178].

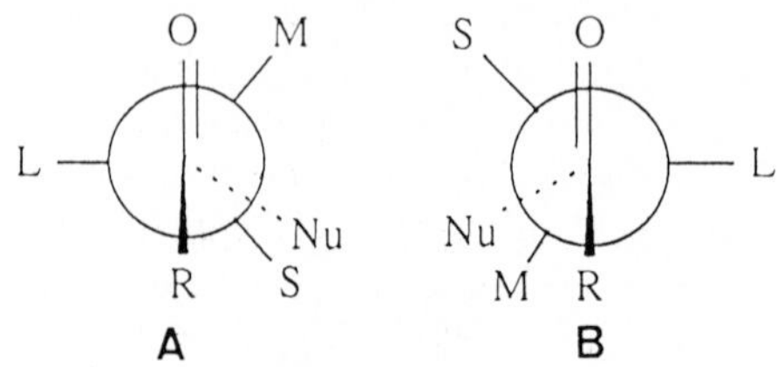

Scheme 6.26

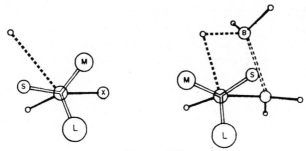

Scheme 6.27

The relationship between the nucleophile trajectory and diastereofacial selectivity was also considered by Heathcock et al. [34]. Contrary to Anh's prediction, these authors found a dramatic increase in diastereofacial selectivity in Lewis acid mediated reactions of chiral aldehydes. This was attributed to the effect of steric bulk of the catalyst (BF_3, $TiCl_4$, etc.), displacing the nucleophile towards the stereogenic C atom. Lodge and Heathcock carried out MNDO calculations for a series of *tert*-butylcarbinol anion structures, where one of the $C-H$ bonds was increasingly elongated over the range of 1.15–2.40 Å (Figure 6.39a). The results confirm that the hydride trajectory deviates from the bisector plane away from the *tert*-butyl group by 7° at a H^-...C distance of about 2.0 Å. The energy cost to bring it back into this plane is about 0.5 kcal mol^{-1}. It was proposed that, if the two carbonyl group ligands have different size, the nucleophile trajectory will be further away from the larger ligand. The magnitude of the displacement from the normal bisector plane (termed the "Flippin-Lodge angle") will be related to the difference in steric bulk on the two sides of this plane (Figure 6.39b). Thus, an increase in the bulk of the groups attached to the carbonyl on the side opposite to the chiral ligand should amplify the diastereofacial preference leading to higher stereoselectivity. This hypothesis fits data reported for $LiAlH_4$ reduction of chiral ketones [177a] (Figure 6.39c) and for alkylations of acetals occurring via additions to chiral oxonium ions (Figure 6.39d).

The "rearside" attack model became equally important in the discussion of stereoselectivity in the aldol condensation. In 1981, Seebach and Golinski proposed a topological rule for aldol-like reactions which says that the H atom, the smaller substituent on the donor component, adopts an *anti* position with respect to the acceptor double bond C=A in the preferred transition state, cf. **A** and **B** in Scheme 6.28 [35]. This preference of **A** over **B** comes about as a result of the increases in steric hindrance when the donor substituent R_2 is jammed against R_1 and H of the acceptor in the non-perpendicular approach **B**.

Eschenmoser et al. have also postulated that the Bürgi-Dunitz trajectory must be taken into account in discussions of the kinetic preference for formation of allose 2,4,6-triphosphate in the hexose series and of ribose 2,4-diphosphate in the pentose series, by aldomerization of glycolaldehyde phosphate in aqueous NaOH solution [37]. In this analysis, however, an additional interaction between the donor and acceptor substituents (b ↔ f in Figure 6.40) is supposed to increase when the approach is non-perpendicular.

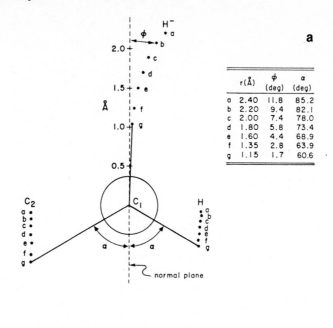

$r(\text{Å})$	ϕ (deg)	α (deg)
a 2.40	11.8	85.2
b 2.20	9.4	82.1
c 2.00	7.4	78.0
d 1.80	5.8	73.4
e 1.60	4.4	68.9
f 1.35	2.8	63.9
g 1.15	1.7	60.6

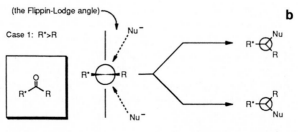

(Flippin, 1983; Lodge, 1987)

Fig. 6.39. a: Attack of hydride on pivalaldehyde [$(CH_3)_3CC(=O)H$], deviation of the attack trajectory from the plane normal to the carbonyl plane and passing through $C=O$
b: definition of the "Flippin-Lodge" angle φ
c: diastereoselectivity for the addition of $LiAlH_4$ to chiral ketones [177a]
d: diastereoselectivity for the addition of a silylenolate to chiral oxonium ion intermediates [34d]

Seebach's argument has been applied to interpret diastereoselectivity in the addition of non-enolate carbanions to aldehydes [179]. Bassindale et al. postulated that addition of $PhCH^-SiMe_3$ to PhCHO proceeds through an early, reactant-like transition state, deemed most consistent with the insensitivity of the product ratio to reaction conditions. In the initial stages of addition, the approach angle is assumed to be at least 109° and the deviation from planarity of the carbonyl is assumed to be minimal (Scheme 6.29). Consequently, one interaction about the incipient bond will be more important than the other two: the major steric influence will result from interactions amongst R^1, R and H. This primary steric effect is a direct consequence of carbanion attack at an angle greater than 90°. The most favored orientation will

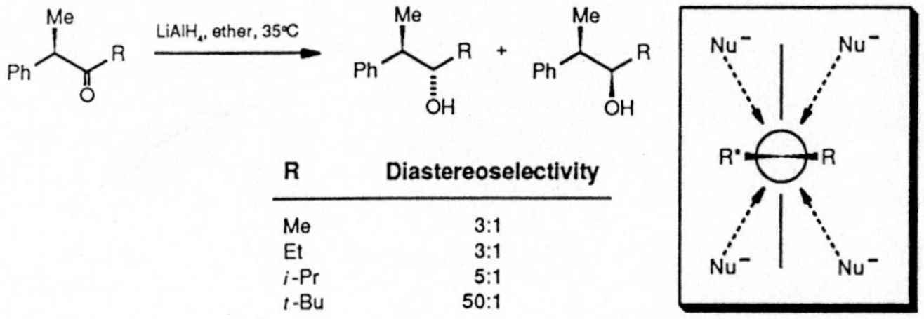

R	Diastereoselectivity
Me	3:1
Et	3:1
i-Pr	5:1
t-Bu	50:1

c

(Cherest, Felkin, Prudent, 1968)

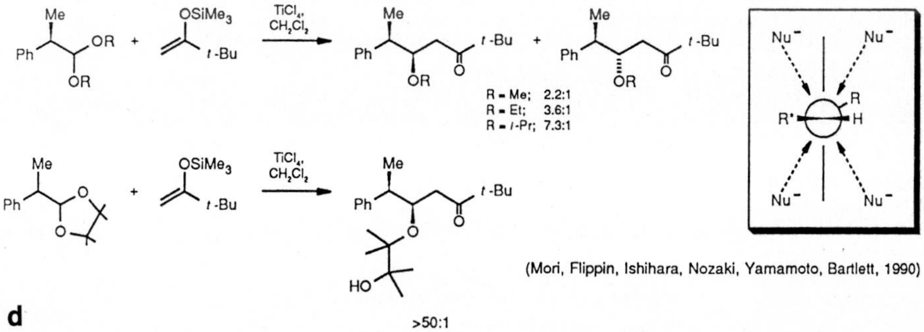

R = Me;   2.2:1
R = Et;   3.6:1
R = *i*-Pr; 7.3:1

(Mori, Flippin, Ishihara, Nozaki, Yamamoto, Bartlett, 1990)

d

>50:1

Fig. 6.39 c, d

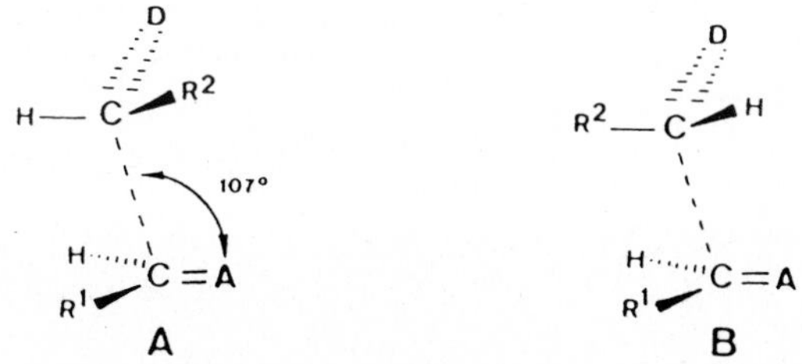

Scheme 6.28

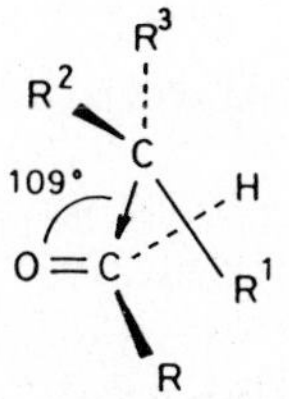

Scheme 6.29

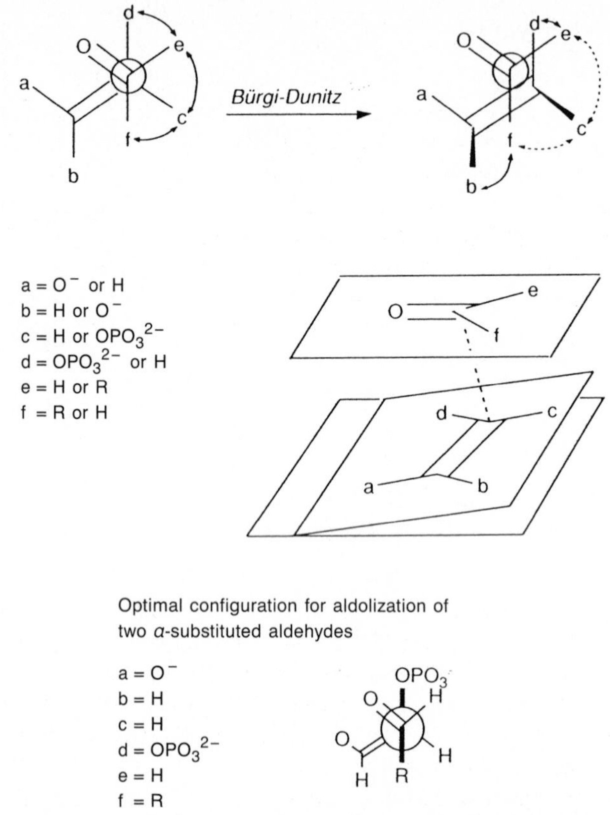

Bürgi-Dunitz

a = O⁻ or H
b = H or O⁻
c = H or OPO_3^{2-}
d = OPO_3^{2-} or H
e = H or R
f = R or H

Optimal configuration for aldolization of
two *a*-substituted aldehydes

a = O⁻
b = H
c = H
d = OPO_3^{2-}
e = H
f = R

Fig. 6.40. Eschenmoser's interpretation of glycolaldehyde phosphate aldomerisation reactions [37]:
"It is generally appreciated... that the Bürgi-Dunitz trajectory... for nucleophilic addition to
C=O groups must be taken into account as steric interactions between reaction center substituents are evaluated. The drawings in (this figure) remind the reader why. While it can be difficult
to weigh the contributions of the four relevant interactions for an aldehyde/ketone-enolate pair,
the problem for the case of an aldehyde/aldehyde-enolate pair turns out to have a unique solution: the one indicated in (this figure), where none of the interacting substituents is juxtaposed
with a non-H-atom partner"

be that in which the primary steric influence is minimized. The steric outcome of
the reaction is then determined by the relative extent of the secondary steric influences: the interactions of R^2 and R^3 with R. It was pointed out that models which
analyze steric interactions for anion attack at 90° fail to account for "erythro-selec-
tivity" in carbanion additions.

Following Anh's and Seebach's analyses, it is now generally accepted that steric
interactions at the "rearside" of carbonyls are more severe than at the "frontside",
i.e. in the directions *gauche* with respect to the carbonyl O [180]. The same argument
has been applied to aromatic derivatives, even though the obtuse angle of attack deviates from colinearity with the π system of the benzene ring [170c, 180e].

6.5.2.3 Directionality and Rate of Intramolecular Reactions of Nucleophilic Addition to Carbonyls

Intramolecular reactions involving 5- or 6-membered cyclic transition states, including corresponding cyclizations, are often faster than their bimolecular counterparts. In 1970, however, Eschenmoser et al. demonstrated that this is not true for S_N2 displacement at tetrahedral C if the cyclic transition state does not allow for a proper alignment of the breaking and forming bonds [181]. Eschenmoser distinguished two situations in such intramolecular reactions (Scheme 6.30), an endocyclic substitution leading to a group transfer, and an exocyclic substitution leading to cyclization. The former cannot occur in 5- or 6-membered rings, since they cannot accommodate the linear S_N2 transition state.

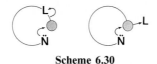

Scheme 6.30

Baldwin extended these considerations to other reactions, including carbonyl additions, in his rules for ring closure: the favored ring closures are those in which the length and nature of the linking chain enables the terminal atoms to achieve the required trajectories to form the final ring bond [182]. The disfavored cases require severe distortions of bond angles and distances to achieve such trajectories.

In the spirit of Baldwin's rules, the "rearside" attack model is often invoked to rationalize the stereochemical course of intramolecular carbonyl additions or to discriminate between various mechanistic options in such processes [183]. However, in contrast to the case of S_N2 displacements, unambiguous experimental support for such conjectures is lacking.

The idea that rates of organic reactions are highly dependent on precise geometrical alignment of the functional groups undergoing the transformation became known as the theory of orbital steering. It was proposed by Koshland et al. to explain differences in lactonization rates by factors of 10^3, found among bicyclic hydroxyacids and in the relative rates of lactonization and thiolactonization in several simple model systems [46]. Assuming that the rate-controlling step of the acid-catalyzed lactonizations is the attack of the hydroxyl group on the carboxylic group, these authors concluded that the differences are due to the differences in orientation of the two groups, described in this instance by the O...C−C angle in the hydroxyacid. It was implied that an angular displacement of merely 10° could cause a decrease in reaction rate by a factor of 10^4. The optimum angle was estimated to be 96°.

Serious objections were raised against Koshland's proposition. For instance, Bruice pointed out that it implies values for the force constants for partly formed bonds that were larger than those for fully formed covalent bonds [184]. One intuitively expects that it should be the other way around, and semi-empirical calculations by Scheiner, Lipscomb and Kleier on methanol addition to formic acid confirmed such expectations [128].

In fact, in 1980, Menger et al. showed that the differences in the attack angle up to 10° have no effect on the rates of lactonization in a model system [185]. Furthermore, the O...C distancs and O...C−C angles, calculated by Houk et al. using the MM2 force field [186], did not show any correlation with experimental rate constants for lactonization of the 15 hydroxyacids studied by Koshland et al. [46], Milstein and Cohen [187], and Menger and Glass [185b].

The second type of intramolecular addition to carbonyl, where the geometry is known in detail along with the rates, is hydride transfer in rigid polycyclic systems [70]. In the series of polycyclic hydroxyketones shown in Scheme 6.4, the rates of hydride transfer increase by a factor of 10^6, as measured by ^{13}C dynamic NMR spectroscopy for the sodium salts in DMSO solutions. In this series, the H...C=O distance in the ground-states decreases: 2.49, 2.40, 2.30, and 2.18 Å; the H...C=O angle also decreases: 99, 98, 96, and 91°, so the ground-state geometry appears to correlate with reactivity. Sherrod and Menger claim, however, on the basis of a semi-empirical AMPAC-AM1 study of the potential energy surface for this reaction, that the major portion of the barrier is here the cost of molecular distortion which brings the H within 1.6 Å of the recipient C atom [188]. The relationship of the ground-state geometry and reactivity might, therefore, not be as simple as the above numbers suggest.

6.5.3 Ground State Geometry of Carbonyl and Related Compounds and Reactivity

Classic structure-reactivity relationships usually involve correlations between logarithms of equilibrium and rate constants. The interest in the relation between reaction rate and structure, understood as ground-state geometry, appears to have been originally stimulated by observations that strained molecules often display enhanced reactivity. A comprehensive investigation in this area was carried out by Rüchardt and Beckhaus and their coworkers, who found excellent correlations between the bond distances and activation energies for homolytic dissociation of substituted ethanes, increasingly strained due to bulky substituents [189]. Attempts to extend this kind of analysis to other unimolecular processes have been made, and similar trends for some solvolytic substitution and elimination reactions have been found, but with less general results [190].

The hypothesis that ground-state geometry can be used to quantify chemical reactivity was also examined for a number of bimolecular processes. Here, the emphasis was invariably on the effect of steric hindrance, which can be extrapolated to the transition state in a fairly straightforward manner. For instance, Seeman et al. evaluated a variety of geometrical parameters and steric congestion models for the methylation of a wide series of alkyl pyridines [191].

Here we are concerned with a different type of structural ground-state effect, termed stereoelectronic, as it is attributed to secondary molecular orbital interactions, stabilizing or repulsive.

6.5.3.1 Valence Angle vs. Bond Length Correlations and Reactivity of Acyl Derivatives towards Elimination

One of the first suggestions that ground-state geometry reveals readiness of a molecule to undergo an elimination reaction was made in the study of α-substituted enamines. The previously described correlations of the valence angles (Section 6.3.2.1) can be interpreted in terms of the increasing keteneiminium character of the enamine fragment with increasing electronegativity of the α-substituent [107]. The diversity of this substitution in the sample does not allow such changes to be correlated directly with the bond distances to the leaving groups, but the comparison of the C−X bond lengths in enamines with those in olefins confirms the expected increase in the ionic character of the bonds to the more electronegative groups: the C−X distances are only 0.01−0.02 Å longer in enamines than in the corresponding $CH_2=CHX$ molecules if $X=CH_3$ or $C\equiv N$, but they are 0.05−0.11 Å longer if X = F, Cl, I. Furthermore, there is a slight rotation of the amino group in the series, which indicates an increasing interaction of the nitrogen lone pair with the C−X bond: the torsion angles between the C=C bond and the assumed lone-pair direction are 33, −17, −6, 17, −7, and 5° for $X=CH_3$, $C\equiv N$, $C\equiv N$, F, Cl, I. Formation of keteneiminium ions does indeed occur readily when α-haloenamines are treated with weak Lewis acids in slightly polar solvents, but no attempts to make quantitative correlations seem to have been made.

Similarly, a qualitative relation between the chemical behavior and the distortion from ideal C_{2v} symmetry was suggested for a series of lithium ester enolates (Scheme 6.13) [108]. Enolate **1**, furthest along the reaction coordinate to ketene, had to be handled at temperatures below −50 °C and decomposed rapidly at temperatures higher than −30 °C. The two other enolates, **2** and **3**, were found to survive in crystalline form at 0 °C and at room temperature, respectively. The decomposition occurs most likely through a ketene-like intermediate, whose transient existence was demonstrated by cleaving the lithium enolate of 2,6-di-*tert*-butyl-4-methylphenyl-2-methylpropanoate at room temperature in the presence of excess *n*-BuLi.

Also, structures of acyl derivatives RC(=O)X indicate higher acylium ion-anion character if X is a good leaving group, readily undergoing elimination in the presence of Lewis acids, than if X is a poor leaving group [50].

6.5.3.2 Ground-State Structure and Reactivity of Acetals: Empirical Potential Energy Surface and Determination of Transition-State Structure for the Spontaneous Hydrolysis of Axial Tetrahydropyranyl Acetals

The link between conformation and reactivity of acetal-like fragments is mainly due to Deslongchamps and his coworkers, who at the beginning of the 70's established stereoelectronic principles governing the breakdown of orthoesters, hemiorthoesters, etc. [51]. The task of verifying Deslongchamps' theory in a somewhat simpler case − the hydrolysis of tetrahydropyranyl acetals (Scheme 6.31) − was undertaken by Kirby and his coworkers. Their extensive efforts to gather both kinetic and structural

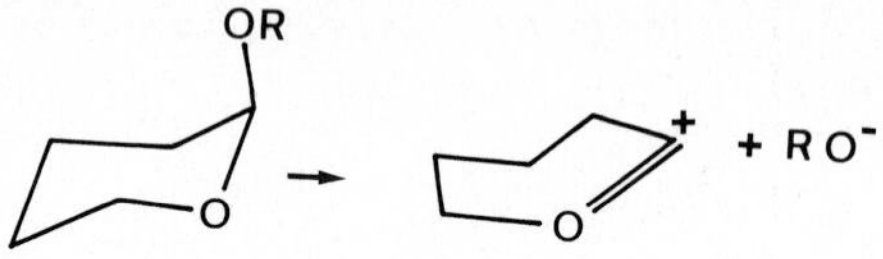

Scheme 6.31

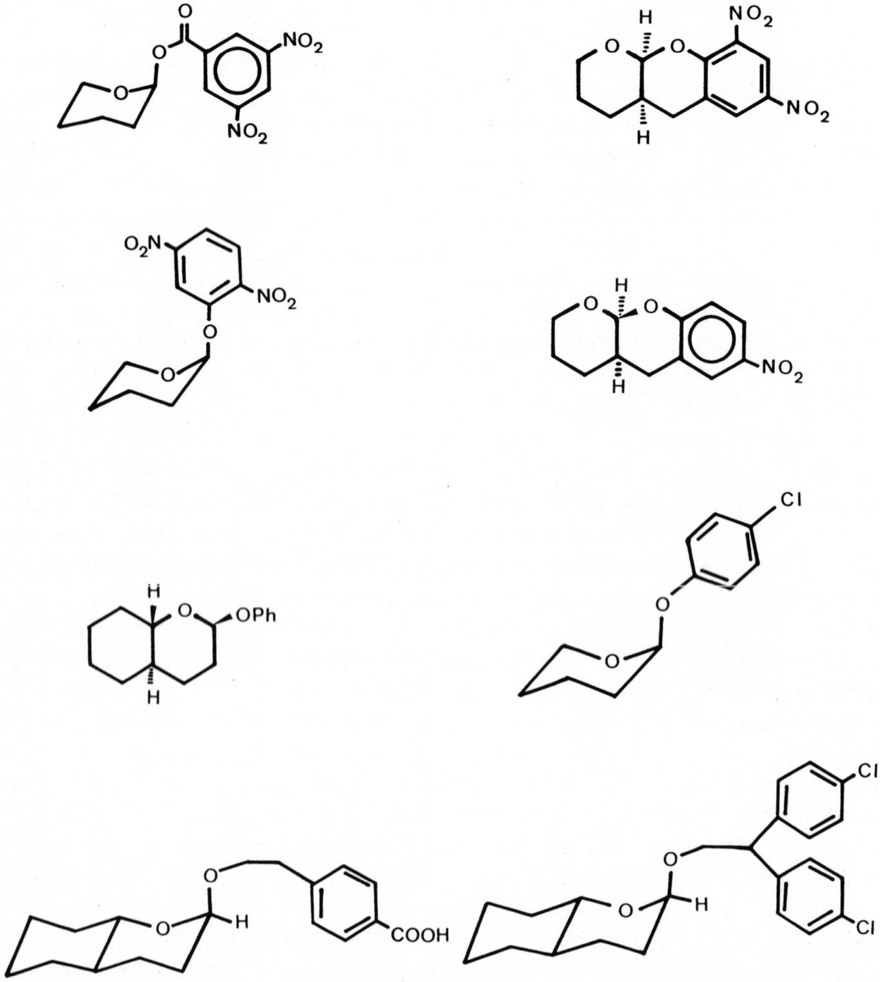

Scheme 6.32

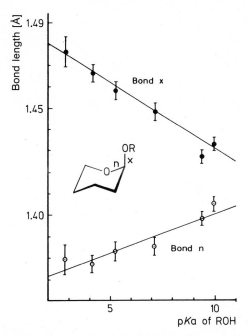

Fig. 6.41. Plot of bond lengths vs. the pK_a of ROH for some of the axial tetrahydropyranyl acetals listed in Scheme 6.32. Error bars represent standard deviations in bond lengths. The equations of the lines drawn are given in [52a]

evidence led to the discovery that the increasing electron demand of the group attached to the exocyclic oxygen not only facilitates hydrolysis but also produces systematic changes in the pattern of the C−O bond lengths [52]. In a series of six axial tetrahydropyranyl aryl acetals, the exocyclic bond lengthens from 1.425 Å to 1.475 Å, and the endocyclic bond shortens from 1.400 Å to 1.375 Å, as the pK_a of the corresponding phenol or carboxylic acid, taken as a measure of leaving-group ability, is lowered from 10 to 3 (Scheme 6.32). Linear regression gives a good fit for the *exo* bond and a slightly poorer one for the *endo* bond (Figure 6.41).

Data for a series of phosphate monoester dianions and phosphate triesters indicate a similar dependence of the P−OR bond length on the pK_a of ROH. In phosphate salts, this bond seems to be about twice as sensitive to the nature of OR as in the esters. Thus, both in acetals and phosphates, stabilization of the incipient electron deficient center is important in influencing the scissile bond distance and its sensitivity to the nature of the leaving group.

In view of the described effects of substitution on heterolytic bond cleavage rate and on the scissile bond length, it is not surprising that this bond distance is correlated with reactivity and that the correlation holds for variation at both ends of the bond. Indeed, Jones and Kirby point out that there are excellent linear correlations of log k_{rel} with intercepts and slopes of the corresponding bond length-pK_a plots for the series: α-glucosides, methoxymethyl acetals, and axial tetrahydropyranyl

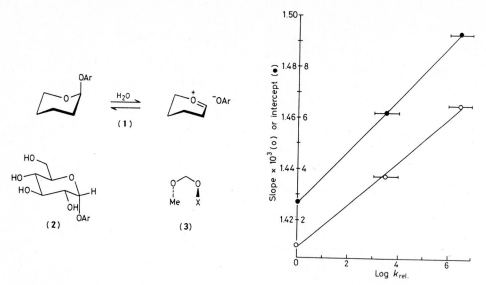

Fig. 6.42. Plots of the sensitivity parameters and intercepts obtained from the bond length-reactivity correlations for compounds **1–3**, against their relative reactivities towards hydrolysis

aryl acetals, which undergo hydrolysis with mean values for relative reactivities $1 : 10^{3.5} : 10^{6.5}$ (Figure 6.42) [52c].

For the spontaneous hydrolysis of the axial tetrahydropyranyl acetals, the rate constants are linearly related to the pK_a of ROH [192]. Consequently, the linear correlation of Figure 6.41 can be translated into a linear relationship between the lengthening of the exocyclic C–O bond and the free energy of activation for its heterolytic cleavage, as shown in Figure 6.43. Jones and Kirby suggested that this implies a linear region of considerable extent on the reaction coordinate, since the set of compounds covers more than 60% of the activation energy. The slope of the correlation line is about 260 kcal mol^{-1} Å^{-1}.

Several assumptions were introduced in order to locate the transition state. First, the transition state is late and thus close in energy to the ion-pair; second, the free energy of the ion-pair is independent of the group RO$^-$ as long as the pH of the medium is high enough for ROH to be fully ionized; third, the rate of recombination of the ion-pair is independent of the basicity of the ion RO$^-$. Consequently, the transition state is assumed to be common for the entire set; the extrapolated C–O bond length is about 1.56 Å, while the pK_a of the conjugate acid of the leaving group at the transition state is around -7, in good agreement with the value expected for a phenolic leaving group, as in the specific acid catalysed hydrolysis of 2-phenoxytetrahydropyran.

To explore the true reaction coordinate, assumed to be represented simply by C–OR bond breaking, the authors proposed a way to relate the data to a standard potential energy curve. The requirement that such a curve has an approximately linear region on the dissociation side of the energy minimum, is fulfilled by the Morse function:

$$V(r - r_0) = D_0 [1 - e^{-\beta(r - r_0)}]^2 .$$ (6.5)

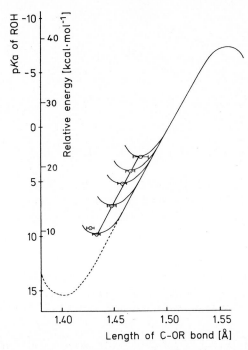

Fig. 6.43. Reaction coordinates for six axial aryl tetrahydropyranyl acetals normalized onto the same scale (see text) and compared with the curve for a "parent" 2-alkoxytetrahydropyran (begins as dashed curve)

D_0 and r_0 were taken to be the free energy of activation for a parent tetrahydro-pyranyl alkyl acetal, and its exocyclic C—O bond length, respectively. The plot of $\ln[(D/D_0)^{1/2} - 1]$ vs. $r - r_0$ gives then the slope, $\beta = 13.9\,\text{Å}^{-1}$ and intercept 0.436. The Morse curve defined by the selected value of D_0 and obtained value of β is shown in Figure 6.44. The experimental data map out a line parallel to this curve but displaced from it by 0.031 Å. The authors concluded that the origin of the observed linear relationship between bond length and reactivity is to be found in the approx-imate linearity of the Morse function at small-to-moderate bond extensions. Since it is an intrinsic property of this function, such linear relationships may, according to Jones and Kirby, turn out to be the rule.

The authors concede that the β value obtained from their model is several times larger than what can be expected on the basis of the standard C—O stretching fre-quency and the reduced mass for a relatively large acetal. The discrepancy was attrib-uted to the fact that the apparent slope, revealed by the one-dimensional model, is a projection of the reaction pathway across an energy hypersurface and depends on the vantage point of the observer, that is, on the choice of the plane of projection.

A different model was proposed by Bürgi and Dunitz [53], who observed that the properties of fractional bonds can be described using simple modifications of in-teratomic potential functions, such as the Morse equation or the general inverse

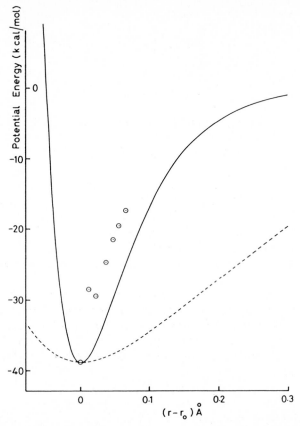

Fig. 6.44. Morse curve (Equation 6.5) calculated by using constants $D_0 = 38.8$ kcal mol^{-1} and $\beta = 13.87$ Å derived (least-squares method; see text) to fit the bond length reactivity data in the way suggested by Figure 6.43. The dashed curve shows the shallower potential function calculated for the gas-phase dissociation into the component atoms of a diatomic molecule joined by a bond with similar properties to that of the $C-OR$ bond of a 2-alkoxytetrahydropyran derived from an alcohol ROH of pK_a 15.5

power potential. For instance, if the Morse equation is rewritten so that the repulsive and attractive terms are explicit, the latter term can be multiplied by a power of the bond number n^q to give

$$V(r-r_0) = D_0[e^{-2\beta(r-r_0)} - 2n^q e^{-\beta(r-r_0)}] \; . \tag{6.6}$$

With the modified equation, the quantities describing the properties of fractional bonds become:

force constant for bond stretching,

$$k(n) = 2\beta^2 D_0 n^{2q} \tag{6.7}$$

equilibrium distance,

$$r_e(n) = r_0 - q\beta^{-1}\ln n = r_0 + [\ln(2\beta^2 D_0)]/2\beta - [\ln k(n)]/2\beta \tag{6.8}$$

dissociation energy,

$$D(n) = D_0 n^{2q} \tag{6.9}$$

and dependence of dissociation energy on bond distance in the ground state,

$$dD(r_e)/dr_e = -2\beta D_0 n^{2q} = -k(n)/\beta \ . \tag{6.10}$$

The equations for $r_e(n)$, $k(n)$ and $D(n)$ derived in this way have the same form as the Pauling definition of fractional bond order [24], Herschbach and Laurie's relationship between stretching force constant and bond distance [193], and Johnston and Parr's expressions for the bond-energy and bond-order relationships [25]. With reasonable values of $\beta \approx 2\,\text{Å}^{-1}$, $D_0 \approx 100\,\text{kcal mol}^{-1}$ and $n = 1$, $dD(r_e)/dr_e$ is about $400\,\text{kcal mol}^{-1}$, thus reproducing the order of magnitude of the experimental slope for acetal hydrolysis.

The energy for a small displacement along the reaction pathway must actually depend on the distance changes for *all* bonds affected; each of these changes can be taken as a linear function of the reaction coordinate Δr_1, the change in the exocyclic C$-$O distance. Thus, in the harmonic approximation, one obtains $2V = k(A\,\Delta r_1)^2$, and in order to yield this expression for small Δr_1 the Morse equation is adapted in the following way:

$$V(\Delta r_1) = D_0[e^{-2\beta'\Delta r_1} - 2e^{-\beta'\Delta r_1}] \tag{6.11}$$

with $\beta' = A\beta$. The value of A is estimated from the bond-length increments for the three most affected acetal bonds to be ca. 8, and $-2\beta' D_0$ then gives $280\,\text{kcal mol}^{-1}$ for the slope of the linear relationship between the free energy of activation and the reaction coordinate Δr_1. This analysis avoids the contradiction of the one-dimensional model based on an unmodified Morse function. A change in the potential has only a small effect on the equilibrium bond distance but a large one on the dissociation energy, as shown in Figure 6.45.

For all its simplicity and appeal, the Morse potential is lacking as a model potential for a chemical reaction in that it does not describe a transition state. In order to obtain an estimate of the transition-state geometry, Jones and Kirby had to introduce several assumptions and thereby a considerable degree of arbitrariness. Recently, this aspect of modeling the hydrolysis of acetals became a focus of a study by Bürgi and Dubler-Steudle [54], who pointed out the contradiction between Jones and Kirby's assumption of a late transition state and their extrapolated C$-$O distance of ca. 1.56 Å for this transition state.

Bürgi and Dubler-Steudle carried out principal component analysis of selected internal coordinates for the acetal fragment in the eight axial tetrahydropyranyl acetals reported by Jones and Kirby. The first principal component accounts for ca. 75% of the total variance, and its eigenvector clearly describes incipient acetal cleavage [194]. A model of the transition-state structure was obtained by linear extrapolation

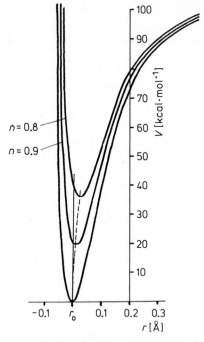

Fig. 6.45. The modified Morse function $V(r-r_0) = D_0\{\exp[-2\beta(r-r_0)] - 2n^q \exp[-\beta(r-r_0)]\}$ (Equation 6.6) with $D_0 = 100$ kcal mol^{-1}, $\beta = 2$ Å^{-1}, $q = 1$, $n = 1$ (outer curve), 0.9, 0.8. Note the sharp decrease in dissociation energy as the equilibrium distance increases (compare Equation 6.9)

along this vector to a length of 1.28 Å for the endocyclic C−O bond (based on the crystal structure of an analog of the oxonium ion reaction intermediate, 2-methoxy-1,7,7-trimethylbicyclo[2.2.1]hept-2-ylium fluoroborate [195]). The corresponding estimate for the exocyclic C−O bond in the transition state is 1.66 Å and may be taken as a lower limit of $r_1^{\ddagger}$. Extrapolation, using the rule of conservation of bond order, Pauling's relationship and $r_2^{\ddagger} = 1.28$ Å, gives an $r_1^{\ddagger}$ value of ca. 1.8 Å, in better agreement with the assumption that the transition state is late.

Bürgi and Dubler-Steudle also proposed a simple way of combining structural, vibrational, and kinetic data to parametrize a potential energy surface for a series of related molecules undergoing the same type of reaction. For the hydrolysis reaction, they developed two models: first a one-dimensional model using the reaction coordinate

$$q = [\Delta r_1^2 + \Delta r_2^2]^{1/2} \tag{6.12}$$

and subsequently a two-dimensional model, allowing for change of both C−O distances along the reaction path, using the reaction coordinates $\Delta r_1 = r_1 - 1.409$ Å and $\Delta r_2 = r_2 - 1.422$ Å, where the numerical constants are bond distances in the reference molecule. In both cases, a cubic polynomial is used to express dependence of energy D on the reaction coordinates, since this is the simplest polynomial showing the necessary extrema, a minimum for the reactant structure and a maximum for the transition-state structure. The effect of small changes in the leaving group is represented by a linear perturbation. Consequently, the one-dimensional model is defined as the following function:

$$V(q) = -D_0^{\ddagger} + (k_q q^2/2 + k_{qq} q^3 + aq) \quad a \leqslant 0 \ . \tag{6.13}$$

The constant k_q is the force constant along q and $D_0^{\ddagger}$ is the experimental free energy of activation of the reference molecule; k_{qq} is an anharmonicity constant, obtained as $-k_q/(3q^{\ddagger})$ from the condition $dV/dq = 0$ for $q = q^{\ddagger}$. A sufficiently small value of a will not change the general appearance of the reaction coordinate; for a perturbed molecule, the new equilibrium will be shifted to $q_e(a)$ and the transition state to $q^{\ddagger}(a)$, whose magnitudes increase and decrease, respectively.

Once values for $D_0^{\ddagger}$ and k_q are selected, activation energies $D^{\ddagger}$, ground-state equilibrium coordinates q_e and transition-state coordinates $q^{\ddagger}$ can be calculated as functions of the perturbation a. The dependence of $D^{\ddagger}$ on q_e is

$$D^{\ddagger} = D_0^{\ddagger} [1 - 2q_e(k_q/6D_0^{\ddagger})^{1/2}]^3 \ . \tag{6.14}$$

This function is plotted in Figure 6.46 and compared with experimental results for which $q_e(\text{obs})$ is taken as $[(r_1 - 1.422)^2 + (r_2 - 1.409)^2]^{1/2}$. The agreement between calculated and experimental values is remarkable. The model leads to 0.55 Å for the distance between the reference equilibrium geometry q_e and the transition state at $q^{\ddagger}$. For the most perturbed molecule in the series, q_e increases by 0.08 Å and $q^{\ddagger}$ decreases by the same amount. The corresponding cubic potential for this molecule is shown in Figure 6.46. It is not known directly how much one bond is shortened and how much the other lengthened, but this information can be obtained by invoking the rule of constant bond order. In this way, one arrives at $r_1^{\ddagger} \approx 1.409 + 0.54 \approx 1.95$ Å and $r_2^{\ddagger} \approx 1.422 - 0.15 \approx 1.27$ Å. Similarly, the right order of magnitude for the anharmonicity constant is obtained.

The success of this simple model suggests that the astonishingly strong dependence of $D^{\ddagger}$ on small changes in ground-state equilibrium geometry may have a very general origin. The perturbation that shifts the ground state affects the reference potential along the entire reaction path: only slightly as far as q_e and $q^{\ddagger}$ are concerned, but substantially in the activation energies.

An analogous cubic polynomial can be used to formulate a two-dimensional potential energy surface as a function of Δr_1 and Δr_2. If these are constrained to be equivalent:

$$V(\Delta r_1, \Delta r_2) = k_{11}[(\Delta r_1)^2 + (\Delta r_2)^2]/2 + k_{12} \Delta r_1 \Delta r_2$$

$$+ k_{111}[(\Delta r_1)^3 + (\Delta r_2)^3] + k_{112}[(\Delta r_1)^2 \Delta r_2 + \Delta r_1 (\Delta r_2)^2] \ . \tag{6.15}$$

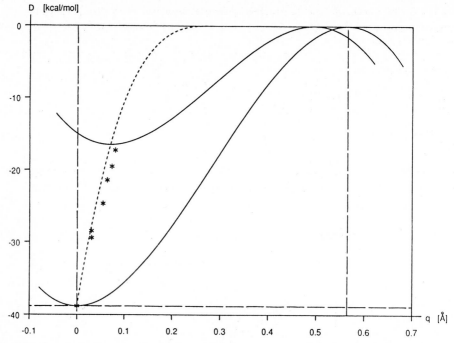

Fig. 6.46. Correlation between change in ground-state structure (q) and activation energy (D) in tetrahydropyranyl acetals (Scheme 6.32). *: experimental values, dotted line: calculated from Equation 6.13; solid lines: energy profiles for slowest and fastest reactions

The linear perturbation has now the form:

$$b(\Delta r_1 - \Delta r_2) \tag{6.16}$$

where the negative sign accounts for the fact that a change of the substituent R lengthens one C−O bond and shortens the other.

Selection of the quadratic force constants can be made from *ab initio* results for related molecules, and the cubic force constants can then be calculated from $D_0^{\ddagger}$ of the reference molecule by assuming that the distances in the transition state obey the rule of constant bond order, while the stretching force constants follow Badger's rule [196] or Hershbach and Laurie's relationship [193]. The structure and force constants of the transition state can thus be estimated. These data are shown in Table 6.3. Finally, an increasing linear perturbation term is added to the energy expression to obtain positions of the new minima, transition states, and the $D^{\ddagger}$ values. The displaced minima are shown in Figure 6.47. Although the model does not invoke bond-order conservation, except in fixing a numerical parameter, the displacements very closely follow the curve of constant bond order and the experimental changes in the ground-state structures. In other words, the anharmonicity built into the model reproduces the non-linear changes in Δr_{1e} and Δr_{2e}. The distances in the transi-

Table 6.3. Numerical constants characterizing the energy surface $V = k_{11}(\Delta r_1^2 + \Delta r_2^2)/2$ $+ k_{12}\Delta r_1 \Delta r_2 + k_{111}(\Delta r_1^3 + \Delta r_2^3) + k_{112}(\Delta r_1^2 \Delta r_2 + \Delta r_1 \Delta r_2^2)$ and calculated transition-state properties (for references see text)

Ground state			
r_{10}	1.422 Å	k_{12}/k_{11}	0.05
r_{20}	1.409 Å	$k_{111} = k_{222}$	-2.18 mdyn Å^{-2}
$k_{11} = k_{22}$	5.34 mdyn Å^{-1}	$k_{112} = k_{122}$	6.70 mdyn Å^{-2}
Transition state			
$r_1^{\ddagger}$	1.95 Å	$k_{12}^{\ddagger}$	-9.1 mdyn Å^{-1}
$r_2^{\ddagger}$	1.27 Å	$k_{22}^{\ddagger}$	10.8 mdyn Å^{-1}
$k_{11}^{\ddagger}$	0 mdyn Å^{-1}	$E_0^{\ddagger}$	38.85 kcal mol^{-1}

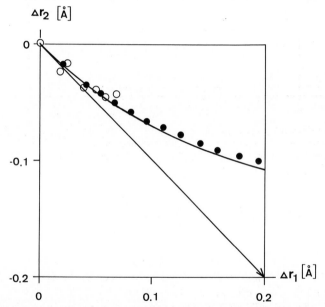

Fig. 6.47. Acetal $C-O$ distances perturbed by variation of the leaving group. Empty circles: observed; filled circles: calculated from Equation 6.15. The solid line is a curve of constant bond order. The diagonal is $\Delta r_1 = -\Delta r_2$

tion state for the reference molecule are the same as those obtained from the one-dimensional model. However, the curve describing the perturbed activation energy $D^{\ddagger}$ as a function of the reaction coordinate is somewhat closer to the experimental data than that obtained from the one-dimensional model.

These simple models do not introduce any assumptions about the nature of the transition state. However, they imply a significant shift in transition-state geometry and energy in response to the change in basicity of the leaving group – modeled by the change in the linear perturbation.

A physical interpretation of the perturbation constants *a* and *b*, in the one- and two-dimensional models, respectively, can be inferred from a description of the spontaneous hydrolysis of acetals in terms of the n, σ^* interaction hypothesis. According to this hypothesis, the process is driven by the double-bond–no-bond resonance that provides stabilization energy *SE* proportional to the overlap of the O(2) *n*- and C(1)–O(1) σ^*orbitals and inversely proportional to the separation of their energy levels

$$SE \sim \frac{S^2(n, \sigma^*)}{\varepsilon(\sigma^*) - \varepsilon(n)} \tag{6.17}$$

Such an interaction increases along the reaction coordinate as elongation of a σ bond lowers its σ^* energy level. Since the dependence of *SE* on the σ^* energy level is inverse, stretching of the C(1)–O(1) bond affects the interaction much more for a low-lying σ^* orbital than for a high-lying one. According to the Bürgi and Dubler-Steudle proposition, the perturbation of the reference potential reflects a change in electronegativity of the substituent at O(1), the alkyl, aryl or acyl moiety R. A more electronegative R lowers the σ^* energy level all along the reaction coordinate, and thereby shifts the fragment into the region of steeper increase in *SE* (Eq. 6.17). The perturbation coefficients *a* and *b* provide a measure of how sharply the n, σ^* interaction increases as the bond distance changes. It follows that *a* and *b* become more negative as the electronegativity of R increases, i.e., as the basicity of the leaving group RO^- decreases. Since basicity and Hammett constants are closely related, the simple MO perturbation argument points to possible connections between the cubic potential model and Hammett-type treatments.

6.5.3.3 Partial Pyramidalization and Intrinsic Face Preference in Diastereofacial Selection

The notion that partial pyramidalization at a trigonal center produces an intrinsic facial preference for the formation of a new bond at such a center was first introduced as a corollary of the frontier molecular orbital (FMO) theory. In 1966, Fujimoto and Fukui reported extended Hückel energies for 2-norbornyl radical in which C(2) had a planar geometry and for three structures in which the attached H was shifted out of the C(1), C(2), C(3) plane by 5° in the *exo* direction and by 5 and 10° in the *endo* direction (Figure 6.48) [27a]. The minimum was found at about 5° in the *endo* direction. This pyramidal distortion was assumed to be associated with mixing of the 2s AO of C(2) with the $2p_z$ unpaired-electron orbital. The authors suggested that the resulting greater extension of the half-occupied hybrid orbital in the *exo* direction favors *exo* attack of a halogen donor. The difference in energy between the two 5° pyramidalized geometries, only about 0.2 kcal mol^{-1}, was expected to increase during bond formation. The argument was subsequently extended to the interpretation of face selectivity in electrophilic additions and in cycloadditions to alkenes such as 2-norbornene, 5-substituted cyclopentadienes, and cyclobutene [27b, 197].

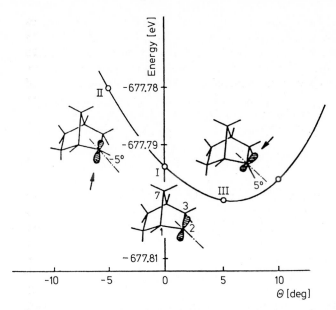

Fig. 6.48. An energy diagram of 2-norbornyl. (θ: the angle between the C–H axis and the C(1)C(2)C(3) plane)

In 1972, Mock considered double-bond reactivity and its relationship to torsional strain, by which he understood the strain imposed on a double bond in medium-ring *trans*-cycloalkenes or by steric compression of large *cis* substituents [28]. He argued that the loss of π overlap due to a "torsion" about the double bond can be partially compensated by rehybridization in these two situations, leading, respectively, to *syn* and *anti* pyramidalization of the double bond; consequently, such bonds will favor different modes of addition (*cis* and *trans*). The proposition was supported by examples of X-ray structures of strained olefins, STO-3G energy calculations for the twisted and pyramidalized ethylene geometries, and by analysis of the out-of-plane vibrational frequencies of ethylene. Mock concluded that small ground-state distortions may produce sizable effects in the transition states.

Subsequently, a number of authors have considered non-planar distortions of olefins and carbonyl groups as a source of diastereofacial selectivity, often emphasizing non-equivalent FMO extension (polarization of the frontier orbitals with respect to the σ plane) or asymmetry of the electrostatic potential, rather than the actual geometry distortion [198, 199]. Interest in the latter, however, was revived by the end of the 70's, stimulated by the early papers on structure correlation as well as several other contributions at that time. One of them was the seminal study by Eschenmoser and Dunitz and their coworkers in 1976, who examined conformations and non-planarity of enamine fragments in a number of crystal structures in search for the origin of face selection in condensations of chiral enamines [200]. Another crucial development appears to be Bartlett and Watson's discovery that large pyramidalizations of the olefinic double bonds in sesquibornene derivatives are associated with unusual chemistry and stereoselectivity [201].

There seem to be two rationales for linking ground-state distortions and face selection. One is due to Morokuma, who proposed that the activation energy for a chemical reaction can be expressed as the sum of an intramolecular deformation energy and an intermolecular interaction energy [202]. For instance, *ab initio* calculations suggested that the deformation energy for ethylene in the *syn* addition of HCl accounts for 40% of the total activation energy. Examination of the non-planar distortions of the ground-state geometry might indicate the direction in which the required deformation is energetically least costly. This argument gained support from results of calculations by Spanget-Larsen and Gleiter [203], Vogel et al. [204], Morokuma et al. [205], and Rastelli et al. [206]. In the empirical domain, systematic investigations of face selection in 3,4-disubstituted cyclobutenes by Gandolfi and Rastelli et al. [207], and independently by Martin et al. [208], led to the conclusion that cycloadditions of such derivatives are governed by the direction of the ground-state bend of the olefinic hydrogens (Scheme 6.33) [209]. Martin et al. pointed out that this bend depends on the angle of the outward splaying of the two substituents [208].

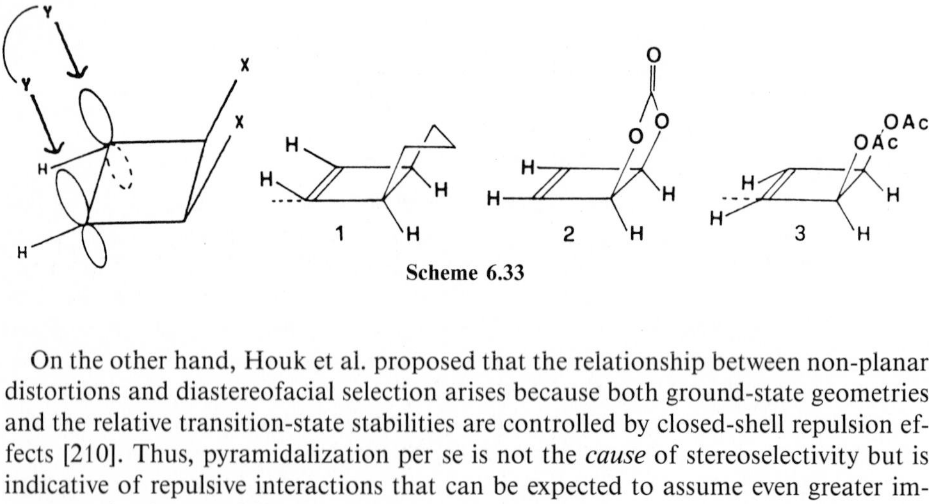

Scheme 6.33

On the other hand, Houk et al. proposed that the relationship between non-planar distortions and diastereofacial selection arises because both ground-state geometries and the relative transition-state stabilities are controlled by closed-shell repulsion effects [210]. Thus, pyramidalization per se is not the *cause* of stereoselectivity but is indicative of repulsive interactions that can be expected to assume even greater importance in the transition state [211].

This concept was adopted by Seebach et al. to rationalize stereoselection in cuprate additions and catalytic hydrogenations of 2,6-disubstituted 1,3-dioxin-4-ones and in alkylations of 1-benzoyl- and 1-(methoxycarbonyl)-2-*tert*-butyl-3,5-dimethyl-4-imidazolidinone enolates [212]. X-ray structures of several dioxinones and silyl enol ethers of imidazolidinones revealed that the reaction centers are slightly pyramidalized in the direction of the observed reagent attack (Figure 6.49). The authors proposed as a general rule that the steric course of attack on a trigonal center can be predicted from the pyramidalization direction. Since pyramidalization of a trigonal center next to a tetrahedral center is caused by the tendency to minimize torsional strain, approach of a reagent from the same direction into which pyramidalization has occurred will minimize strain even more. In contrast, approach from the opposite

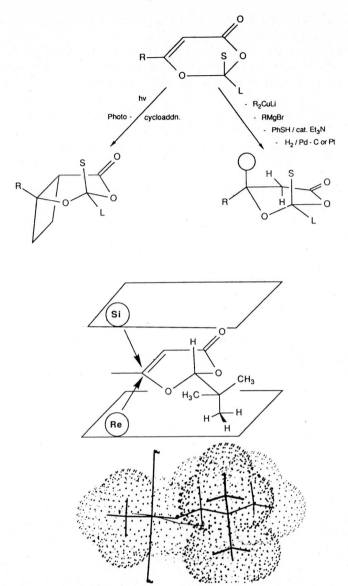

Fig. 6.49. Stereoselectivity of photocycloaddition, nucleophilic addition to and hydrogenation of dioxinones

site will increase torsional strain (Figure 6.50). The energy difference between the two transition states will be larger than the one between the oppositely pyramidalized ground states. Bürgi and Dubler-Steudle pointed out that Seebach's interpretation of face selectivity in reactions of dioxinones can be readily cast in terms of the cubic potential model, by treating pyramidalization towards a reagent as a negative pertur-

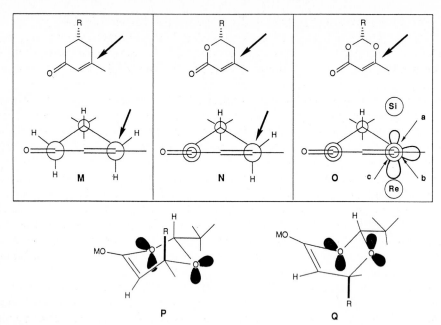

Fig. 6.50. Minimization of torsional strain in addition reactions to trigonal centers (for **O** the product is **P** not **Q**)

bation ($aq < 0$ in Equation (6.3), leading to a decrease in activation energy), and the one away from a reagent as a positive (leading to an increase in activation energy) perturbation [54].

The search for experimental evidence of a link between ground-state pyramidalization and face selectivity was only recently extended to ketones [72, 213, 214]. Preliminary results are contradictory [213, 214], and more work seems needed to reach consensus in this area.

6.6 Concluding Remarks

We have attempted here to recapitulate development of the notion that molecular geometry of carbonyl derivatives can be analyzed in terms of partial progress of their addition and elimination reactions. This analysis has undergone a remarkable transmutation from a qualitative concept born in natural product chemistry to a formalized model of empirical parameterization of the potential energy surfaces for such reactions. As we have also amply shown, the ideas issuing from the related structural studies were widely incorporated into the groundwork of thinking about reactivity and asymmetric synthesis in organic chemistry.

The major issues raised in these studies were (I) the question of local symmetries of the species along the reaction path for a nucleophilic addition to a trigonal center and (II) the correlations between the fractional bond order, the pyramidalization of the trigonal center, and the bond distance during the breaking and forming of bonds.

In regard to the first question, two pictures of the paths for $sp^2 \rightleftharpoons sp^3$ transformations emerged early on. Approach of nucleophiles to the trigonal centers carrying three equivalent ligands was shown to proceed along the normal to the ligand plane and to cause such centers, regardless of their nature, to deform along essentially identical paths, maintaining an approximate C_{3v} symmetry throughout. On the other hand, approach of nucleophiles to trigonal centers such as carbonyl C, i.e. sp^2 centers carrying non-equivalent ligands was postulated to occur along a path significantly deviating from the normal to the carbonyl plane. In particular, in the case of ketones, this deviation ("rearside" attack, non-perpendicular attack) was assumed to result from the tendency of the bond angle between the nucleophile and the carbonyl O to remain constant during the attack, close in value to the standard tetrahedral angle even at fairly large distances.

The evidence, brought originally in support of this distinction, seems now less compelling in the light of new crystallographic and computational data. Our reassessment of the data on the geometry of intramolecular nucleophile-electrophile contacts led to the conclusion that the large obtuse angle of the nucleophile approach to a keto group ($\alpha = 112°$), found in the early studies to be constant over a wide range of N...C=O distances, is unique to the transannular interaction in the 5-azacyclooctanone ring. No other fragment reported to contain intramolecular N...C=O or O...C=O interactions shows such a high α angle: the available mean α-values range from 93 to 102°. Furthermore, we have found that the intermolecular contacts of neutral N and O nucleophiles with carbonyls appear to occur along the normal to the plane of the trigonal C atom, that is, the mean α is invariably close to 90°, at least at distances of $2.8-3.2$ Å.

Similarly, reexamination of the patterns of acetal distortions suggests that, contrary to initial conclusions, the $O-C-O$ angle does not remain constant along the reaction coordinate for the expulsion of one of the O ligands (spontaneous breakdown of acetals). This has been noted in the analysis of structural variance in the Kirby-Jones set of acetals and now been confirmed by our analysis of distortions in glycosides. In the fragments where either geometry or electronic bias make one of the two O ligands a preferred leaving group, i.e. foster expulsion of one RO^- rather than the other, the $O-C-O$ angle correlates quite well with the difference in the $C-O$ bond distances $\Delta d = d_1 - d_2$.

Computational results generally support the notion of nonperpendicular approach in additions of anionic nucleophiles to aldehydes and ketones. This finding, however, has to be qualified in two ways. First, the extent of deviation from the perpendicular approach direction clearly depends on the flexibility of the basis set: it is very large with the minimal and small split-valence basis sets, known to exaggerate anisotropy of the charge distribution and thereby contributions of the coulombic interactions. Second, available information concerning the effects of Lewis acid catalysis, solvation and dielectric medium on the transition structures, suggests that the gas-phase results offer a poor approximation of the transition-state geometries for additions that occur in solution. Consequently, the *ab initio* data we have now can hardly be considered the last word on the matter.

In this situation, we are inclined to take the iconoclastic point of view that the perpendicular approach model is probably an adequate starting point for the discussion of the transition-state geometries for nucleophilic addition to trigonal centers. The questions whether, when, and to what extent, deviations from this approach geometry occur, remain to be answered.

In regard to the second problem, correlation of the fractional bond order and pyramidalization, the newer crystallographic and computational evidence consistently supports the original model, albeit a modification of its quantitative aspects is clearly necessary. Similarly, the simple relations invoking bond-order conservation during bond breaking and forming proved useful also in the more recent studies.

As a final note, we comment on the development of structure correlation analysis in the context of the development of MO calculations of potential energy hypersurfaces. We have seen how the early structure correlation studies inspired and complemented the first *ab initio* investigations of reaction paths. Over the following two decades, the growth in the capacity of both machines and computational techniques transformed the latter field into an autonomous and fast expanding area of MO theory. At the same time, the structure correlation approach to the modeling of potential energy surfaces was shown to offer a unique empirical alternative to MO calculations. This aptly illustrates, in our view, the status of structure correlation analysis as an indispensable source of inspiration and a tool of verification in the study of bonding and reactivity in organic chemistry.

Acknowledgement

The author expresses his gratitude to Prof. K. B. Wiberg of Yale University for granting access to computing facilities.

References

[1a] Smiles, J., *Pharm. J.* **1867**, *8*, 595
[1b] Hesse, O., *Ann. Chem. Suppl.* **1872**, *8*, 261
[2] Onda, M., Takahashi, H., in: *The Alkaloids*, A. Brossi (ed.) Academic Press, San Diego, **1988**, Vol. 34, p. 181
[3] Perkin Jr., W. H., *J. Chem. Soc.* **1916**, *109*, 815
[4a] Voss, A., Gadamer, J., *Arch. Pharm.* **1910**, *248*, 43
[4b] Pyman, F. L., *J. Chem. Soc.* **1913**, *103*, 817

[5]   Decker, H., Klauser, O., *Chem. Ber.* **1904**, *37*, 520
[6]   Gadamer, J., *Arch. Pharm.* **1920**, *258*, 148
[7]   Kermack, W.O., Robinson, R., *J. Chem. Soc.* **1922**, *121*, 427
[8]   Anet, F.A.L., Bailey, A.S., Robinson, Sir Robert, *Chem. Ind.* **1953**, 944
[9a] Mottus, E.H., Schwarz, H., Marion, L., *Can. J. Chem.* **1953**, *31*, 144
[9b] Anet, F.A.L., Marion, L., *Can. J. Chem.* **1954**, *32*, 453
[10] Huisgen, R., Wieland, H., Eder, H., *Ann. Chem.* **1949**, *561*, 193
[11] Bailey, A.S., Robinson, R., *Nature* **1948**, *161*, 433
[12] Leonard, N.J., *Rec. Chem. Prog.* **1956**, *17*, 243
[13] Tulinsky, A., van der Hende, J.H., *J. Am. Chem. Soc.* **1967**, *89*, 2905
[14] Wunderlich, J.A., *Acta Cryst.* **1967**, *B23*, 846
[15] Hall, S.R., Ahmed, F.R., *Acta Cryst* **1968**, *B24*, 337
[16] Hall, S.R., Ahmed, F.R., *ibid.*, 346
[17a] Birnbaum, K.B., Klasek, A., Sedmera, P., Snatzke, G., Johnson, L.F., Santavy, F., *Tetrahedron Lett.* **1971**, 3421
[17b] Birnbaum, K.B., *Acta Cryst.* **1972**, *B28*, 2825
[18] Mooney Slater, R.C.L., *Acta Cryst.* **1959**, *12*, 187
[19] Slater, J.C., *Acta Cryst.* **1959**, *12*, 197
[20] Hirschfelder, J., Diamond H., Eyring, H., *J. Chem. Phys.* **1937**, *5*, 695
[21] Bent, H., *Chem. Rev.* **1968**, *68*, 587
[22] Bürgi, H.B., Dunitz, J.D., Shefter, E., *Cryst. Struct. Comm.* **1973**, *3*, 667
[23] Bürgi, H.B., Dunitz, J.D., Shefter, E., *J. Am. chem. Soc.* **1973**, *95*, 5065
[24] Pauling, L., *J. Am. Chem. Soc.* **1947**, *695*, 542
[25a] Johnston, H.S., *Adv. Chem. Phys.* **1960**, *3*, 131
[25b] Johnston, H.S., Parr, C., *J. Am. Chem. Soc.* **1963**, *85*, 2544
[26a] Bürgi, H.B., *Inorg. Chem.* **1973**, *12*, 2321
[26b] Bürgi, H.B., *Angew. Chem.* **1975**, *87*, 461; *Int. Ed. Engl.* **1975**, *14*, 460
[27a] Fujimoto, H., Fukui, K., *Tetrahedron Lett.* **1966**, 5551
[27b] Inagaki, S., Fukui, K., *Chem. Lett.* **1974**, 509
[28a] Mock, W., *Tetrahedron Lett.* **1972**, 483
[28b] Mock, W., *Bioorg. Chem.* **1975**, *4*, 270
[29] Dunitz, J.D., *X-Ray Analysis and the Structure of Organic Molecules,* Cornell, London, **1979**
[30] Baldwin, J.E., *J. Chem. Soc. Chem. Comm.* **1976**, 738
[31] Liotta, C.S., Burgess, E.M., Eberhardt, W.H., *J. Am. Chem. Soc.* **1984**, *106*, 4849
[32] Nguyen Trong Anh, Eisenstein, O., *Nouv. J. Chim.* **1977**, *1*, 61
[33] Houk, K.N., Paddon-Row, M.N., Rondan, N.G., Wu, Y.-D., Brown, F.K., Spellmeyer, D.C., Metz, J.T., Li, Y., Loncharich, R.J., *Science* **1986**, *231*, 1108 and references therein
[34a] Heathcock, C.H., Flippin, L.A., *J. Am. Chem. Soc.* **1983**, *105*, 1667
[34b] Lodge, E.P., Heathcock, C.H., *J. Am. Chem. Soc.* **1987**, *109*, 2819
[34c] Mori, I., Bartlett, P.A., Heathcock, C.H., *J. Am. Chem. Soc.* **1987**, *109*, 7199
[34d] Mori, I., Ishihara, K., Flippin, L.A., Nozaki, K., Yamamoto, H., Bartlett, P.A., Heathcock, C.H., *J. Org. Chem.* **1990**, *55*, 6107
[34e] Heathcock, C.H., *Aldrichim. Acta* **1990**, *23*, 99
[35a] Seebach, D., Golinski, J., *Helv. Chim. Acta.* **1981**, *64*, 1413
[35b] Brook, M.A., Seebach, D., *Can. J. Chem.* **1987**, *65*, 836
[36] Nguyen Trong Anh, Bui Trong Thanh, *Nouv. J. Chim.* **1986**, *10*, 681
[37] Müller, D., Pitsch, S., Kittaka, A., Wagner, E., Wintner, C.E., Eschenmoser, A., *Helv. Chim. Acta* **1990**, *73*, 1410
[38a] Evans, D.A., Nelson, J.V., Taber, T.R., *Top. Stereochem.* **1982**, *13*, 1 and references therein
[38b] Oare, D.A., Heathcock, C.H., *Top. Stereochem.* **1991**, *20*, 87 and references therein
[39] Laudon, G.M., *Organic Chemistry*, 2nd Edn., Benjamin Cummings, Menlo Park, CA, **1988**, p. 775
[40a] Hodgkin, D.C., Maslen, E.N., *Biochem. J.* **1961**, *79*, 393
[40b] Abrahamsson, S., Hodgkin, D.C., Maslen, E.N., *Biochem. J.* **1963**, *86*, 514
[41] Prelog, V., personal communication

[42a] Bürgi, H.B., Lehn, J.M., Wipff, G., *J. Am. Chem. Soc.* **1974**, *96*, 1956
[42b] Bürgi, H.B., Dunitz, J.D., Lehn, J.M., Wipff, G., *Tetrahedron* **1974**, *30*, 1563
[43]   Wintner, C.E., *J. Chem. Educ.* **1987**, *64*, 587
[44]   Bader, R.F.W., MacDougall, P.J., Lau, C.D.H., *J. Am. Chem. Soc.,* **1984**, *106*, 1594
[45]   Lee, Ch., Yang, W., Parr, R.G., *J. Mol. Struct. (Theochem.)* **1988**, *163*, 305
[46a] Storm, D.R., Koshland, D.E., Jr., *Proc. Natl. Acad. Sci. USA* **1970**, *66*, 445
[46b] Storm, D.R., Koshland, D.E., Jr., *J. Am. Chem. Soc.* **1972**, *94*, 5805, 5815
[46c] Dafforn, A., Koshland, D.E., Jr., *Biochem. Biophys. Res. Commun.* **1973**, *52*, 779
[47]   Rosenfield Jr., R.E., Dunitz, J.D., *Helv. Chim. Acta* **1978**, *61*, 2176
[48a] Wijnberg, J.B.P.A., Speckamp, W.N., Schoemaker, H.E., *Tetrahedron Lett.* **1974**, 4073
[48b] Bailey, D.M., Johnson, R.E., *J. Org. Chem.* **1970**, *35*, 3574
[49]   Bürgi, H.B., Dunitz, J.D., *Acc. Chem. Res.* **1983**, *16*, 153
[50]   Ferretti, V., Dubler-Steudle, K.C., Bürgi, H.B., in: *Accurate Molecular Structures,* Domenicano, A., Hargittai, I. (eds.) Oxford, London, **1992**, pp. 412–436
[51a] Deslongchamps, P., *Tetrahedron* **1975**, *31*, 2463
[51b] Deslongchamps, P., *Pure Appl. Chem.* **1975**, *43*, 351
[51c] Deslongchamps, P., *Stereoelectronic Effects in Organic Chemistry*, Pergamon, New York, **1983**
[51d] Deslongchamps, P., Pothier, N., *Can. J. Chem.* **1990**, *68*, 597
[52a] Briggs, A.J., Glenn, R., Jones, P.G., Kirby, A.J., Ramaswamy, P., *J. Am. Chem. Soc.* **1984**, *106*, 6200
[52b] Jones, P.G., Kirby, A.J., *J. Am. Chem. Soc.* **1984**, *106*, 6207
[52c] Jones, P.G., Kirby, A.J., *J. Chem. Soc. Chem. Comm.* **1986**, 444
[53]   Bürgi, H.B., Dunitz, J.D., *J. Am. Chem. Soc.* **1987**, *109*, 2924
[54]   Bürgi, H.B., Dubler-Steudle, K.C., *J. Am. Chem. Soc.* **1988**, *110*, 7291
[55]   Allen, F.H., Kennard, O., Taylor R., *Acc. Chem. Res.* **1983**, *16*, 146
[56a] McIver, J.W., Komornicki, A., *J. Am. Chem. Soc.* **1972**, *94*, 2625
[56b] Dewar, M.J.S., *Chem. Brit.* **1975**, *11*, 97
[56c] Flanigan, M.C., Komornicki, A., McIver, J.W., *Semiempirical Methods of Electronic Structure Theory*, in: *Modern Theoretical Chemistry.* Vol. 8, Segal, G.A. (ed.) Plenum Press, New York, **1977**
[57]   Pulay, P., *Applications of Electronic Structure Theory*, in: *Modern Theoretical Chemistry,* Vol. 4. Schaefer, H.F. (ed.) Plenum, New York, **1977**
[58]   Binkley, J.S., Pople, J.A., Hehre, W.J., *J. Am. Chem. Soc.* **1980**, *102*, 939
[59]   Birnbaum, G.I., *J. Am. Chem. Soc.* **1974**, *96*, 6165
[60a] Kaftory, M., Dunitz, J.D., *Acta Cryst.* **1975**, *B31*, 2912
[60b] Kaftory, M., Dunitz, J.D., *Acta Cryst.* **1975**, *B31*, 2914
[60c] Kaftory, M., Dunitz, J.D., *Acta Cryst.* **1976**, *B32*, 1
[60d] Bye, E., Dunitz, J.D., *Acta Cryst.* **1978**, *B34*, 3245
[61]   Schweizer, W.B., Procter, G., Kaftory, M., Dunitz, J.D., *Helv. Chim. Acta* **1978**, *61*, 2783
[62]   Spanka, G., Boese, R. Rademacher, P., *J. Org. Chem.* **1987**, *52*, 3362
[63]   Bürgi, H.B., Dunitz, J.D., Shefter, E., *Acta Cryst.* **1974**, *B30*, 1517
[64]   Chadwick, D.J., Dunitz, J.D., *J. Chem. Soc. Perkin II* **1979**, 276
[65]   Brennan, R.G., Prive, G.G., Blonski, W.J.P., Hruska, F.E., Sundaralingam, M., *J. Am. Chem. Soc.* **1983**, *105*, 7737
[66]   Chadwick, D.J., Whittleton, S.N., *J. Chem. Res.* **1984**, *8*, 398
[67]   Cossu, M., Bachmann, C., Yao N'Guessan, T., Viani, R., Lapasset, J., Aycard, J.P., Bodot, H., *J. Org. Chem.* **1987**, *52*, 5313
[68]   Irngartinger, H., Reimann W., Garner, P., Dowd, P., *J. Org. Chem.* **1988**, *53*, 3046
[69]   Murray-Rust, J., Murray-Rust, P., Parker, W.C., Tranter, R.L., Watt, C.I.F., *J. Chem. Soc. Perkin II* **1979**, 1496
[70a] Cernik, R., Craze, G.A., Owen, S.M., Watt, I., *J. Chem. Soc. Perkin II* **1982**, 361
[70b] Cernik, R.V., Craze, G.A., Owen, S.M., Watt, I., Whittleton, S.N., *J. Chem. Soc. Perkin II* **1984**, 685
[71]   Bye, E., *Acta Chem. Scand.* **1974**, *B28*, 5

[72a]   Cieplak, A. S., Bürgi, H. B., in preparation. The data were retrieved using the CSD Version 4.5 July 1991. Stereoprojections of the distributions of θ_c vs. ω_1 and ω_2 were generated using the WÜRFEL program developed by Drs. W. Hummel and J. Hauser of the Laboratorium für chemische und mineralogische Kristallographie, University of Berne

[72b]   The torsion angles were defined as $\omega_1 = \omega(C4-C3-C2-C1)$ and $\omega_2 = \omega(C6-C5-C2-C1)$ for the fragment $(C8\,C7\,C4)C3-C2(=O1)-C5-C6-C9-N10$. The data were retrieved using the CSD, Version 4.6 January 1992

[73]   Cieplak, A. S., *J. Am. Chem. Soc.* **1985**, *107*, 271

[74a]   Bye, E., *Acta Chem. Scand.* **1976**, *30B*, 323

[74b]   Barlein, F., Mostad, A., *Acta Chem. Scand.* **1981**, *35B*, 613

[74c]   Portoghese, P. S., Poupaert, J. H., Larson, D. L., Grontas, W. C., Meitzner, G. D., Swenson, D. C., Smith, G. D., Duax, W. L., *J. Med. Chem.* **1982**, *25*, 684

[74d]   Gerardy, B. M., Poupaert, J. H., Vandervorst, D., Dumont, P., Declerq, J. P., van Meersche, M., Portoghese, P. S., *J. Labelled Comp. Radiopharm.* **1985**, *22*, 5

[74e]   Singh, P., Moreland, C. G., *Acta Cryst.* **1989**, *45C*, 1469

[75a]   Childs, R. F., Kostyk, M. D., Lock, C. J. L., Mahendran, M., *J. Am. Chem. Soc.* **1990**, *112*, 8912

[75b]   Childs, R. F., Mahendran, M., Zweep, S. D., Shaw, G. S., Chadda, S. K., Burke, N. A. D., George, B. E., Faggiani, R., Lock, C. J. L., *Pure Appl. Chem.* **1986**, *58*, 111

[76]   Cieplak, A. S., unpublished
       All the structures containing a given combination of the two functional groups (C=O and a N or O nucleophile) were retrieved using the CSD, Version 4.6 January 1992. From these large samples, the fragments containing intermolecular electrophile-nucleophile contacts ($d_1 < 3.2$ Å) were retrieved using the GSTAT91 facility INTER FROM *i* TO *j* EXT. The carbonyl ligands were not specified. For *tert*-aliphatic amines, $N = 6$ fragments were obtained from the sample of 773 structures; for *tert*-anilines $N = 5/255$; for trigonal *tert*-amines $N = 28/2064$; for oximes and hydrazones $N = 161/974$; for triazoles $N = 15/96$; for cyanides $N = 90/769$; for azides $N = 15/44$; and for nitro compounds $N = 161/1004$. The α distributions appear to be normal about their means

[77]   Dennen, R. S., *J. Mol. Spectr.* **1969**, *291*, 163

[78]   Jeffrey, G. A., Pople, J. A., Binkley, J. S., Vishveshara, S., *J. Am. Chem. Soc.* **1978**, *100*, 373

[79]   Wolfe, S., Whangbo, M.-H., Mitchell, D. J., *Carbohydr. Res.* **1979**, *69*, 1

[80]   Fuchs, B., Schleifer, L., Tartakovsky, E., *Nouv. J. Chim.* **1984**, *8*, 275

[81]   Cosse-Barbi, A., Dubois, J. E., *J. Am. Chem. Soc.* **1987**, *109*, 1503

[82]   Wiberg, K. B., Murcko, M. A., *J. Am. Chem. Soc.* **1989**, *111*, 4821

[83]   Irwin, J. J., Ph. D. Thesis No. 9397, ETH Zürich, **1991**

[84a]   Juaristi, E., Cuevas, G., *Tetrahedron* **1992**, *48*, 5019

[84b]   Beckhaus, H.-D., Dogan, B., Verevkin, S., Hadrick, J., Rüchardt, C., *Angew. Chem.* **1990**, *102*, 313

[84c]   Montagnani, R., Tomasi, J., *Int. J. Quant. Chem.* **1991**, *39*, 851

[84d]   Koritsanszky, T., Strumpel, M. K., Buschman, J., Luger, P., Hansen, N. K., Pichon-Pesme, V., *J. Am. Chem. Soc.* **1991**, *113*, 9148

[84e]   Bertolasi, V., Ferretti, V., Gilli, G., Marchetti, P., D'Angeli, F., *J. Chem. Soc. Perkin II* **1990**, 2135

[85a]   Altona, C., Knobler, C., Romers, C., *Acta Cryst.* **1963**, *16*, 1217

[85b]   Altona, C., Romers, C., *Acta Cryst.* **1963**, *16*, 1225

[85c]   Altona, C., Ph. D. Thesis, University of Leiden, **1964**

[86a]   Caserio, M. C., Souma, Y., Kim, J. K., *J. Am. Chem. Soc.* **1981**, *103*, 6712

[86b]   Caserio, M. C., Shih, P., Fisher, C. L., *J. Org. Chem.* **1991**, *56*, 5517

[87a]   Bennett, A. J., Sinnott, M., *J. Am. Chem. Soc.* **1986**, *108*, 7287

[87b]   Sinnott, M., *Adv. Phys. Org. Chem.* **1988**, *24*, 113

[88]   Perrin, C. L., Nunez, O., *J. Am. Chem. Soc.* **1987**, *109*, 522

[89a]   Huber, R., Knierzinger, A., Krawczyk, E., Obrecht, J. P., Vasella, A., *Pure Appl. Chem.* **1985**, 255

[89b]   Bernet, B., Krawczyk, E., Vasella, A., *Helv. Chim. Acta* **1985**, *68*, 2299

[89c]   Huber, R., Vasella, A., *Tetrahedron* **1990**, *46*, 33

[90a] Ratcliffe, A. J., Mootoo, D. R., Andrews, C. W., Fraser-Reid, B., *J. Am. Chem. Soc.* **1989**, *111*, 7661

[90b] McPhail, D. R., Lee, J. R., Fraser-Reid, B., *J. Am. Chem. Soc.* **1992**, *114*, 1905

[90c] Andrews, C. W., Fraser-Reid, B., Bowen, J. P., *J. Am. Chem. Soc.* **1991**, *113*, 8293

[91]   Geokjian, P. G., Wu, T.-Ch., Kishi, Y., *J. Org. Chem.* **1991**, *56*, 6412

[92]   Cieplak, A. S., unpublished
       The data were retrieved from the CSD, Version 4.6 January 1992

[93]   Zobetz, E., *Z. Kristallogr.* **1982**, *160*, 81

[94a] Zobetz, E., *Z. Kristallogr.* **1988**, *185*, 518

[94b] Zobetz, E., *Z. Kristallogr.* **1990**, *191*, 45

[95a] Murray-Rust, P., Bürgi, H. B., Dunitz, J. D., *J. Am. Chem. Soc.* **1975**, *97*, 595

[95b] Murray-Rust, P., Bürgi, H. B., Dunitz, J. D., *Acta Cryst.* **1978**, *B34*, 1793

[96]   Baur, W. H., *Trans. Am. Cryst. Assoc.* **1970**, *6*, 129

[97]   Brown, I. D., Brown, M. C., Hawthorne, F. C., BIDICS-1981, *Bond Index to the Determinations of Inorganic Crystal Structures*. Institute for Materials Research. McMaster Univ., Hamilton, Ontario, Canada, **1982**

[98]   Procter, G., Britton, D., Dunitz, J. D., *Helv. Chim. Acta* **1981**, *64*, 471

[99]   Wallis, J. D., Dunitz, J. D., *J. Chem. Soc. Chem. Comm.* **1984**, 672

[100] Baxter, P. N. W., Connor, J. A., Povey, D. C., Wallis, J. D., *J. Chem. Soc. Chem. Comm.* **1991**, 1135

[101] Earl, J. C., Leake, E. W., Le Fever, R. J. W., *Nature* **1947**, *160*, 366

[102a] Earl, J. C., Mackney, A. W., *J. Chem. Soc.* **1935**, 899

[102b] Eade, R. A., Earl, J. C., *J. Chem. Soc.* **1946**, 591

[103] Baker, W., Ollis, W. D., Poole, V. D., *J. Chem. Soc.* **1949**, 307

[104] Thiessen, W. E., Hope, H., *J. Am. Chem. Soc.* **1967**, *89*, 5977

[105] Gieren, A., Lamm, V., *Acta Cryst.* **1978**, *B34*, 3248

[106] Cieplak, A. S. (unpublished) A search of the CSD, Version 4.6 January 1992, yielded 23 structures containing the fragment depicted in Scheme 6.10. The total sample of sydnone fragments (26) included 20 fragments of *RFAC $<$ 0.05 and 16 fragments of *AS = 1. The latter reveal good correlations of β_1 with d_1 ($r = 0.826$), with β_2 ($r = -0.705$) and α ($r = -0.762$); for definition of symbols see Scheme 6.14

[107] Van Meerssche, M., Germain, G., Declercq, J. P., Colens, A., *Acta Cryst.* **1979**, *B35*, 907

[108] Seebach, D., Amstutz, R., Laube, T., Schweizer, W. B., Dunitz, J. D., *J. Am. Chem. Soc.* **1985**, *107*, 5402

[109] Nørskov-Lauritsen, L., Bürgi, H. B., Hofmann, P., Schmidt, H. R., *Helv. Chim. Acta* **1985**, *68*, 76

[110] Borthwick, P. W., *Acta Cryst.* **1980**, *B36*, 628

[111] Davidson, E. R., Feller, D., *Chem. Rev.* **1986**, *86*, 681

[112a] Dykstra, C. E., Arduengo, A. J., Fukunaga, T., *J. Am. Chem. Soc.* **1978**, *100*, 6008

[112b] Eisenstein, O., Procter, G., Dunitz, J. D., *Helv. Chim. Acta* **1978**, *61*, 2538

[112c] Houk, K. N., Rondan, N. G., Schleyer, P. v. R., Kaufmann, E., Clark, T., *J. Am. Chem. Soc.* **1985**, *107*, 2821

[113a] Nguyen, M. T., Hegarty, A. F., *J. Am. Chem. Soc.* **1984**, *106*, 1552

[113b] Leung-Toung, R., Tidwell, T. T., *J. Am. Chem. Soc.* **1990**, *112*, 1042

[114a] Nguyen, M. T., Hegarty, A. F., *J. Mol. Struct. (Theochem.)* **1987**, *150*, 319

[114b] Kaufmann, E., Sieber, S., Schleyer, P. v. R., *J. Am. Chem. Soc.* **1989**, *111*, 4005

[115] Nguyen, M. T., Ha T.-K., Hegarty, A. F., *J. Phys. Org. Chem.* **1990**, *3*, 697

[116] Bayly, C. I., Grein, F., *Can. J. Chem.* **1988**, *66*, 149

[117] Stone, A. J., Erskine, R. W., *J. Am. Chem. Soc.* **1980**, *102*, 7186

[118a] Williams, I. H., Maggiora, G. M., Schowen, R. L., *J. Am. Chem. Soc.* **1980**, *102*, 7832

[118b] Williams, I. H., Spangler, D., Femec, D. A., Maggiora, G. M., Schowen, R. L., *J. Am. Chem. Soc.* **1980**, *102*, 6620

[119] Madura, J. D., Jorgensen, W. L., *J. Am. Chem. Soc.* **1986**, *108*, 2517

[120] Bayly, C. I., Grein, F., *Can. J. Chem.* **1988**, *67*, 176

[121] Bernardi, F., Olivucci, M., Poggi, G., Robb, M. A., Tonachini, G., *Chem. Phys. Lett.* **1988**, *144*, 141

[122]   Houk, K. N., personal communication to W. Jorgensen
[123]   Howard, A. E., Kollman, P. A., *J. Am. Chem. Soc.* **1988**, *110*, 7196
[124]   Volatron, F., Eisenstein, O., *J. Am. Chem. Soc.* **1987**, *109*, 1
[125]   Li, Y., Paddon-Row, M. N., Houk, K. N., *J. Org. Chem.* **1990**, *55*, 481
[126]   Yamabe, H., Kato, H., Yonezawa, T., *Bull. Chem. Soc. Jpn.* **1971**, *44*, 611
[127]   Minato, T., Fujimoto H., Fukui, K., *Bull. Chem. Soc. Jpn.* **1978**, *51*, 1621
[128]   Scheiner, S., Lipscomb, W. N., Kleier, D. A., *J. Am. Chem. Soc.* **1976**, *98*, 4770
[129]   Shokhen, M. A., Tikhimirov, D. A., Yeremeyev, A. V., *Tetrahedron* **1983**, *39*, 2975
[130]   Williams, I. H., Spangler, D., Maggiora, G. M., Schowen, R. L., *J. Am. Chem. Soc.* **1985**, *107*, 7717
[131]   Williams, I. H., *J. Am. Chem. Soc.* **1987**, *109*, 6299
[132]   Kozmutza, C., Evleth, E. M., Kapuy, E., *J. Mol. Struct. (Theochem.)* **1991**, *233*, 139
[133]   Yamataka, H., Nagase, S., Ando, T., Hanafusa, T., *J. Am. Chem. Soc.* **1986**, *108*, 601
[134]   Sevin, A., Maddaluno, J., Agami, C., *J. Org. Chem.* **1987**, *52*, 5611
[135]   Chang, J.-W. A., Taira, K., Urano, S., Gorenstein, D. G., *Tetrahedron* **1987**, *43*, 3863
[136]   The conformational dependence was attributed to the kinetic α-effect, interpreted in terms of the n,σ* interaction hypothesis. This explanation was originally proposed by Baddeley, G., *Tetrahedron Lett.* **1973**, 1645. See also: Cieplak, A. S., *J. Am. Chem. Soc.* **1981**, *103*, 4540
[137]   Dewar, M. J. S., McKee, M. L., *J. Am. Chem. Soc.* **1978**, *100*, 7499
[138]   Eisenstein, O., Schlegel, H. B., Kayser, M. M., *J. Org. Chem.* **1982**, *47*, 2886
[139]   Bonaccorsi, R., Palla, P., Tomasi, J., *J. Mol. Struct.* **1982**, *87*, 181
[140]   Nagase, S., Uchibori, Y., *Tetrahedron Lett.* **1982**, 2585
[141]   Kaufmann, E., Schleyer, P. v. R., Houk, K. N., Wu, Y.-D., *J. Am. Chem. Soc.* **1985**, *107*, 5560
[142]   Bachrach, S. M., Streitwieser Jr., A., *J. Am. Chem. Soc.* **1986**, *108*, 3946
[143 a] Wong, S. S., Paddon-Row, M. N., *J. Chem. Soc. Chem. Comm.* **1990**, 456
[143 b] Wong, S. S., Paddon-Row, M. N., *Aust. J. Chem.* **1991**, *44*, 765
[144]   Bestmann, H. J., *Pure Appl. Chem.* **1980**, *52*, 771
[145]   Wu, Y.-D., Houk, K. N., *J. Am. Chem. Soc.* **1987**, *109*, 908
[146]   Wong, S. S., Paddon-Row, M. N., *J. Chem. Soc. Chem. Comm.* **1991**, 327
[147]   Frenking, G., Kohler, K. F., Reetz, M. T., *Tetrahedron* **1991**, *47*, 9005
[148]   Dorigo, A., Morokuma, K., *J. Am. Chem. Soc.* **1989**, *111*, 4635
[149]   Sheldon, J. C., *Aust. J. Chem.* **1981**, *34*, 1189
[150]   Cieplak, A. S., Wiberg, K. B., in preparation
[151]   Wu, Y.-D., Tucker, J. A., Houk, K. N., *J. Am. Chem. Soc.* **1991**, *113*, 5018
[152]   Wu, Y.-D., Houk, K. N., Paddon-Row, M. N., *Angew. Chem., Int. Ed. Engl.* **1992**, *31*, 1019
[153]   Frenking, G., Kohler, K. F., Reetz, M. T., *Angew. Chem.* **1991**, *103*, 1167
[154]   Alagona, G., Scrocco, E., Tomasi, J., *J. Am. Chem. Soc.* **1975**, *976*, 6976
[155]   Aleksandrov, S. L., Antonov, V. K., *Mol. Biol.* **1984**, *18*, 1576
[156]   Weiner, S. J., Chandra Singh, U., Kollman, P. A., *J. Am. Chem. Soc.* **1985**, *107*, 2219
[157]   Maraver, J. J., Marcos, E. S., Bertran, J., *J. Mol. Struct. (Theochem.)* **1987**, *150*, 251
[158]   Yamabe, S., Minato, T., *J. Org. Chem.* **1983**, *48*, 2972
[159]   Blake, J. F., Jorgensen, W. L., *J. Am. Chem. Soc.* **1987**, *109*, 3856
[160]   Oie, T., Loew, G. H., Burt, S. K., Binkley, J. S., McElroy, R. D., *J. Am. Chem. Soc.* **1982**, *104*, 6169
[161]   Williams, I. H., Spangler, D., Femec, D. A., Maggiora, G. M., Schowen, R. L., *J. Am. Chem. Soc.* **1983**, *105*, 31
[162]   Yu, H.-A., Karplus, M., *J. Am. Chem. Soc.* **1990**, *112*, 5706
[163 a] Miertus, S., Scrocco, E., Tomasi, J., *Chem. Phys.* **1981**, *55*, 117
[163 b] Bonaccorsi, R., Cimiraglia, R., Tomasi, J., Miertus, S., *J. Mol. Struct. (Theochem.)* **1983**, *94*, 11
[164 a] Wong, M. W., Frisch, M., Wiberg, K. B., *J. Am. Chem. Soc.* **1991**, *113*, 4776
[164 b] Wong, M. W., Wiberg, K. B., Frisch, M. J., *J. Chem. Phys.* **1991**, *95*, 8991
[164 c] Wong, M. W., Wiberg, K. B., Frisch, M. J., *J. Am. Chem. Soc.* **1992**, *114*, 523, 1645
[165]   Cieplak, A. S., Wiberg, K. B., *J. Am. Chem. Soc.* **1992**, *114*, 9226

[166a] Weygand, F., Kinkel, K.G., Tietjen, D., *Chem. Ber.* **1950**, *83*, 394
[166b] Brewster, J.H., Fusco, A.M., *J. Org. Chem.* **1963**, *28*, 501
[166c] Vaughan, W.R., Goetschel, C.T., Goodrow, M.H., Warren, C.L., *J. Am. Chem. Soc.* **1963**, *85*, 2282
[166d] Bloomfield, J.J., Lee, S.L., *J. Org. Chem.* **1967**, *32*, 3919
[166e] Taub, D., Girotra, N.N., Hoffsommer, R.D., Kuo, C.H., Slates, H.L., Weber, S., Wendler, N.L., *Tetrahedron* **1968**, *24*, 22443
[166f] Cross, B.E., Stewart, J.C., *Tetrahedron Lett.* **1968**, 3589
[166g] Birckelbaw, M.E., Le Quesne, P.W., Wocholski, C.K., *J. Org. Chem.* **1970**, *35*, 558
[166h] Bailey, D.M., Johnson, R.E., *J. Org. Chem.* **1970**, *35*, 3574
[166i] Winterfeld, E., Nelke, J.M., *Chem. Ber.* **1970**, *103*, 1174
[166j] Burke, D.E., Le Quesne, P.W., *J. Org. Chem.* **1971**, *36*, 2397
[166k] Hubert, J.C., Steege, W., Speckamp, W.N., Huisman, H.O., *Synth. Comm.* **1971**, *1*, 103
[166l] Wijnberg, J.B.P.A., Speckamp, W.N., Schoemaker, H.E., *Tetrahedron Lett.* **1974**, 4073
[167]   Suess, R., *Helv. Chim. Acta* **1977**, *60*, 1650
[168]   Wijnberg, J.B.P.A., Schoemaker, H.E., Speckamp, W.N., *Tetrahedron* **1978**, *34*, 179
[169a] Kayser, M.M., Morand, P., *Can. J. Chem.* **1978**, *56*, 1524
[169b] Kayser, M.M., Morand, P., *Tetrahedron Lett.* **1979**, 695
[170a] Eid, Jr. C.N., Konopelski, J.P., *Tetrahedron* **1991**, *47*, 975
[170b] Schwartz, C.E., Curran, D.P., *J. Am. Chem. Soc.* **1990**, *112*, 9274
[170c] Duncia, J.V., Chiu, A.T., Carini, D.J., Gregory, G.B., Johnson, A.L., Price, W.A., Wells, G.J., Wong, P.C., Calabrese, J.C., Timmermans, P.B.M.W.M., *J. Med. Chem.* **1990**, *33*, 1312
[170d] Oppenlander, T., Schonholzer, P., *Helv. Chim. Acta* **1989**, *72*, 1792
[170e] Trost, B.M., Yang, B., Miller, M.L., *J. Am. Chem. Soc.* **1989**, *111*, 6482
[170f] Stanetty, P., Frohlich, H., Sauter, F., *Monatsh. Chem.* **1986**, *117*, 69
[171a] Morand, P., Salvator, J., Kayser, M.M., *J. Chem. Soc. Chem. Comm.* **1982**, 458
[171b] Kayser, M.M., Salvador, J., Morand, P., Krishnamurthy, H.G., *Can. J. Chem.* **1982**, *60*, 1199
[171c] Kayser, M.M., Salvador, J., Morand, P., *Can. J. Chem.* **1983**, *61*, 439
[172]   Kayser, M.M., Wipff, G., *Can. J. Chem.* **1982**, *60*, 1192
[173a] Das, J., Kubela, R., MacAlpine, G.A., Stojanac, Z., Valenta, Z., *Can. J. Chem.* **1979**, *57*, 3808
[173b] Liotta, D., Saindane, M., Barnum, C., *J. Org. Chem.* **1981**, *46*, 3369
[173c] Liotta, D., Saindane, M., Sunay, U., Jamison, W.C.L., Grosman, J., Phillips, P., *J. Org. Chem.* **1985**, *50*, 3241
[174a] Knight, D.W., Pattenden, G., *J. Chem. Soc. Perkin I* **1979**, 62
[174b] Kayser, M.M., Morand, P., *Can. J. Chem.* **1980**, *58*, 2484
[174c] Kayser, M.M., Breau, L., Eliev, S., Morand, P., Ip, H.S., *Can. J. Chem.* **1986**, *64*, 104
[175a] McAlees, A.J., McCrindle, R., Sneddon, D.W., *J. Chem. Soc. Perkin I* **1977**, 2030, 2037
[175b] Makhlouf, M.A., Rickborn, B., *J. Org. Chem.* **1981**, *46*, 4811
[175c] Krepelka, J., Holubek, J., *Coll. Czech. Chem. Comm.* **1981**, *46*, 2123
[175d] Krishnamurthy, S., Vreeland, W.B., *Heterocycles* **1982**, *18*, 265
[175e] Narasimhan, S., *Heterocycles* **1982**, *18*, 131
[175f] Braun, M., Veith, R., Moll, G., *Chem. Ber.* **1985**, *118*, 1058
[175g] Soucy, C., Favreau, D., Kayser, M.M., *J. Org. Chem.* **1987**, *52*, 129
[176]   Kayser, M.M., Morand, P., *J. Org. Chem.* **1979**, *44*, 1338
[177a] Cherest, M., Felkin, H., Prudent, N., *Tetrahedron Lett.* **1968**, 2199
[177b] Cherest, M., Felkin, H., *Tetrahedron Lett.* **1968**, 2205
[178]   Midland, M.M., Kwon, Y.C., *J. Am. Chem. Soc.* **1983**, *105*, 3725
[179]   Bassindale, A.R., Ellis, R.J., Lau, J.C-Y., Taylor, P.G., *J. Chem. Soc. Perkin II* **1986**, 593
[180a] DeShong, P., Li, W., Kennington, Jr. J.W., Ammon, H.L., *J. Org. Chem.* **1991**, *56*, 1364
[180b] Roush, W.R., *J. Org. Chem.* **1991**, *56*, 4151
[180c] Paquette, L.A., Moorhoff, C.M., Maynard, G.D., Hickey, E.R., Rogers, R.D., *J. Org. Chem.* **1991**, *56*, 2449
[180d] Roy, R., Rey, A.W., *Can. J. Chem.* **1991**, *69*, 62

[180e] Davies, S.G., Goodfellow, C.L., *J. Chem. Soc. Perkin I* **1990**, 393

[180f] Alcaide, B., Rodriguez-Lopez, J., Monge, A., Perez-Garcia, V., *Tetrahedron* **1990**, *46*, 6799

[180g] Alexakis, A., Sedrani, R., Normant, J.F., Mangeney, P., *Tetrahedron: Asymmetry* **1990**, *1*, 283

[180h] Jefford, C.W., Jaggi, D., Boukouvalas, J., *Tetrahedron Lett.* **1987**, *28*, 4037

[180i] Auvray, P., Knochel, P., Normant, J.F., *Tetrahedron Lett.* **1986**, *27*, 5091

[180j] Nagao, Y., Ikeda, T., Inoue, T., Yagi, M., Shiro, M., Fujita, E., *J. Org. Chem.* **1985**, *50*, 4072

[181] Tenud, L., Farooq S., Seibl, J., Eschenmoser, A., *Helv. Chim. Acta* **1970**, *53*, 2059

[182] Baldwin, J.E., *J. Chem. Soc. Chem. Comm.* **1976**, 734

[183a] Sato, M., Sakaki, J., Sugita, Y., Yasuda, S., Sakoda, H., Kaneko, C., *Tetrahedron* **1991**, *47*, 5689

[183b] Mulzer, J., Kirstein, H.M., Buschmann, J., Lehmann, C., Luger, P., *J. Am. Chem. Soc.* **1991**, *113*, 910

[183c] Snyder, J.R., Serianni, A.S., *Carboh. Res.* **1988**, *184*, 13

[183d] Tietze, L.F., Brumby, T., Pretor, M., Remberg, G., *J. Org. Chem.* **1988**, *53*, 813

[183e] Anwar, S., Davis, A.P., *Tetrahedron* **1988**, *44*, 3761

[183f] Lavallee, J.-F., Deslongchamps, P., *Tetrahedron Lett.* **1988**, *29*, 6033

[183g] Rubin, M.B., Inbar, S., *J. Org. Chem.* **1988**, *53*, 3355

[183h] Williams, R.M., Tomizawa, K., Armstrong, R.W., Dung, J.-S., *J. Am. Chem. Soc.* **1987**, *109*, 4031

[183i] Wroblewski, A., *Tetrahedron* **1986**, *42*, 3595

[183j] Akiba, K., Nakatani, M., Wada, M., Yamamoto, Y., *J. Org. Chem.* **1985**, *50*, 63

[183k] Baldwin, J.E., Beckwith, P.L.M., Wallis, J.D., Orrell, A.P.K., Prout, K., *J. Chem. Soc. Perkin II* **1984**, 53

[183l] Han, W.C., Takahashi, K., Cook, J.M., Weiss, U., Silverton, J.V., *J. Am. Chem. Soc.* **1982**, *104*, 318

[184] Bruice, T.C., Brown, A., Harris, D.O., *Proc. Natl. Acad. Sci. USA* **1971**, *68*, 658

[185a] Menger, F.M., *Tetrahedron* **1983**, *39*, 1013

[185b] Menger, F.M., Glass, L.E., *J. Am. Chem. Soc.* **1980**, *102*, 5404

[186] Dorigo, A.E., Houk, K.N., *J. Am. Chem. Soc.* **1987**, *109*, 3698

[187] Milstein, S., Cohen, L.A., *J. Am. Chem. Soc.* **1972**, *94*, 9158

[188] Sherrod, M.J., Menger, F.M., *Tetrahedron Lett.* **1990**, *31*, 459

[189] Rüchardt, C., Beckhaus, H.-D., *Angew. Chem. Int. Ed. Engl.* **1980**, *19*, 429

[190a] Allen, A.D., Kanagasabapathy, V.M., Tidwell, T.T., *J. Am. Chem. Soc.* **1983**, *105*, 596

[190b] Kaftory, M., Apeloig, Y., Rappoport, Z., *J. Chem. Soc. Perkin II* **1985**, 29

[190c] Kanagasabapathy, V.M., Sawyer, J.F., Tidwell, T.T., *J. Org. Chem.* **1985**, *50*, 503

[190d] Schneider, H.-J., Buchheit, U., Schmidt, G., *J. Chem. Res.* **1987**, *8*, 92

[191] Seeman, J.I., Viers, J.V., Schug, J.C., Stovall, M.D., *J. Am. Chem. Soc.* **1984**, *106*, 143

[192] Kirby, A.J., Martin, R.J., *J. Chem. Soc. Chem. Comm.* **1978**, 172

[193a] Herschbach, D.R., Laurie, V.W., *J. Chem. Phys.* **1961**, *35*, 458

[193b] Johnston, H.S., *J. Am. Chem. Soc.* **1964**, *86*, 1643

[194] Results of a covariance analysis on five internal coordinates; conclusion are practically the same if more variables are introduced

[195] Montgomery, L.K., Grendze, M.P., Huffman, J.C., *J. Am. Chem. Soc.* **1987**, *109*, 4749

[196] Badger, R.M., *J. Chem. Phys.* **1934**, *2*, 128

[197] Inagaki, S., Fujimoto, H., Fukui, K., *J. Am. Chem. Soc.* **1976**, *98*, 4054

[198a] Anh, N.T., Eisenstein, O., Lefour, J.-M., Dan, M.-T.H., *J. Am. Chem. Soc.* **1973**, *95*, 6146

[198b] Klein, J., *Tetrahedron Lett.* **1973**, 4307

[198c] Klein, J., *Tetrahedron* **1974**, *30*, 3349

[198d] Liotta, C.L., *Tetrahedron Lett.* **1975**, 519, 523

[198e] Ashby, E.C., Boone, J.R., *J. Org. Chem.* **1976**, *41*, 2890

[198f] Paquette, L.A., Hertel, L.W., Gleiter, R., Bohm, M., *J. Am. Chem. Soc.* **1978**, *100*, 565

[198g] Hoffmann, R.W., Hauel, N., *Tetrahedron Lett.* **1979**, 4959

[198h] Eisenstein, O., Klein, J., Lefour, J.-M., *Tetrahedron* **1979**, *35*, 225

[198i] Giddings, M.R., Hudec, J., *Can. J. Chem.* **1981**, *59*, 459

[198j] Burgess, E.M., Liotta, C.L., *J. Org. Chem.* **1981**, *46*, 1703

[198k] Ito, S., Kahehi, A., *Bull. Chem. Soc. Jpn.* **1982**, *55*, 1869

[198l] Gleiter, R., Paquette, L. A., *Acc. Chem. Res.* **1983**, *16*, 328

[198m] Washburn, W. N., Hillson, R. A., *J. Am. Chem. Soc.* **1984**, *106*, 4575

[198n] Pradhan, S. K., *Tetrahedron* **1986**, *42*, 6351

[198o] Nasipuri, D., Saha, A., *Ind. J. Chem.* **1990**, *29B*, 471, 474

[198p] Ishida, M., Beniya, Y., Inagaki, S., Kato, S., *J. Am. Chem. Soc.* **1990**, *112*, 8980

[198q] Kurita, Y., Takayama, C., *Tetrahedron* **1990**, *46*, 3789

[198r] Ohwada, T., *J. Am. Chem. Soc.* **1992**, *114*, 8818

[199a] Kahn, S. D., Pau, C. F., Chamberlin, A. R., Hehre, W. J., *J. Am. Chem. Soc.* **1987**, *109*, 650

[199b] Broughton, H. B., Green, S. M., Rzepa, H. S., *J. Chem. Soc. Chem. Comm.* **1992**, 998

[199c] Pudzianowski, A. T., Barrish, J. C., Spergel, S. H., *Tetrahedron Lett.* **1992**, *33*, 293

[200] Brown, K. L., Damm, L., Dunitz, J. D., Eschenmoser, A., Hobi, R, Kratky C., *Helv. Chim. Acta* **1978**, *61*, 3108

[201a] Bartlett, P. D., Blakeney, A. J., Kimura, M., Watson, W. H., *J. Am. Chem. Soc.* **1980**, *102*, 1383

[201b] Paquette, L. A., Carr, R. V. C., Bohm, M. C., Gleiter, R., *J. Am. Chem. Soc.* **1980**, *102*, 1186, 7218

[201c] Watson, W. H., Galloy, J., Bartlett, P. D., Roof, A. A. M., *J. Am. Chem. Soc.* **1981**, *103*, 2022

[201d] Pinkerton, A. A., Schwarzenbach, D., Stibbard, J. H., Carrupt, P.-A., Vogel, P., *J. Am. Chem. Soc.* **1981**, *103*, 2095

[201e] Bartlett, P. D., Roof, A. A. M., Winter, W. J., *J. Am. Chem. Soc.* **1981**, *103*, 6520

[202] Nagase, S., Morokuma, K., *J. Am. Chem. Soc.* **1978**, *100*, 1666

[203a] Spanget-Larsen, J., Gleiter, R., *Tetrahedron Lett.* **1982**, *23*, 2435

[203b] Spanget-Larsen, J., Gleiter, R., *Tetrahedron* **1983**, *39*, 3345

[204a] Carrupt, P.-A., Vogel, P., *J. Mol. Struct. (Theochem.)* **1985**, *124*, 9

[204b] Carrupt, P.-A., Vogel, P., Mison, P., Eddaif, A., Pellisier, N., Faure, R., Loiseleur, H., *Nouv. J. Chim.* **1986**, *10*, 277

[205] Koga, N., Ozawa, T., Morokuma, K., *J. Phys. Org. Chem.* **1990**, *3*, 519

[206] Rastelli, A., Cocchi, M., Schiatti, E., Gandolfi, R., Burdisso, M., *J. Chem. Soc. Faraday* **1990**, *86*, 783

[207a] Burdisso, M., Gandolfi, R., Pevarello, P., Poppi, A. L., Rastelli, A., *Tetrahedron Lett.* **1985**, *26*, 4653

[207b] Burdisso, M., Gamba, A., Gandolfi, R., Toma, L., Rastelli, A., Schiatti, E., *J. Org. Chem.* **1990**, *55*, 3311

[207c] Burdisso, M., Gandolfi, R., Rastelli, A., *Tetrahedron Lett.* **1991**, *32*, 2659

[208] Landen, H., Margraf, B., Martin, H.-D., *Tetrahedron Lett.* **1988**, *29*, 6593

[209] Caramella, P., Marinone Albini, F., Vitali, D., Rondan, N. G., Wu, Y.-D., Schwartz, T. R., Houk, K. N., *Tetrahedron Lett.* **1984**, *25*, 1875

[210a] Houk, K. N., Rondan, N. G., Brown, F. K., Jorgensen, W. L., Madura, J. D., Spellmeyer, D. C., *J. Am. Chem. Soc.* **1983**, *105*, 5980

[210b] Houk, K. N., Rondan, N. G., Brown, F. K., *Isr. J. Chem.* **1983**, *23*, 3

[210c] Jeffrey, G. A., Houk, K. N., Paddon-Row, M. N., Rondan, N. G., Mitra, J., *J. Am. Chem. Soc.* **1985**, *107*, 321

[211] Houk, K. N., Wu, Y.-D., Mueller, P., Caramella, P., Paddon-Row, M. N., Nagase, S., Mazzocchi, P. H., *Tetrahedron Lett.* **1990**, *31*, 7289

[212a] Seebach, D., Zimmermann, J., Gysel, U., Ziegler, R., Ha, T.-K., *J. Am. Chem. Soc.* **1988**, *110*, 4763

[212b] Seebach, D., Maetzke, T., Petter, W., Klötzer, B., Plattner, D. A., *J. Am. Chem. Soc.* **1991**, *113*, 1781

[213] Amarendra Kumar, V., Venkatesan, K., Ganguly, B., Chandrasekhar, J., Khan, F. A., Mehta, G., *Tetrahedron Lett.* **1992**, *33*, 3069

[214] Laube, T., Hollenstein, S., *J. Am. Chem. Soc.* **1992**, *114*, 8812

7 Reaction Paths for Nucleophilic Substitution (S$_N$2) Reactions

Hans-Beat Bürgi and Valery Shklover

7.1 Introduction

There are two general classes of reactions to change coordination at a reaction center: addition/elimination and substitution. The words used to identify and further characterize these reactions are traditionally different in organic and in inorganic chemistry. In order to have a common basis for discussing analogous processes at such diverse reaction centers as Cd, Pb, Sn, Ge, Si, C, Al, and B, we start by giving a condensed summary of the respective classifications.

Addition and elimination are simply different aspects of the same transformation followed either in the forward or backward directions. In the process the coordination number of the reaction center is increased or decreased by one. The transition state of both reactions is characterized by a partial bond between the reactants. In Chapter 6 several examples of addition/elimination reactions have been discussed, primarily those involving carbonyl carbon and other three-coordinate atoms as reaction centers. The inorganic analogs given there illustrate the relationship between the organic transformations and the more general concept of addition and elimination of Lewis bases to and from Lewis acids.

In a substitution reaction the coordination number does not change. However, the reaction may take two quite different courses. Either it proceeds by way of an intermediate, i.e. across two transition states, or it involves only a single transition state. The two-step reaction may be considered as a sequence of an addition and an elimination step, or as the reverse sequence. In the nomenclature of Langford and Gray [1] these are the associatively activated process with an intermediate of increased coordination number (A) and the dissociatively activated process with an intermediate of decreased coordination number (D). In the chemistry of four-coordinate carbon the latter is called an S$_N$1 reaction (whereas the former is of no practical importance). In the one-step substitution two variants are distinguished: one in which the new bond is largely formed in the transition state, with little or no weakening of the bond to be broken; and the other in which both the old and the new bond are very weak in the transition state. These are the associatively and dissociatively activated interchange processes (I$_a$, I$_d$), respectively. In the chemistry of four-coordinate carbon both variants are called S_N2 reactions. Here the distinction is made by characterizing the corresponding transition states as either tight or loose.

The above summary shows that mechanistic parlance implies energetic and structural aspects. As discussed in Chapter 5, structure correlation does not allow detailed energetic deductions to be made and can usually not distinguish one step from multistep mechanisms, or early from late transition states. Other aspects of mechanism, however, can be addressed by structure correlation methods, e.g. the detailed structural changes necessary to deform the ground state of the reactants into a transition state, and the structural characteristics of the latter such as looseness or tightness.

On this general background we shall discuss ligand interchange reactions at four-coordinate main group and d^{10}-transition elements. We start with Cd, proceed to Sn and Si, digress to Pb and Ge for purposes of comparison, regain the main alley with a short comparison of Al and B, and end up with structural aspects of S_N2 reactions at C.

7.2 The XCdS₃Y-Fragment

The Cd(II) ion forms a variety of trigonal-bipyramidal complexes with different degrees of distortion of the coordination polyhedron. X-ray structural data on a subset of such complexes have been analyzed [2]. In these complexes, thioglycolate ligands are bound equatorially to Cd via the S atoms, the axial X and Y ligands are I, S or O atoms. To compare quantitatively the geometries in different Cd-complexes, the following parameters were used (Figure 7.1): (1) the differences Δx and Δy between the observed distances $d(\text{Cd}-\text{X})$ and $d(\text{Cd}-\text{Y})$ and appropriate reference distances, e.g. sums of covalent radii, (2) the deviation Δz of Cd from the plane of

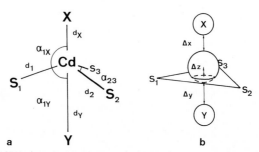

Fig. 7.1. Geometric parameters for describing distorted trigonal bipyramidal XCdS₃Y fragments. a: Distances and angles. b: Deviations Δx and Δy from sums of covalent radii and deviation Δz of Cd from S_3 plane

the three equatorial S atoms or (3) the average angles $\alpha_x = \langle \alpha(\text{XCdS}) \rangle$ and $\alpha_y = \langle \alpha(\text{YCdS}) \rangle$, which are respectively larger or smaller than 90° when Δz is positive or negative.

From the scatterplot of Δx, Δy and Δz in Figure 7.2, $\Delta z \approx 0$ Å when $\Delta x \approx \Delta y \approx 0.32$ Å, i.e. the Cd atom acquires the strictly trigonal bipyramidal geometry when the axial distances are about 0.3 Å larger than the sum of covalent radii or about equal to the sum of ionic radii. No significant changes of the equatorial Cd–S distances could be detected.

Logarithmic equations can be fitted to the experimental data in Figure 7.2:

$$\Delta x = d(\text{X, obs}) - d(\text{X, ref}) = -a \log [(\Delta z + \Delta z_{\text{max}})/2 \Delta z_{\text{max}}]$$

$$\Delta y = d(\text{Y, obs}) - d(\text{Y, ref}) = -a \log [(-\Delta z + \Delta z_{\text{max}})/2 \Delta z_{\text{max}}] \ .$$

$$(7.1)$$

Alternatively, Δz can be expressed in terms of α_X or α_Y. This leads to

$$\Delta x = -a \log [(\cos \alpha_X + \cos \alpha_{X,\text{max}})/(2 \cos \alpha_{X,\text{max}})]$$

$$\Delta y = -a \log [(\cos \alpha_Y + \cos \alpha_{X,\text{max}})/(2 \cos \alpha_{X,\text{max}})] \ .$$

$$(7.2)$$

Δz_{max} was chosen as $d(\text{Cd} - \text{S}_{\text{eq}})/3$, corresponding to $\alpha_{X,\text{max}} = 109.47°$. The value of the constant $a = 1.05$ Å was determined by a least-squares calculation on the observed values of Δx, Δy and Δz. With these constants, Equations (7.1) and (7.2) satisfy the condition that the function passes through the points [$\Delta x = 0$, $\alpha_X = 109.47°$] and [$\Delta y = \infty$, $\alpha_Y = 70.53°$]. Equations 7.1 and 7.2 were inspired by Pauling's formula relating differences in interatomic distances Δd to the logarithm of a bond number n [3, 4]

$$\Delta d = -a \log n \ .$$

$$(7.3)$$

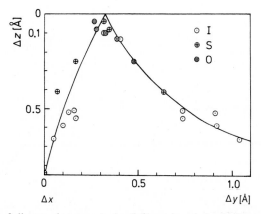

Fig. 7.2. Scatterplot of distance increments Δx (left arm) and Δy (right arm) versus deviation of Cd from the S_3 plane in XCdS$_3$Y fragments (see Figure 7.1). Solid curve, see text

If the arguments of the logarithmic functions (7.1) and (7.2) are identified with bond numbers n_X and n_Y, the bond strengths of ligands X and Y can be expressed entirely in terms of structural parameters (Δz, α_X or α_Y). Furthermore, all points along the correlation curve obey the invariance condition

$$n_X + n_Y = 1 \ . \tag{7.4}$$

This may be interpreted to mean that the total bonding is approximately constant for all compounds considered here. Equation (7.4) may be rewritten to obtain the dependence of Δx on Δy

$$10^{-\Delta x/a} + 10^{-\Delta y/a} = 1 \ . \tag{7.5}$$

The hyperboloid curve implied in Equation (7.5) is shown in Figure 7.3 together with the experimental data.

Equations (7.1) or (7.2) and (7.5), together with a single adjustable constant $a = 1.05$ Å, reproduce all essential aspects of the empirical correlations. They describe the continuous transition between two limiting situations: initially Y is far away from an S_3CdX tetrahedron, below the S_3 triangle and opposite to X. As the distance Y...Cd decreases, the distance Cd–X increases, and the SCdX-angles also decrease until, for equal Y–Cd and Cd–X distances, the SCdY and XCdS angles equal 90°. Further shortening of the distance Y–Cd lengthens the Cd...X distance and inverts the S_3Cd fragment until, at large Cd...X distance, an S_3CdY has formed. This sequence of events derived from a series of crystal structures describes exactly the structural changes associated with an organic S_N2 reaction: attack of a nucleophile Y opposite to a leaving group X, formation of a new bond to Y, breaking of the bond to X, and inversion of configuration at the reaction center

$$Y + S_3CdX \ \rightarrow \ Y \ldots S_3Cd \ldots X \ \rightarrow \ YCdS_3 + X \ . \tag{7.6}$$

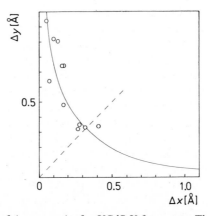

Fig. 7.3. Correlation plot of Δx versus Δy for $XCdS_3Y$ fragments. The solid line shows a curve of constant sum of bond orders (see Equations (7.4) and (7.5))

The static pictures seen in crystal structures and ordered into a continuous series may therefore be interpreted as a model of the reaction path of an S$_N$2 or associative ligand interchange reaction. From the above discussion, it remains an open question whether the five-coordinate species would be an intermediate or transition state if such a reaction occurred in solution. However, the numerical value of the constant *a* characterizes the transition state: tight if it is small and loose if it is large.

7.3 Tin Compounds with Coordination Numbers Four to Six

The study on Cd complexes surveyed in the preceding section was based on only four crystal structures [2]. Eight years later, Britton and Dunitz [5] published an analogous study on tin complexes. Their work was based on systematic searches in the Cambridge Structural Database (CSD; [6] and Chapter 3). It could rely on a much larger and broader selection of structural data (186 entries, 250 examples of Sn(IV) atoms).

7.3.1 The XSnC$_3$Y Fragment

For this fragment two limiting situations have been analyzed; one in which the electronegative X and Y substituents are arranged *trans* to each other and another in which X and Y are arranged in *cis* positions.

The structural data for *trans*-XSnC$_3$Y fragments are summarized in Figures 7.4 and 7.5. Figure 7.4 correlates CSnX and YSnC angles with deviations Δx and Δy from standard X–Sn and Sn–Y distances for C$_3$SnX and XSnC$_3$Y fragments. The solid line is based on Equation (7.2) with $a = 1.20$ Å and $\alpha_{max} = 109.5°$. Figure 7.5 correlates Δx versus Δy, the solid curve being based on Equation (7.5). The distribution of data points in this figure is symmetrical about the diagonal. This is because each of the two axial distances has been labeled once as Δx and once as Δy (see Chapter 2). The two figures are analogous to Figures 7.2 and 7.3 described earlier for Cd complexes and so is their general interpretation: they may be taken to map an S$_N$2 pathway with inversion of configuration at Sn.

Note that for large Δx- and Δy-values (>0.6 Å, Figure 7.4) the simple model (Equation (7.2)) does not describe the data well; also at $\Delta x \approx \Delta y \approx 0.35$ Å where the X–Sn–C angle is ca. 90°, the data cluster is very broad (Figure 7.4) and similarly along the diagonal $\Delta x = \Delta y$ (Figure 7.5). The shortcomings of the simple bond-order

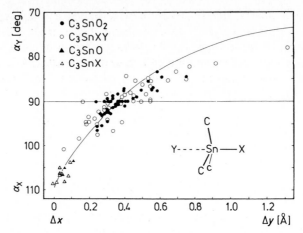

Fig. 7.4. CSnX and YSnC angles α_X and α_Y vs. Δx and Δy for $XSnC_3Y$ fragments. The curve is based on Equation (7.2) with $a = 1.20$ Å and $\alpha_{max} = 109.5°$

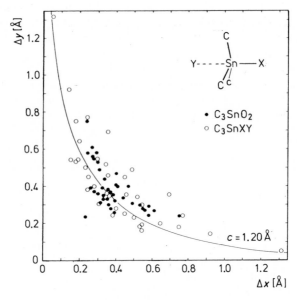

Fig. 7.5. Δx vs. Δy for $XSnC_3Y$ fragments. The solid curve is Equation (7.5) with $a = 1.20$ Å

model may be partly intrinsic and partly due to an inappropriate choice of standard covalent X−Sn and Sn−Y distances [4, 5]. In spite of this, it seems clear that the numerical constant is larger for Sn than for Cd, i.e. the trigonal bipyramidal structure is somewhat looser for Sn than for Cd.

When X and Y are part of the same bidentate, chelating ligand, they must occupy *cis*-positions of the trigonal bipyramid. In principle, such structures could serve as models for mapping a ligand exchange with retention of configuration at Sn. However, at the time the study was done, only seven examples of the *cis*-arrangement were known, too few to map out the path in detail [5] (see, however, Section 7.4.6).

7.3.2 The X$_2$SnC$_2$Y$_2$ Fragment

For most of the 45 X$_2$SnC$_2$Y$_2$ fragments analyzed in [5] ligands X are approximately in *trans*-position to ligands Y; the two C atoms are also transoid. The structural data are discussed in two steps: first, the deviations Δx and Δy from standard distances and the corresponding XSnC and CSnY angles α_X and α_Y will be looked at for each transoid X$-$SnC$_2$ ... Y subfragment independently. In a second step, the correlation between the two X$-$SnC$_2$ ... Y fragments will be analyzed.

The correlation of α_X, α_Y vs. Δx, Δy is given in Figure 7.6 and that of Δx vs. Δy in Figure 7.7. The smooth curves are based on Equations (7.2) and (7.5) with $a = 1.20$ Å (for details, see [5]). The two figures show that the formation of a Y$-$Sn bond, the breaking of the *trans*-Sn$-$X bond, and the changes of the corresponding XSnC and CSnY angles α_X and α_Y, are correlated in very much the same way as for the trigonal bipyramidal X$-$SnC$_3$$-$Y arrangement.

Correlation between the X$-$SnC$_2$$-$Y and X'$-SnC_2$$-$Y' subfragments may be investigated on the basis of α_X (CSnX) and α'_X (CSnX') alone, since there is a relationship between these and the other structural parameters characterizing the

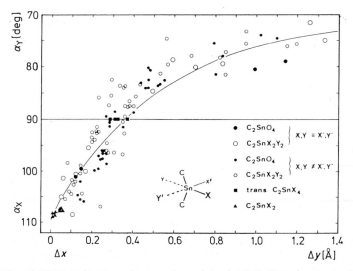

Fig. 7.6. XSnC and CSnY angles α_X and α_Y vs. Δx and Δy for X$_2$SnC$_2$Y$_2$ fragments. The curve is the same as in Figure 7.4

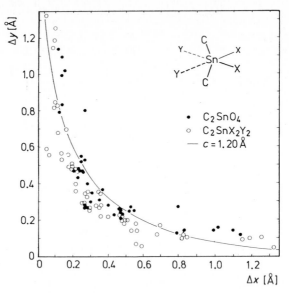

Fig. 7.7. Δx vs. Δy for $X_2SnC_2Y_2$ fragments. The curve gives Equation (7.5) with $a = 1.20$ Å

separate fragments (Figure 7.6). The corresponding scatterplot, shown in Figure 7.8, shows a concentration of points along the diagonal $\alpha_X = \alpha'_X$, corresponding to $X_2SnC_2Y_2$ fragments with approximate C_{2v} symmetry. Proceeding along the solid line from the triangles close to the center of the plot towards its bottom left corner corresponds to the simultaneous strengthening of the Sn–Y and Sn–Y′ bonds with concomitant weakening of the X–Sn and X′–Sn bonds until, at $\alpha_X = \alpha'_X = 90°$, an octahedral coordination of Sn is reached. Upon continuing to $\alpha_X = \alpha'_X \approx 70°$ (not shown in Figure 7.8), the initial XX′SnC$_2$ ... YY′ fragment is transformed into XX′ ... SnC$_2$YY′. It is not known whether the corresponding reaction, in which two pairs of bonds are simultaneously formed and broken, can actually occur in solution or in the gas phase; on entropic grounds it would seem unlikely since it requires a three-body collision. However, if this structure correlation means anything, the corresponding path should correspond to a fairly well defined valley in the potential energy hypersurface and map the symmetric addition of two ligands Y and Y′ to a tetrahedral XX′SnC$_2$ molecule to give an octahedral intermediate or transition state. This course of events has been referred to as an S$_N$3 reaction, where the 3 indicates that the rate law for such a reaction is expected to depend on three concentrations, [XX′SnC$_2$], [Y] and [Y′] [7].

There are at least two other processes that have to be considered in any discussion of double substitution reactions. One is the consecutive addition of Y and Y′, followed by the consecutive elimination of X and X′. The first half of this process is shown as a dotted line in Figure 7.8. The data points deviating from the S$_N$3 path are rather scarce, "deviations ... seem to have a rather unsystematic nature and it is difficult to draw any conclusions from them" [5]. The second alternative to S$_N$3 would be

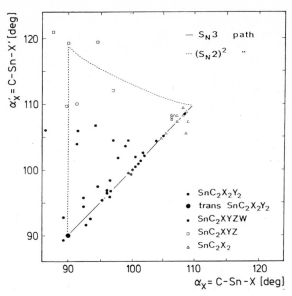

Fig. 7.8. X′SnC vs. XSnC angles (α_X vs. α'_X). The S$_N$3 pathway, i.e. simultaneous addition of Y and Y′, follows the solid line. Consecutive addition of Y and Y′ is mapped approximately by the dotted line

two consecutive addition/elimination steps, i.e. two S$_N$2 reactions, one after the other. These would have to follow the path shown in Figures 7.4 and 7.5.

7.3.3 Analogies with Pb(IV)

Pb(IV) is more readily reduced to Pb(II) than Sn(IV), and consequently Pb(IV) compounds are less common than Sn(IV) ones. Correlations of Δx, Δy, α_X and α_Y are less well defined for Pb than for Sn, but they are generally similar [5]. There is one important difference, however: the model of constant bond order (Equations (7.2) and (7.5)) requires $a = 1.66(10)$ Å to reproduce the observed correlations. This value is significantly larger than for Sn(IV).

7.4 Silicon Compounds with Coordination Numbers Four and Five

The structure correlations described so far for Cd, Sn and Pb were based on whatever structural data happened to be available at the time the work was done. In comparison, the compounds **1**–**19** discussed in this section were made expressly to investigate specific factors influencing the bonding in the $O-SiC_3-X$ fragment (X = oxygen or halogen) [8–15].

7.4.1 The OSiC$_3$X Fragment

The molecules discussed in this section are structurally quite similar. They all show, as a common feature, an $\overline{N-C=O}\ldots Si-C$ five-membered ring incorporating an amide moiety and a five-coordinated silicon atom. The three carbon substituents at Si are always equatorial, the amide oxygen and the second electronegative ligand (Cl or O) are always *trans*-axial. The distances $d(Si-O)$ and $d(Si-X)$ and the deviations Δz of Si from the plane defined by the three equatorial carbon atoms are given in Table 7.1. The filled symbols in Figure 7.9 show the correlation betwen $d(Si-O)$ and Δz; it is relatively smooth, well defined, and of sufficient quality for a more stringent test of the simple-minded constant bond-order models used in preceding sections.

Table 7.1. Bond lengths $d(Si-O)$ and $d(Si-X)$ (Å) in the axial $O-Si-X$ moieties of **1**–**19**, and deviations Δz (Å) of the Si atom from the equatorial C_3-plane towards O

	X	$d(Si-O)$	$d(Si-X)$	Δz	REFCODE
1	I	1.749(2)	3.734(1)	0.348(1)	FOGBIM
2	Br	1.800(4)	3.122(2)	0.218(2)	FUFXAF 10
3	Cl	1.954(2)	2.307(2)	−0.058(2)	FUSYIB 10
4[a]	F	2.316 to	1.665 to	−0.238 to	GEGDOL 10
		2.461(8)	1.630(7)	−0.305(3)	
5	I	1.830(3)	3.030(1)	0.090	[15]
6[a]	Br	1.978 to	2.507 to	−0.052 to	[15]
		1.982(6)	2.467(3)	−0.069	
7[a]	Cl	2.027(6)	2.259(4)	−0.134	JOFFOZ
		2.077(6)	2.229(4)	−0.160	
8	Cl	1.950(2)	2.315(2)	−0.055(1)	GEGDAX 10
9	Cl	1.989(2)	2.313(1)	−0.074(1)	GEGDEB
10	Cl	2.050(2)	2.284(1)	−0.205(1)	[15]
11	Cl	2.435	2.146	−0.251	GEGDIF
12	Cl	1.788(1)	2.624(1)	0.178(1)	GILYIJ
13	Cl	1.879(1)	2.432(1)	0.078(1)	FUBPOH
14	Cl	1.918(3)	2.348	0.029	CMSIAT
15	Cl	1.975(1)	2.294(1)	−0.077	[15]
16[a]	Cl	2.142	2.223	−0.187	[15]
		2.219	2.191	−0.205	
17	O	1.753(2)	2.784(2)	0.300(1)	[15]
18	O	2.228(2)	1.778(2)	−0.204(1)	[15]
19	O	2.367(2)	1.711(2)	−0.296(1)	[15]

[a] Crystals of **4**, **6**, **7** and **16** contain 4, 3, 2 and 2 crystallographically independent molecules, respectively. For entries **4** and **6** only ranges of values are given

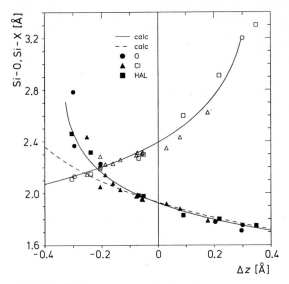

Fig. 7.9. Correlation of Si−O (filled symbols) and Si−X distances (empty symbols) vs. deviation Δz of Si from the C_3 plane towards O in the $OSiC_3X$ fragment of molecules **1−19** (circles: X = OR; triangles: X = Cl; squares: X = F, Br, I). Dashed and solid curves, see text

As a starting point for a correlation curve, one may assume, as before, a regular C_3SiO-tetrahedron with $\Delta z_{max} = 0.62$ Å (from $d(Si−C, ref) = 1.86$ Å) and $d(Si−O, ref) = 1.64$ Å [16]. With $a = 0.96$ Å and Equation (7.1) the dashed curve in Figure 7.9 is obtained. It fits the experimental data (filled symbols) quite well when $\Delta z > 0$, but not when $\Delta z < 0$, as was also seen in Figure 7.4 for the $XSnC_3Y$ fragments. Of course, there is no reason why a C_3SiO-tetrahedron built into a five-membered lactam ring should show local C_{3v} symmetry with bond angles close to 109.5°. Indeed, it is likely that in an almost planar five-membered ring the bond angles at $C(sp^2)$, $N(sp^2)$ and $O(sp^2)$ atoms are larger and those at Si smaller than the average value of 108° in a planar pentagon; angle strain will then be distributed about equally among the C, N, O, and Si atoms. Thus, the ring constraint forces the ring angles at C, N, and O to be larger than 108°, the one at Si to be below 108°, and Δz cannot reach its reference value of 0.62 Å. If the data in Figure 7.9 are used to redefine the numerical constants in Equation (7.1) (linear least-squares fit, $d(Si−O, ref) = 1.73$ Å, $\Delta z_{max} = 0.35$ Å and $a = 0.644$ Å), the solid curve is obtained; it accounts quite well for all data points.

A plot of the distances $d(Si−Cl)$ against Δz is also found in Figure 7.9 (empty symbols). From a least-squares calculation analogous to that for $d(Si−O)$, the constants used to obtain the solid line are $d(Si−Cl, ref) = 2.10$ Å, $\Delta z_{max} = 0.35$ Å and $a' = 0.975$ Å. The distances $d(Si−X)$ (X = O, F, Br, I) may also be included in the correlation if they are appropriately corrected for the differences in standard radii (Cl: 2.07(1) Å; O: 1.645(12) Å; F: 1.588(14) Å; Br: 2.28 Å [16]; I: 2.50 Å [17]). All data are reasonably reproduced by the solid curve. The two correlations taken together lead to

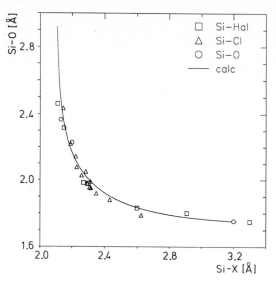

Fig. 7.10. Correlation of Si−O vs. Si−X distances in the OSiC$_3$X fragment of molecules **1−19** (circles: X = OR; triangles: X = Cl; squares: X = F, Br, I). The solid curve is Equation (7.5); see text

$$10^{-\Delta x/a} + 10^{-\Delta y/a'} = 1 \qquad\qquad (7.6)$$

which is analogous to Equation (7.5). The comparison of the data and the model in Figure 7.10 shows that the rule of constant bond order is followed if the difference between Si−O and Si−X bonds is taken into account by a proper choice of a and a'.

The correlations shown in Figures 7.9 and 7.10 are relatively smooth compared with those shown earlier. This is because the data refer to a set of very similar molecules. The original model has been adapted to the specific features of the fragment investigated, here the angle strain in the N−C=O ... Si−C ring; furthermore, the difference in the behavior of different bonds, here the Si−O and Si−Cl bonds, could be studied in detail. It is unlikely, however, that the correlation curves obtained here hold without modification for other connectivities, e.g. a set of molecules lacking the five-membered N−C=O ... Si−C ring. This is demonstrated by the results of Klebe [18, 19], who retrieved 46 independent molecular fragments with different molecular connectivities, but always containing five-coordinate Si, from the CSD [6]. His analysis is based on slightly different geometrical parameters, namely A_2'' symmetry displacement coordinates [20], specifically the umbrella deformation of the equatorial SiR$_3$-fragment, expressed in terms of axial-equatorial bond angles, and the difference $\Delta x - \Delta y$ of axial distance increments. His correlation curve also maps the transition from 4+1 to 1+3+1 coordination unambiguously (Figure 7.11, bottom right to top left), but it shows far more scatter than Figures 7.9 and 7.10. This is due, however, not to the choice of geometrical parameters, but to the diversity of connectivities included in Klebe's analysis.

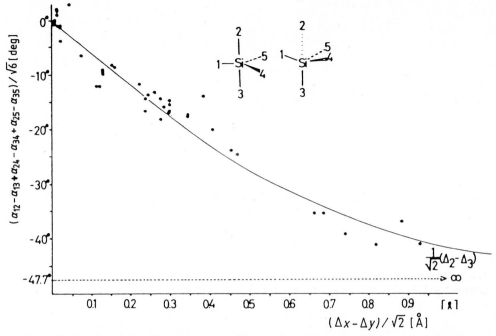

Fig. 7.11. Correlation of A_2'' symmetry displacement coordinates involving angles (vertical) and distances at Si (horizontal). The solid curve maps the expansion of the coordination at Si from $4+1$ (bottom right) to $1+3+1$ (top left)

7.4.2 Influence of Peripheral Substituents

The collection of molecules **1–19** allows several comparisons. If, for example, the six-membered ring in **1–3** is replaced by an acetyl group at nitrogen, as in **5–7**, $d(\mathrm{Si-O})$ lengthens by ca. 0.1 Å and $d(\mathrm{Si-X})$ shortens substantially (Table 7.1), due, presumably, to a lowering of the electron density at the coordinating oxygen atom and therefore of its nucleophilicity. This observation is substantiated by comparing **14** or **15** with **7**, all of which have only one ring. In the more electron-rich **14** or **15** $d(\mathrm{Si-O})$ is shorter and $d(\mathrm{Si-Cl})$ is longer than in **7**. An analogous difference is found between **12** and **13**. The size and substitution of the ring fused to the five-membered lactam may also play a role, as seen from **8, 3, 9, 10** and **11**. The Si–O distance increases continuously as the ring size decreases from seven to five and as the number of substituents on the annellated five-ring increases.

In the derivatives **17–19** the chloride ion is replaced by organic anions leading to a $\mathrm{O-SiC_3-O}$ fragment. From the Si–O distances and from the degree of pyramidality at Si, the relative nucleophilicity of these oxygen ligands is found to be $\mathrm{CF_3SO_3^- < N-C=O < C_6H_5COO^- < C_6H_5O^-}$, as expected (Figure 7.12).

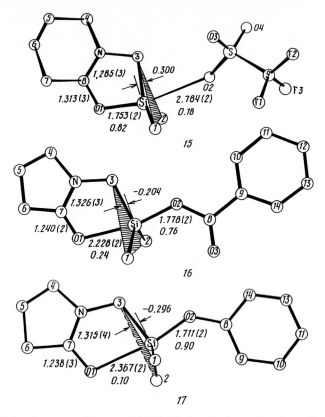

Fig. 7.12. Molecular structures of compounds **17–19** showing inversion of configuration at Si induced by different oxygen substituents (right) and different annellated rings (left)

For the *N*-acetyl compounds **5–7**, activation barriers have been determined by NMR for the degenerate isomerization in which the two oxygen atoms exchange place. In the process, the Si–O (partial) bond is broken and each of the two acetyl groups rotates about the respective C–N bond before the new Si–O bond is formed. The barriers are $\Delta G^{\pm} = 8.5$, 9.3 and 10.4 kcal mol^{-1} for **5**, **6** and **7**, respectively [21], highest for the longest Si–O distance. In Chapter 5 (Section 5.5) it was demonstrated that small bond-distance changes of a few hundreths of an Å may be correlated with significant changes in activation energies. The similarity of the barriers, the relatively large difference in d(Si–O) (1.83–2.05 Å), and the sign of the correlation between distance and energy, all seem to contradict these arguments. The explanation is probably that the rate-limiting step of the isomerization reaction is internal rotation about one of the C–N bonds in the β-diketonate system and not the breaking of the Si–O bond. As a consequence, the upper limit of the energy necessary to break the Si–O bonds in **5–7** can be estimated to be less than 10 kcal mol^{-1}.

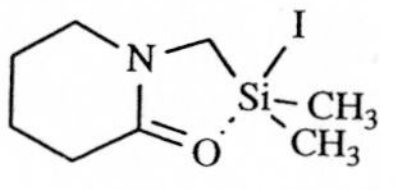

1

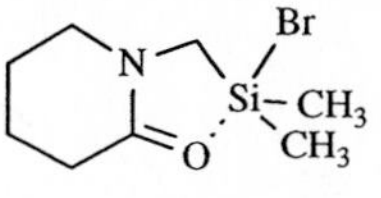

2

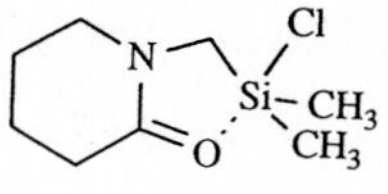

3

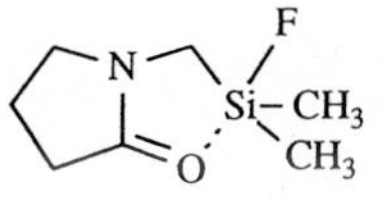

4

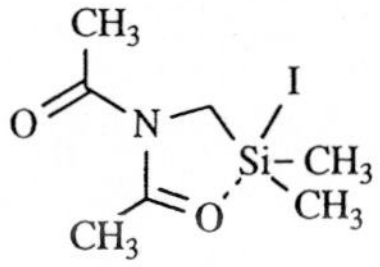

5

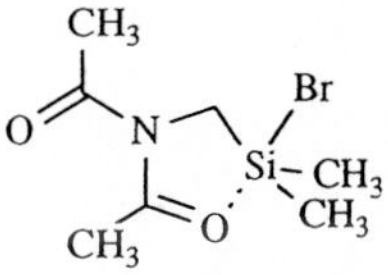

6

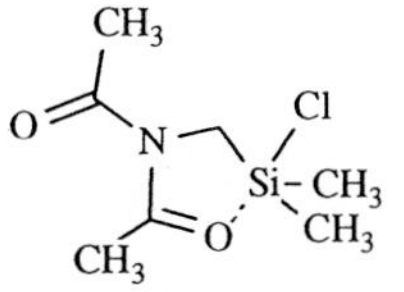

7

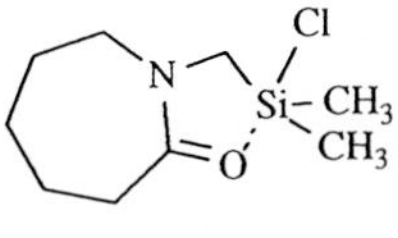

8

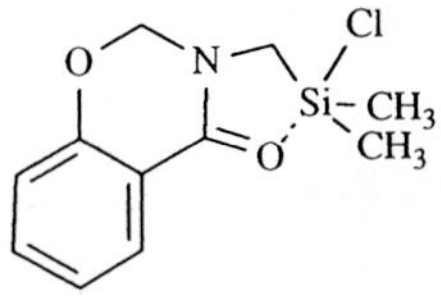

9

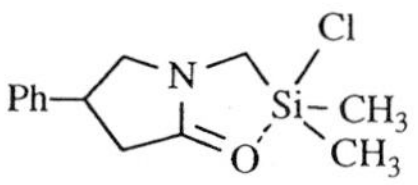

10

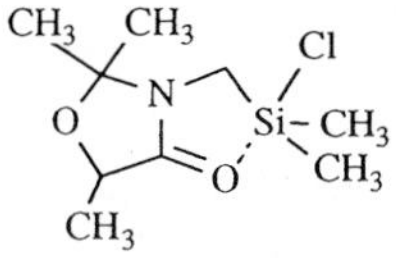

11

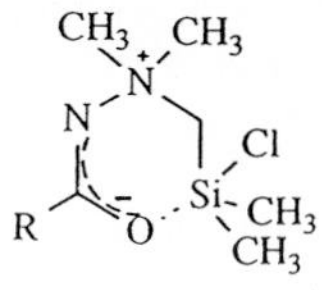

R= -C_6H_4OCH_3

12

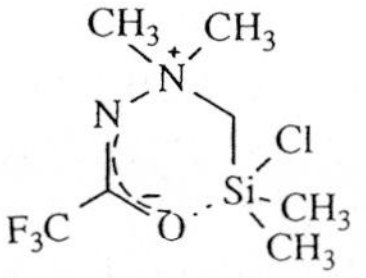

13

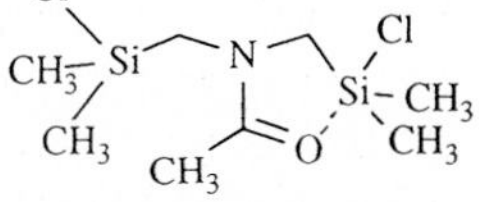

14

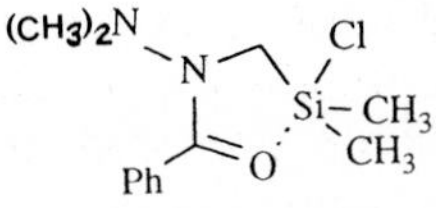

15

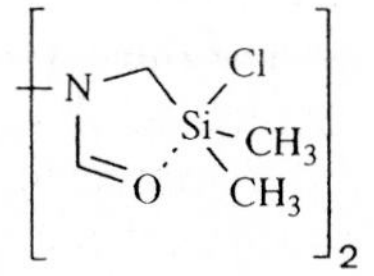

16

17

18

19

20

21

22

23

24

25

26

27

28

29

30

31

32

33

7.4.3 Analogies between Si and Ge, the XGeR$_3$Y Fragment

The series of germanium compounds available for comparison with **1–19** fall into three classes: (1) **20** and **21** are lactams comparable to the Si-molecules **10** and **8**. (2) In **22–24** the axial oxygen atom has been replaced by a nitrogen atom. (3) The largest series, **25–33**, has two equatorial chlorine atoms instead of methyl groups. In itself, each series is too small for meaningful correlations to be made on the basis of a single type of distance, as was done for the Si−O and Si−Cl contacts (Table 7.2). The authors of the original study [22] therefore looked for reference distances in appropriate four-coordinated compounds and correlated approximately corrected distances with Δz. For Ge−Cl bonds in **20–24** they chose 2.17 Å from ClGe(CH$_3$)$_3$ [23], in **25–33** they used 2.13 Å from CH$_3$GeCl$_3$ [24]. A reference Ge−O distance of 1.86 Å was chosen from CF$_3$C(=O)OGe(C$_6$H$_5$)$_3$ [25] for **20–21**, and of 1.75 Å from GeO$_4^{4-}$ for **25–33** [26]. The reference value for Ge−N bonds is 1.92 Å from [(CH$_3$)$_3$Ge−N(CH$_3$)C(=S)]$_2$ [27]. The resulting corrected distances are plotted against Δz in Figure 7.13 and against each other in Figure 7.14. The regression curves according to Equations (7.1) and (7.5) were calculated with d(Ge−Cl, ref) = 2.13 Å, Δz_{max} = 0.6 Å, a = 0.67 Å and d(Ge−O, ref) = 1.68 Å, Δz_{max} = 0.6 Å, a' = 0.98 Å. They account reasonably well for most of the experimental observations. The data for **20–24**, corrected for the differences in assumed standard distances as given above, are included in the figures, but not in the determination of a and a'.

Table 7.2. Distances d(Ge−Cl) and d(Ge−X) (Å), in the axial Cl−Ge−X moieties of **20–33** and deviations Δz (Å) of the Ge atom from the equatorial plane towards Cl

	X	d(Ge−Cl)	d(Ge−X)	Δz	REFCODE
ClGe(C$_3$)X					
20	O	2.324	2.311	0.192	JIPZEN
21	O	2.354	2.194	0.154	JIPZIR
22[a,b]	N	2.301	2.508	−	BZAMGE
	N	2.327	2.479	−	
23	N	2.458	2.148	0.003	JIPZOX
24	N	2.566	2.064	− 0.048	JIPZUD
ClGe(CCl$_2$)X					
25	O	2.134	3.228	0.577	BOSBIU01
26	O	2.140	3.075	0.533	DADKEY
27	O	2.160	2.790	0.520	JIRLOL
28	O	2.253	2.166	0.223	BOSBEQ
29	O	2.264	2.123	0.201	CUBYUT
30	O	2.253	2.140	0.190	CUBZAA
31[a]	O	2.239	2.092	0.200	KOJFEU
	O	2.252	2.080	0.175	
32	O	2.181	2.507	0.428	CUBMUH
33	O	2.169	2.770	0.476	DISGOB

[a] The crystal contains two crystallographically independent molecules
[b] Only the bond distances and some angles have been published for **22**

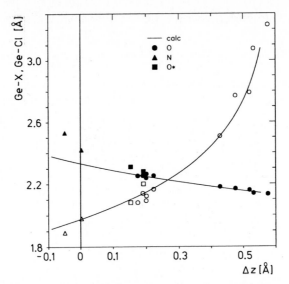

Fig. 7.13. Correlation of Ge−Cl distances (filled symbols) and Ge−X distances (empty symbols) vs. deviation Δz of Ge from R_3 plane towards Cl in the $ClGeR_3X$ fragment of compounds **20−33** (circles: $R_3 = CCl_2$, X=O; triangles: $R_3 = C_3$, X=N; squares: $R_3 = C_3$, X=O). Solid curves, see text

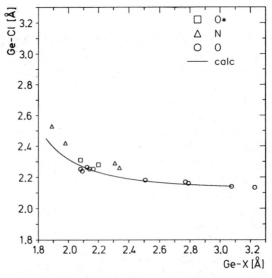

Fig. 7.14. Correlation of Ge−Cl vs. Ge−X distances in the $ClGeR_3X$ fragment of molecules **20−33** (circles: $R_3 = CCl_2$, X=O; triangles: $R_3 = C_3$, X=N; squares: $R_3 = C_3$, X=O). Solid curve, see text

   The effects of substitution are not unlike those observed for silicon. Electron donation shortens the Ge−O contact, viz. **25−29**. Note that the ketone derivatives **32** and **33** show Ge−O distances intermediate between the carboxylic acid or ester compounds **25−27** and the amides **28−31**. The larger anellated ring in **21** as compared with **20** is accompanied by a shorter Ge−O bond, as observed for the Si-analogs **8** and **10**.

7.4.4 Silatranes, the $NSiO_3X$ Fragment

The name "metallatrane" applies to molecules **34**, which may be considered as five-coordinate metal complexes of triply deprotonated triethanolamine. We use the term for analogous triphenolamine complexes **35** as well. Unlike the molecules described before, silatranes have three electronegative equatorial O-atoms. Because of the tricyclic nature of the molecules, the amine nitrogen atom is forced into an axial position. The Si−N distance may thus serve to probe the nature of the second axial ligand X. The atrane family of compounds is even more akin than the lactams discussed above because differences in connectivity are essentially confined to the ligand X.

A search in the CSD ([6], version of October 1992) produced the data collected in Table 7.3. The Si−C distances in the $NSiO_3C$ fragments range from 1.85 to 1.96 Å; they tend to be short for bulky aromatic substituents on Si, intermediate for alkyl and unhindered aromatic substituents, and long for CH_2Y substituents where Y is $N(CH_3)_3$, $OC(=O)R$, $S(CH_3)_2$, or Cl. The Si−N distances magnify this trend, but in the opposite direction, they change from 2.42 to 2.05 Å. In the $NSiO_3O$-series, axial Si−O distances cover a somewhat broader range, from 1.65 to 1.83 Å, the Si−N distances shorten from 2.19 to 1.96 Å. The shortest Si−O distance is observed for the electron-rich, poor leaving group $(CH_3)_3C-O^-$ of high basicity; the longest distance for the good leaving group $(CH_3)_2O$ of low basicity. Judging from the N−Si distances, the fluoro and chloro derivatives indicate a leaving group ability of F^- and Cl^- comparable with that of a carboxylic acid anion.

In the phenylsilatrane with the more rigid ligand (**35**, M = Si, X = C_6H_5), the Si−N distance is 2.34 Å, whereas in the more flexible **34** (M = Si, X = C_6H_5) this distance decreases to ca. 2.14 Å. In the process, the Si−C distance lengthens from 1.85 to 1.90 Å. Similarly, for the corresponding chloromethyl derivatives, Si−N shortens from 2.26 to 2.11 Å, and Si−C lengthens from 1.86 to 1.92 Å. In the series of compounds **34** with M = Si and X = CH_3, CH_2Cl, $CHCl_2$, an inductive effect of chlorine is observed; Si−N shortens from 2.17 to 2.11 to 2.06 Å, Si−C lengthens from 1.89 to 1.92 to 1.93 Å.

    34   **35**

Table 7.3. Geometry of $N - SiO_3 - X$ fragment in silatranes **34**, **35** (X = C, O, F, Cl; distances in Å, angles in deg.)

X [a]	$d(Si-X)$	$d(Si-N)$	⟨cos (OSiX)⟩	REFCODE
8-Dimethylamino-1-naphthyl-	1.892 [c]	2.420	101.5	KEWMOO
Phenyl- [b]	1.853	2.344	100.1	PNPOSI
1-(2-Dimethylaminoethyl)phenyl	1.896	2.282	98.9	KEWMII
Chloromethyl- [b]	1.864	2.256	98.1	FICPIQ
1,2-(1-Disilatranyl) ethane	1.870	2.230	98.4	KAGYEW
Cyclopropyl-	1.943	2.227	98.3	VAWGOP
1-(3-Thiocyanato) propyl-	1.881	2.209	97.7	CASSAQ
1-(3-Mercapto)propyl-	1.871	2.178	97.3	VEBJUH
Methyl-	1.870	2.174	97.2	MSILTR
1-(3-Hydroxy)propyl-	1.869	2.173	97.3	JEDZEX
p-Tolyl-	1.887	2.171	97.2	DEKGIJ
	1.892	2.167	97.0	
1-(3-Cyano)propyl-	1.885	2.164	96.8	BAZCEK10
Phenyl-	1.908	2.155	96.8	PNEOSI11
Phenyl-	1.894	2.132	96.4	PNEOSI02
Tricarbonyl-η^5-{(*tert*-butoxy carbonyl methyl) cyclohexadienyl silatrane}-manganese	1.899	2.127	96.0	VIFZOZ
N-(1-(1-Silatranyl) ethyl) pyrrolidone monohydrate	1.939	2.126	95.8	FOWNUA
Benzoyloxymethyl-	1.955	2.123	95.5	BAZXEF
1-(1-(2-Oxoperhydroazepino)ethyl)-	1.909	2.122	96.0	FEFSUE
(2-Pyridyl carboxy) oxomethylsilatrane	1.905	2.119	95.7	JAKRUI
3-Nitrophenyl-	1.904	2.116	95.8	NNESIL
Chloromethyl-	1.921	2.111	95.3	CMSILT
Tricarbonyl-(η^5-phenyl-silatrane)-chromium	1.906	2.108	95.3	JESBIS
Trimethyl-(1-silatranylmethyl)am-monium iodide	1.926	2.083	94.6	DILKOY
Dichloromethyl-	1.926	2.063	94.3	CIPBUY
1-(Dimethylsulfoniomethyl)-Silatrane iodide	1.930	2.046	94.0	CINTEY
1-*tert*-Butoxy-	1.659 [d]	2.189	97.0	GERJES
Ethoxy-	1.651	2.165	93.6	TALVUX
	1.665	2.139	96.1	
1-*p*-Tolyloxy-	1.676	2.107	95.3	DOKKUJ
1-*m*-Chlorophenoxy-	1.691	2.079	94.7	BIDSIQ
O-Methoxy benzoyloxy-	1.724	2.069	94.0	JAKSAP
1-Ethoxysilatrane trifluoroacetic acid	1.710	2.051	93.5	TALWEI
1-Silatranyl-dimethyloxonium tetra-fluoroborate	1.830	1.964	90.6	TALWAE
Fluoro-	1.622 [e]	2.042	94.1	CIZSUZ
Chloro-	2.153	2.023	93.3	CLSITR

[a] Usually only a description of the substituent X is given. For complicated molecules the full name of the compound is given. The complexes are of type **34** unless indicated otherwise
[b] Molecule of type **35**;   [c] X = C;   [d] X = O;   [e] X = F, Cl

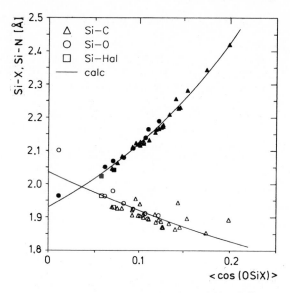

Fig. 7.15. Correlation of Si−X distances (empty symbols) and Si−N distances (filled symbols) vs. ⟨cos (OSiX)⟩ in silatranes **34**, **35** (triangles: X = C(aryl), C(alkyl); circles: X = OR; squares: X = F, Cl). Solid curves, see text

Correlation of the N−Si, Si−X distances and the OSiX angles shows a picture which by now should be familiar to the reader (Figure 7.15). Si−O, Si−F and Si−Cl distances have been corrected on the basis of $d(\text{Si}-\text{C}) = 1.888(23)$ Å [16] and the reference distances given in Section 7.4.1; they are included in the figure as well. The scatterplot of Si−N distances vs. OSiX angles is very smooth, as expected for identical molecular fragments, whereas the Si−X vs OSiX correlation reflects the diversity of carbon-, oxygen-, and halogen substituents. The curves used to fit the data (Figure 7.15) are based on Equation (7.2) with: $\alpha_{max} = 109.47°$, $d(\text{Si}-\text{N}, \text{ref}) = 1.505$ Å, $a = 1.244$ Å, and $d(\text{Si}-\text{C}, \text{ref}) = 1.729$ Å, $a' = 1.02$ Å.

7.4.5 Germatranes, the NGeO₃X Fragment

Fewer data are available for germatranes (Table 7.4). The general trends are similar to those observed for the silatranes (compare Figure 7.16 to Figure 7.15): The correlation between the Ge−N distances and XGeO-angles is smooth. The constants used to obtain the curve in Figure 7.16 are: $\alpha_{max} = 109.47°$, $d(\text{Ge}-\text{N}, \text{ref}) = 1.69$ Å, $a = 1.01$ Å. The Ge−X distances follow the expected trend, but their scatter does not warrant a correlation curve to be drawn. The reference distances used to compare Ge−C with Ge−O and Ge−Br distances are: 1.89(1) [24], 1.75 [26], and 2.276(2) Å [28], respectively.

Table 7.4. Geometry of NGeO₃X-fragment in germatranes **34** (X = C, O, Br; distances in Å, angles in deg)

X	$d(Ge-X)$	$d(Ge-N)$	$\langle \cos(OGeX) \rangle$	REFCODE
Digermatranylmethane	2.014 to 1.894[b]	3.317 to 2.258	99.7 to 99.0	BODPIT [e]
Ethyl-	1.966	2.243	97.0	EGERTR10
1-Naphthyl-	1.935	2.237	98.2	NPGERT
1-*tert*-Butyl-	1.972	2.236	98.6	BUWBUQ
N-(1-(1-Silatranyl)ethyl) pyrrolidone	1.980	2.223	97.8	DADKAU
Iodomethyl-	1.967	2.192	96.7	BABLIZ
N-((1-Germatranyl)methyl)-p-Chloro benzamide	1.959	2.184	97.1	FINMIY
5-Ethoxy carbonyl-2-furyl-	1.937	2.164	96.3	CIKFUX
1-(−)-Menthoxy-	1.768[c]	2.149	95.8	SISBAX
Hydroxy-	1.788	2.146	95.3	DUBZIJ
Bromo-	2.359[d] 2.357	2.098 2.084	93.6 94.5	BUWCUR

[a] see footnote [a] of Table 7.3
[b] X = C
[c] X = O
[d] X = Br
[e] contains 4 crystallographically independent molecules

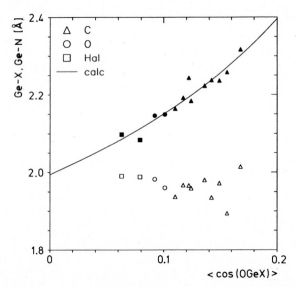

Fig. 7.16. Correlation of Ge−N distances (filled symbols) and Ge−X distances (empty symbols) vs. $\langle \cos(OGeX) \rangle$ in germatranes **34** (triangles: X = C(aryl), C(alkyl); circles: X = OR, squares: X = Br). Solid curve, see text

7.4.6 Inversion versus Retention of Configuration at Si

Experimentally, substitution reactions at Si may proceed with inversion or with retention (**36**, **37**, Nu: nucleophile, L: leaving group); in both cases a five-coordinate intermediate seems to be involved [29, 30]. *Ab initio* calculations indicate that L and Nu are both apical in the intermediate leading to inversion, whereas the mechanism for retention involves attack of Nu opposite to a substituent other than L, a Berry rearrangement bringing L into apical opposition, followed by departure of L [31]. The preferred mechanism depends on the propensity of L and Nu to be in apical position. The results of Deiters and Holmes [31] indicate that for L=Cl$^-$, the inversion mechanism **36** is favored by 20 to 25 kcal mol^{-1} relative to retention **37**, whereas this preference is reduced to ca. 10 kcal mol^{-1} if L = H$^-$, OH$^-$, F$^-$, SH$^-$. This trend is said to correlate with the preference for inversion observed in solution when L = Cl$^-$ and with the observation of partial or complete retention for the other leaving groups (or their chemical analogs, such as RO$^-$, RS$^-$) [29, 30]. The lactam derivatives **1–19** all relate to **36**, since the electronegative ligands O and X are always in apical position. They demonstrate primarily the accessibility of the five-coordinate intermediate (as do many other five-coordinate Si-species) and the subtle interplay of the nature of R$_1$, R$_2$, R$_3$, L, and Nu in determining its structure.

36

37

Overall, the five-coordinate intermediate **36** seems to be quite flexible (see Section 7.4.2), at least along the addition/elimination coordinate (Figures 7.9 – 7.16). Similar comments apply to the alkoxysilatranes and the silatranylhalogenides **34**, (M=Si, X=OR⁻, F⁻, Cl⁻); both families show only electronegative atoms in apical positions.

Alkyl- and arylsilatranes **34** and **35** (M = Si, X = alkyl, aryl), however, pertain to the retention mechanism **37**. The polycyclic skeleton forces the electronegative oxygens into equatorial positions, leaving the apical positions to the less electronegative substituent atom N and to the alkyl or aryl groups, which are poor leaving groups. The N−Si and Si−C distances, as well as the CSiO angles, clearly demonstrate the feasibility of five-coordinate intermediates akin to those postulated in **37**. Dichloromethylsilatrane, for example (Table 7.3), shows a N−Si distance of 2.06 Å, only 0.14 Å longer than the opposite Si−C distance. The CSiO angle of 94° also shows that this molecule is very close to a trigonal bipyramid.

Two more families of molecules, the silylnitronates **38** [32] and the silylcarboxylate esters **39** [33], may also be considered as incipient stages of the five-coordinate intermediates postulated in **37**. The O . . . Si distances are 2.83, 2.80, 2.83 and 2.80 Å, somewhat longer than the N . . . Si distances discussed above, but markedly shorter than the sum of the *van der Waals* radii, 1.52 and 2.10 Å for O and Si respectively [34]. The relevant O . . . Si−C angles are 154 and 150° for **38**, but have not been determined reliably for **39**. The activation barrier for a supposedly intramolecular process in which the Si−O bond is switched from one oxygen atom to the other has been measured by NMR to be ca. 10 kcal mol⁻¹ for compound **40** [32].

7.5 Nucleophilic Substitution at First Row Atoms

7.5.1 An Alatrane and a Boratrane

What can be learned from the sila- and germatranes about the S_N2 reaction path of first-row atoms such as carbon and boron? For these elements, stable five-coordinate molecules are unknown. The structural difference between corresponding second- and first-row molecules is nicely illustrated by the alatrane **42** and the boratrane **41** [35 – 37]. The $NAlO_3N$ fragment in **42** is close to trigonal bipyramidal with an Al−N(pyridine) distance of 1.992(6) Å, an Al−N(amine) distance of 2.153(6) Å, and an OAlN(pyridine)-angle of 94.7(3)°. Corresponding values for the boratrane **41** are 1.631(4) Å, 2.816(4) Å, and 103.7(2)°, indicating an almost tetrahedral NBO_3 fragment with a distant fifth ligand. In **42** the NC_3 fragment is pyramidalized towards Al, but in **41** away from B (CNAl angle 103.0(4)°, CNB angle 81.8(2)°).

In tetrahydrofurane solution at about room temperature the boratrane **41** is in equilibrium with **43** (stereoplot) and pyridine. In **43** both the BO_3 and NC_3

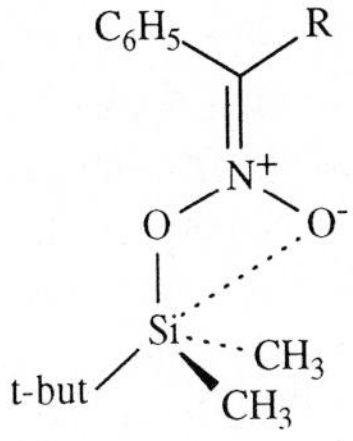

38

39

40

R=H,C₆H₄

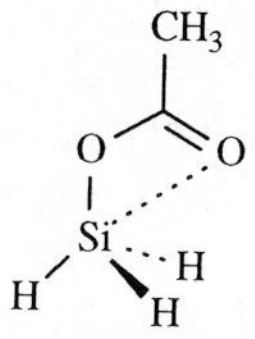

(s) and (g)

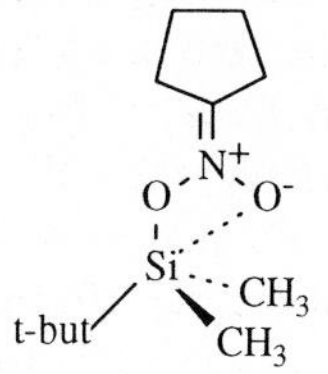

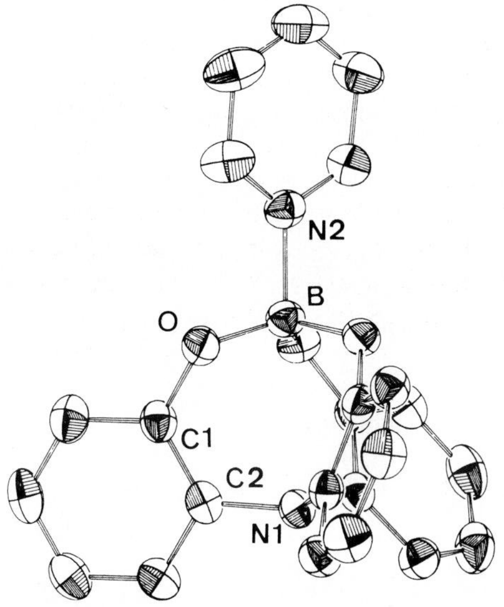

41

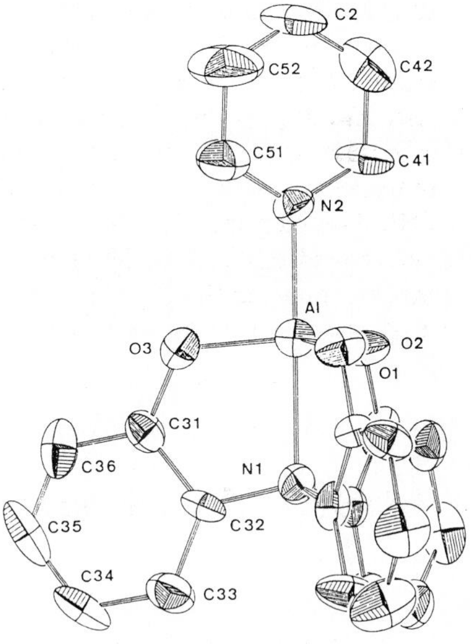

42

fragments have inverted to form an intramolecular bond of length 1.681(5) Å [36]. The reaction enthalpy is 9.5(1.0) kcal mol^{-1}, the entropy 24.5(3.3) cal mol^{-1} K^{-1}, strongly positive, as expected for a dissociation reaction producing two molecules from one. The corresponding activation parameters are $\Delta H^{\ddagger} = 14.9(1.0)$ kcal mol^{-1} and $\Delta S^{\ddagger} = 2.7(3.3)$ cal mol^{-1} K^{-1} [35]. Note that the activation entropy is close to zero. This implies that in the transition state the pyridine is still (partially) bonded to boron, i.e. the transition state is a five-coordinate species. There remains the question of how much the $B-N(py)$ distance has lengthened in the transition state. In Chapter 5, Section 5.5, a model has been introduced which allows one to estimate this lengthening from ground-state structures and activation parameters; from the above structural and kinetic data the lengthening is found to be ca. 0.3 Å. This is roughly the same as the difference between the X−M distance in the tetrahedral X−MR$_3$ fragments and the trigonal bipyramidal X−MR$_3$−Y fragments discussed in preceding sections for M = Pb, Sn, Ge, Si. It thus seems justified to assume that the correlated changes between distances and angles shown in several figures for the heavier group IV elements are a good model for the reaction path of an S$_N$2 reaction at boron. The more important case of nucleophilic substitution at four-coordinate carbon will be taken up next.

7.5.2 Carbon, the XCR$_3$Y Fragment

Among the attempts to synthesize five-coordinate carbon species with a linear XCR$_3$X$^-$ structure only those by Martin and his coworkers will be mentioned here [38−40]. Eight molecules of type **44** with different counter ions and in different

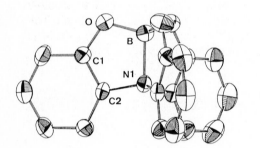

43

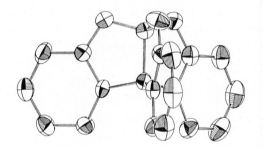

44   45

solvents were investigated by ^{1}H-NMR. The degenerate "bell clapper" rearrangement of **44**, inverting the tetrahedral carbon atom and switching the C–S bonds, was found to proceed with $\Delta H^{\neq}$ between 10 and 20 kcal mol^{-1}, thus indicating that the symmetric five-coordinate species corresponds to a transition state or intermediate [38, 40]. In a similar study on three molecules of type **45**, Forbus and Martin interpreted their ^{19}F- and ^{1}H-NMR spectra in terms of *symmetric* structures [40]! One of the compounds was crystalline, but attempts to grow crystals suitable for X-ray analysis were unsuccessful. To the best of our knowledge, the problem still awaits a conclusive resolution.

In the absence of experimental structural information on five-coordinate XCR$_3$Y species we have only calculations to go on. *Ab initio* studies on S$_N$2-reactions are plentiful indeed. An exhaustive review of the results would be beyond the scope of this chapter, however. Here we concentrate on recent results obtained by Shi and Boyd who have studied sixteen different XCH$_3$Y$^-$ species both at the Hartree-Fock level and with corrections for correlation energy (second-order Møller-Plesset perturbation calculations; [41]). The large number of systems calculated at the same level makes it possible to apply structure correlation methods to the structures of stationary points on the *ab initio* energy surface, i.e. to structures of ion-molecule complexes such as Y$^-$ ... CH$_3$X or YCH$_3$ ... X$^-$ and to the corresponding transition states [Y ... CH$_3$ ... X]$^-$. The HCH$_3$X$^-$ species given in [41] serve as an example (X = H, NH$_2$, OH, F, NC, CCH, CN, SH, Cl, all calculated at the MP2 level). The "axial" H–C distances and the "equatorial" HCX-angles (or rather their cosines) are plotted in the usual way in Figure 7.17. The data corresponding to HCH$_3$ ... X$^-$

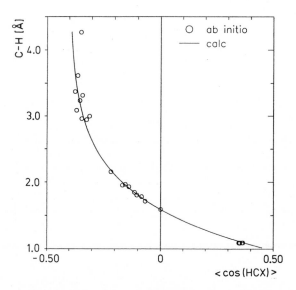

Fig. 7.17. Correlation of C–H distances vs. ⟨cos (HCX)⟩ in the molecules HCH$_3$X$^-$. The structural data have been obtained by ab-initio calculations on H$^-$ ... CH$_3$X and HCH$_3$ ... X$^-$ molecule-ion pairs (beginning and end region) and on [H ... CH$_3$ ... X]$^-$ transition states (middle region) [41]. For definitions of X see Table 7.5; solid curve, see text

are all very similar (C−H: 1.09 Å, HCH: 110.5 °) and are given as two closely spaced points. The structures corresponding to H⁻ ... CH₃X show a large spread, and only a feeble trend may be discerned: for decreasing HCX angles the H ... C distances tend to decrease. The calculated transition-state structures are the most interesting part of the figure. They cover the H ... C distance range between 1.59 and 2.15 Å; the HCX angles vary between 90 and 103 ° and are correlated to the distances in the familiar way. Note that the [H ... CH₃ ... X]⁻ transition states are all early, i.e. closer to H⁻ ... CH₃X than to HCH₃ ... X⁻. This reflects the exo-energetic nature of the reaction, up to 90 kcal mol⁻¹.

The curve in Figure 7.17 has the form of Equation (7.2) with d(C−H, ref) = 1.062 Å, α_{max} = 113.7°, a = 1.745 Å; it fits the *ab initio* data well. The numerical constants have been determined by least-squares calculation from the data in Table 7.5 (excluding H⁻ ... CH₃H with d(H ... C) = 4.264 Å). The general form

Table 7.5. Structural parameters of HCH₃X fragments from ab initio calculations [41]. Except for HCH₃H, the entries come in groups of three with the ion-molecule complex H⁻ ... CH₃X first, the transition state [H ... CH₃ ... X]⁻ second, and the molecule-ion pair HCH₃ ... X⁻ last (distances in Å, angles in deg)

HCH₃X⁻	d(CH)	α(HCX)
H ... CH₃H	4.264	110.4
H ... CH₃ ... H	1.589	90.0
H ... CH₃NH₂	3.616	111.3
H ... CH₃ ... NH₂	1.712	93.9
HCH₃ ... NH₂	1.092	68.8
H ... CH₃OH	3.314	110.0
H ... CH₃ ... OH	1.803	96.0
HCH₃ ... OH	1.092	68.6
H ... CH₃F	2.996	108.1
H ... CH₃ ... F	1.928	98.0
HCH₃ ... F	1.091	68.8
H ... CH₃NC	2.958	110.2
H ... CH₃ ... NC	1.964	98.9
HCH₃ ... NC	1.089	69.4
H ... CH₃CCH	3.372	112.0
H ... CH₃ ... CCH	1.780	94.8
HCH₃ ... CCH	1.090	69.4
H ... CH₃CN	3.086	111.6
H ... CH₃ ... CN	1.845	96.6
HCH₃ ... CN	1.089	69.6
H ... CH₃SH	3.235	110.7
H ... CH₃ ... SH	1.952	99.6
HCH₃ ... SH	1.090	69.7
H ... CH₃Cl	2.943	108.9
H ... CH₃ ... Cl	2.152	102.5
HCH₃ ... Cl	1.089	69.5

of the correlation curve is the same as that discussed throughout this chapter for experimentally observed ground-state structures. From this we conclude that the principle of structure-structure correlation (Chapter 5, Section 5.4) and the basis on which it stands can be extended to all stationary points along the reaction coordinate (see below).

The correlation described above implies that the reaction path for a specific S_N2 reaction is expected to follow the curves given by Equation (7.1) or (7.2), and in particular, the hyperboloid curve of constant bond order, Equation (7.5), relating the changes of $d(C-H)$ and $d(C-X)$. A Hartree-Fock calculation of several points along the S_N2 path of the reaction $Cl^- + CH_3Cl \rightarrow [Cl\ldots CH_3 \ldots Cl]^- \rightarrow ClCH_3 \ldots Cl^-$ [44] seems to contradict this expectation. A plot of the two $Cl-C$ distances (Figure 7.18) shows that initially only the long $Cl\ldots C$ distance shortens (to ca. 2.6 Å) without concomitant lengthening of the short $C-Cl$ distance; then it is mainly the short $C-Cl$ distance which lengthens until the symmetric transition state is reached at $d(C-Cl) = 2.383$ Å. This course of events is distinctly different from that implied by Equation (7.5) (see Figures 7.3, 7.5, 7.7, 7.10 for examples), as there is a kink in the curve relating the two $C-Cl$ distances. One could conclude that the structure correlation method for tracing the reaction path is not applicable in this case, or that the results of the calculation are questionable. For comparison, the results of the Hartree-Fock calculations by Shi and Boyd [41], obtained with a similar basis set are included in Figure 7.18 as empty circles. They agree with the values of Chandrasekhar et al. [44]. The situation is different if corrections for elec-

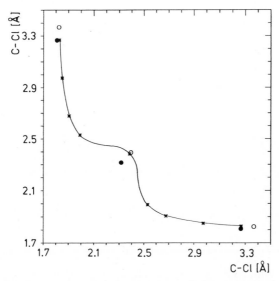

Fig. 7.18. Correlation of $C-Cl$ distances in $ClCH_3Cl^-$ calculated *ab initio* along the reaction path for an S_N2-exchange of Cl^- (stars and solid curve) [44]. The first and last points represent molecule-ion complexes, the middle point corresponds to the transition state. Results of a calculation with a similar basis set are also shown (open circles: Hartree-Fock calculation, filled circles: second order Møller-Plesset perturbation calculation)

tron correlation are included in the calculations (MP 2, filled circles in Figure 7.18). Although the point for the ion-molecule complex falls on the Hartree-Fock path, the point for the transition state indicates Cl−C distances shortened by ca. 0.09 Å! The kink in the Hartree-Fock path is thereby reduced, although not completely removed. This indicates the importance of electron correlation effects in these XCH_3Y species. It remains to be seen whether or not better calculations will remove the kink completely.

To summarize, two points can be made:

(1) When discussing activation parameters in terms of transition-state structures, we generally rely on the laws, rules, regularities, i.e. electronic and steric effects, used to classify and interpret the measurable details of ground-state structure. The finding that transition-state structures are correlated in very much the same way as are ground-state structures would seem to justify this practice.

(2) Structure correlations based on ground-state structures have been shown to map minimum energy paths for a variety of bond-breaking, bond-making and conformational interconversion reactions. The idea is that the environment of a given fragment perturbs its structure preferentially along the shallow directions of its energy surface. If the energy in the transition-state region is relatively low, the perturbations are often sufficient to map a complete path from reactant to product (principle of structure-structure correlation, Chapter 5, Section 5.4). If this energy is relatively high, the perturbations may be insufficient to push the ground-state structure close to a transition-state geometry. It has been shown, however, that the perturbations responsible for small changes in ground-state structure also affect activation energy in a predictable way. Corresponding changes in transition-state geometry can only be estimated (principle of structure-energy correlation, Chapter 5, Section 5.5.4). Our analysis of transition-state structures derived from *ab initio* calculations completes the picture; it indicates that changes in transition-state structure do occur in a way that is consistent with the general ideas of structure-structure and structure-energy correlations.

7.6 Conclusions

Large parts of this chapter, at least many of its illustrations, may appear somewhat monotonous. All those correlations among the distances and angles in XMR_3Y fragments look very much the same, regardless of the nature of the atoms involved. However, this monotony attests to the generality of the observed trends and therefore to a common feature of the Cd, Pb, Sn, Ge, Si, C, Al, and B fragments investigated. In fact, this generality started to become apparent already 20 years ago [45], when it could be shown that the distances in I−I−I, S−S−S, Cl−Sb−Cl, O−Mo−O and

O−H−O systems all follow the hyperboloid correlation illustrated in Figures 7.3, 7.5, 7.10 and 7.14 of this chapter.

What then is the feature common to the various fragments from across the periodic table? One unifying characteristic is a three-center X−M−Y system held together by two two-electron bonds, i.e. three-center four-electron bonding. In all examples a central Lewis-acid M is complexed to two Lewis bases X and Y in a linear fashion. The nature of X, Y and M merely modulates the energetic details along a common deformation path. For the heavier elements the path tends to show an energy minimum for the symmetric X−M−Y arrangement, whereas for the lighter ones the path shows energy minima for the asymmetric situations XM . . . Y or X . . . MY and a saddle point close to the symmetric one.

The examples presented in this chapter also illustrate a development of the structure correlation method itself. Initially, it was applied to whatever representatives of a specific fragment happened to be available (Cd). Later, directed searches in the CSD [6] led to sometimes surprising new types of correlations (Sn). More recently, compounds have been synthesized in a planned way and their structures determined, so that a specific structural feature or reactivity problem could be studied (Si, B). We have also seen that the methods and interpretations of structure correlation can be applied to results of quantum-chemical calculations. Combinations of kinetic, mechanistic, and computational studies together with structure correlations, are just beginning to illuminate as yet poorly understood problems of chemical reactivity and selectivity, e.g. the factors differentiating between substitutions proceeding with retention or inversion at Si.

Acknowledgments

The authors thank Drs. A. Bieniok, D. De Ridder and V. Malogajski for help with CSD searches and with the drawings. This work was supported by the Schweizerischer Nationalfonds.

References

[1] Langford, C. H., Gray, H. B., *Ligand Substitution Processes*, Benjamin, New York, **1965**
[2] Bürgi, H. B., *Inorg. Chem.* **1973**, *12*, 2321−2325
[3] Pauling, L., *J. Am. Chem. Soc.* **1947**, *69*, 542−553
[4] Pauling, L., *The Nature of the Chemical Bond*, Cornell University Press, Ithaca, N.Y., **1960**
[5] Britton, D., Dunitz, J. D., *J. Am. Chem. Soc.* **1981**, *103*, 2971−2979

[6] Allen, F. H., Bellard, S., Brice, M. C., Cartwright, B. A., Doubleday, A., Higgs, H., Hummelink, T., Hummelink-Peters, B. G., Kennard, O., Motherwell, W. D. S., Rodgers, J. R., Watson, D. G., *Acta Cryst.* **1979**, *B35*, 2331–2339

[7] See also Chapter 8, Section 8.4.4, where a third-order reaction in Co(III)-chemistry is discussed

[8] Macharashvili, A. A., Shklover, V. E., Struchkov, Yu. T., Baukov, Yu. I., Kramarova, E. P., Oleneva, G. I., *J. Organomet. Chem.* **1987**, *327*, 167–172

[9] Macharashvili, A. A., Baukov, Yu. I., Kramarova, E. P., Oleneva, G. I., Pestunovich, V. A., Struchkov, Yu. T., Shklover, V. E., *Zh. Strukt. Khim.* **1987**, *28*, 114–119

[10] Macharashvili, A. A., Shklover, V. E., Struchkov, Yu. T., Oleneva, G. I., Kramarova, E. P., Shipov, A. G., Baukov, Yu. I., *Chem. Comm.* **1988**, 683–685

[11] Macharashvili, A. A., Shklover, V. E., Struchkov, Yu. T., Voronkov, M. G., Gostevsky, B. A., Kalikhman, I. D., Bannikova, O. B., Pestunovich, V. A., *J. Organomet. Chem.* **1988**, *340*, 23–29

[12] Macharashvili, A. A., Shklover, V. E., Struchkov, Yu. T., Gostevsky, B. A., Kalikhman, I. D., Bannikova, O. B., Voronkov, M. G., Pestunovich, V. A., *J. Organomet. Chem.* **1988**, *356*, 23–30

[13] Macharashvili, A. A., Shklover, V. E., Struchkov, Yu. T., Pestunovich, V. A., Baukov, Yu. I., Kramarova, E. P., Oleneva, G. I., *Zh. Strukt. Khim.* **1988**, *29*, 121–124

[14] Macharashvili, A. A., Shklover, V. E., Chernikova, N. Yu., Antipin, M. Yu., Struchkov, Yu. T., Baukov, Yu. I., Oleneva, G. I., Kramarova, E. P., Shipov, A. G., *J. Organomet. Chem.* **1989**, *359*, 13–20

[15] Shklover, V. E., Macharashvili, A. A., Pestunovich, V. A., Baukov, Yu. I., *Internal Report*, Institute of Organo-Element Compounds, Russian Academy of Science, **1988**

[16] Allen, F. H., Kennard, O., Watson, D. G., Brammer, L., Orpen, A. G., Taylor, R., *J. Chem. Soc. Perkin II* **1987**, S1–S19. See also Appendix A to this book

[17] No standard distances available in [16]. Reference values have been estimated according to $d(Si-X) = r(X) + r(Si) - 0.09 |EN(X) - EN(Si)|$ where EN are Pauling's electronegativity values. See Schomaker, V., Stevenson, D. P., *J. Am. Chem. Soc.* **1941**, *63*, 37

[18] Klebe, G., *J. Organomet. Chem.* **1985**, *293*, 147–159

[19] Klebe, G., *Struct. Chem.* **1990**, *1*, 597–616

[20] Murray-Rust, P., Bürgi, H. B., Dunitz, J. D., *Acta Cryst.* **1979**, *A35*, 703–713. See also Chapter 5, Section 5.3.2

[21] Kalikhman, I. D., Bannikova, O. B., Belousova, L. I., Gostevsky, B. A., Liepinsh, E., Vyazankina, O. A., Vyazankin, N. S., Pestunovich, V. A., *Metalloorg. Khim.* **1988**, *1*, 683–688

[22] Tandura, S. N., Gurkova, S. N., Gusev, A. I., Alexeev, N. V., *Zh. Strukt. Khim.* **1985**, *26*, 136–139

[23] Li, Y. S., Durig, J. R., *Inorg. Chem.* **1973**, *12*, 306–309

[24] Drake, J. E., Hencher, J. L., Shen, Q., *Can. J. Chem.* **1977**, *55*, 1104–1110

[25] Glidewell, C., Liles, D. C., *J. Organomet. Chem.* **1983**, *243*, 291–297

[26] Hesse, K. F., Eysel, W., *Acta Cryst.* **1987**, *B37*, 429–431; Sidorkin, V. F., Pestunovich, V. A., Balakhchi, G. K., Voronkov, M. G., *Izv. AN SSSR* **1985**, 622–625

[27] Halder, T., Schwarz, W., Weidlein, J., *J. Organomet. Chem.* **1983**, *246*, 29–48

[28] Sonza, G. G. B., Wieser, J. D., *J. Mol. Struct.* **1975**, *25*, 442

[29] Corriu, R. J. P., Guerin, C., *J. Organomet. Chem.* **1980**, *198*, 231–320

[30] Corriu, R. J. P., Guerin, C., *Adv. Organomet. Chem.* **1982**, *20*, 265–312

[31] Deiters, J. A., Holmes, R. R., *J. Am. Chem. Soc.* **1987**, *109*, 1686–1692; ibidem, 1692–1696

[32] Colvin, E. W., Beck, A. K., Bastani, B., Seebach, D., Kai, Y., Dunitz, J. D., *Helv. Chim. Acta* **1980**, *63*, 697–710

[33] Barrow, M. J., Cradock, S., Ebsworth, E. A. V., Rankin, D. W. H., *J. Chem. Soc. Dalton* **1981**, 1988–1993

[34] Bondi, H., *J. Phys. Chem.* **1964**, *68*, 441–451

[35] Müller, E., Bürgi, H. B., *Helv. Chim. Acta* **1987**, *70*, 499–510

[36] Müller, E., Bürgi, H. B., *Helv. Chim. Acta* **1987**, *70*, 511–519

[37] Müller, E., Bürgi, H. B., *Helv. Chim. Acta* **1987**, *70*, 520–532

[38] Basalay, R. J., Martin, J. C., *J. Am. Chem. Soc.* **1973**, *95*, 2563–2571

[39] Martin, J.C., Basalay, R.J., *J. Am. Chem. Soc.* **1973**, *95*, 2572–2578

[40] Forbus, T.R., Jr., Martin, J.C., *J. Am. Chem. Soc.* **1979**, *101*, 5057–5059

[41] Shi, Z., Boyd, R.J., *J. Am. Chem. Soc.* **1990**, *112*, 6789–6796. (Basis set 6-31++G** for non-hydrogen atoms, 6-31 G for hydrogen atoms [42, 43])

[42] Clark, T., Chandrasekhar, J., Spitznagel, G.W., Schleyer, P.v.R., *J. Comput. Chem.* **1983**, *4*, 294–301

[43] Hariharan, P.C., Pople, J.A., *Theor. Chim. Acta* **1973**, *28*, 213–222

[44] Chandrasekhar, J., Smith, S.F., Jorgensen, W.L., *J. Am. Chem. Soc.* **1985**, *107*, 154–163

[45] Bürgi, H.B., *Angew. Chem.* **1975**, *87*, 461–475; *Angew. Chem. Int. Ed. Engl.* **1975**, *14*, 460–473

8 Ligand Rearrangement and Substitution Reactions of Transition Metal Complexes

Thomas Auf der Heyde

8.1 Introduction

Most structure correlation studies have been performed on systems that might classically be termed "organic"; far fewer have analyzed "inorganic" systems containing metals, particularly transition metals. A citation search for three key papers in the structure correlation area [1, 2] at the end of 1989, for instance, yielded about 460 citations, of which just under 70 were in papers dealing with transition metal compounds. Presumably, this preponderance of studies on organic systems reflects the fact that there are a greater number of organic systems offering meaningful comparisons of experimental/computational information with the results of structure correlation analyses, than there are inorganic systems. This is so because reaction mechanisms for organic systems are more fully understood and characterized, and also these systems are generally easier to model computationally, than are those containing metal atoms.

Nevertheless, several successful, intriguing, and skillful examinations of metal-organic systems have revealed information that might not have been accessible by any other means. This chapter reviews about 40 such studies. The material was largely collected by way of the citation search mentioned above, by a follow-up search to the middle of 1991, and through the author's own routine searches for pertinent papers − assisted by helpful colleagues throughout the world. Many of the studies revealed in this way are communications in which attempts were made to place a given structure in the context of some deformation pathway, but without rigorously searching for structural correlations. In order to avoid a review merely of structure communications, we include here only studies that explicitly report geometric reaction coordinates − either in the form of analytic expressions, or graphically − derived from a correlation analysis, even where this may have been done on the basis of as few as three or four structures. Excluded are papers in which the focus is on rationalizing particular chemical characteristics of a molecule in terms of its specific structural features. Moreover, in reporting the various reaction paths, emphasis will be placed on the geometric characteristics as they have been determined from the solid state; mention of solution studies will only be made in cases where the mechanistic information is not generally available from an advanced inorganic chemistry textbook. In any event, experimental/computational support for the results of a structure correlation analysis is invariably discussed in each individual report. Although this chapter principally reviews structure correlation studies of transition metal complexes, some studies of non-transition metals are included for the sake of comparison and completeness.

Throughout this chapter, reference will be made to techniques and approaches described elsewhere in this book, and a certain familiarity with these topics will be assumed: Methods of representing molecular conformation, and different coordinate systems (Chapter 1), ways of dealing with symmetry aspects (Chapter 2), data retrieval from the Cambridge Structural Database (CSD; Chapter 3) [3], and multivariate statistical techniques such as principal component analysis (PCA) and cluster analysis (CA; Chapter 4).

The papers reviewed here can be divided into two broad categories: (i) those that are fairly non-specific in their choice of data, studying basically all complexes with a common stoichiometry ML_n, where M = metal, L = a coordinated ligand atom, and $n = 3, 4, 5, \ldots, 12$, and (ii) those that set out to examine specific reaction paths − such as metal cluster rearrangements, geared rotations, ring whizzing, etc. − and which are fairly selective in data retrieval. We shall discuss the papers in the above order.

8.2 Reaction Paths of ML_n Coordination Complexes

8.2.1 ML_3 Complexes

Ligands of the type 2,11-bis(di-R-phosphinomethyl)benzo[c]-phenanthrene (bpbp) form trigonal complexes with metals such as Cu, Ag, Pd, Au, etc. Two extreme conformations exist for these complexes: in the P (parallel) conformer the CH_2-P vectors point to the same side of the mean bpbp plane, whereas in the A (anti-parallel) conformer they point to opposite sides of this plane.

bpbp =

An early discussion of the Au(I), Ag(I), and Cu(I) complexes with R = phenyl considered the possibility of interpreting correlated changes in the observed bond distances and angles in terms of a low-energy pathway leading from bent three-coordinate to linear two-coordinate complexes, or of the reverse reaction [4]. A later publication, incorporating 19 independent fragments that were symmetry expanded (according to C_2 and C_s symmetry for A and P conformers, respectively) did not pursue this possibility, but instead investigated conformational preferences and changes of the complexes [5]. It was shown that A-type conformers are conformationally more flexible than P-type ones, being far less resistant to changes in metal or counterion. A variable temperature NMR experiment on an A-type complex seemed to support the proposition of fluxionality. Principal component analysis (PCA) on the A conformers revealed two components (or factors), the first associated only with the deformation from C_2 symmetry, the second with a distortion of the hydrocarbon backbone of the ligand, such that the tetracyclic moiety becomes more non-planar as the $P-M-P$ angle increases.

8.2.2 ML$_4$ Complexes

8.2.2.1 Polyhedral Isomerization

Although there is a wealth of complexes of the ML$_4$ type, only the $CuCl_4^{2-}$ fragment has received any attention. In their first paper on polytopal rearrangement reactions, Muetterties and Guggenberger [6] very cursorily remark that the various conformations adopted by this fragment in different crystal environments may be used to map a $T_d \rightarrow D_{4h}$ deformation that can occur via either of two mechanisms. The first is a tetrahedral compression along an S_4 axis, the second a digonal twist about this axis, in which the angles subtended at the metal are not constrained. A smooth curve, based on 15 data points describing this rearrangement, illustrates that the average bond distance lengthens sigmoidally as the fragment distorts away from T_d to D_{4h}, perhaps due to the greater steric strain in the square-planar conformer [7, 8].

8.2.2.2 Conformational Interconversions in Metal-Phosphine Complexes

Gearing motions (geared rotations) of PR$_3$ ligands (where R = various alkyl and phenyl substituents) in complexes of the type [XM(PR$_3$)$_3$] and [XYM(PR$_3$)$_2$] (M = Rh(I), Pt(II); X, Y = various anions) have been investigated in two studies by Bürgi et al. [9, 10]. The former study incorporated 15 fragments, whose conformations were expressed in terms of torsion angles about the $P-M$ bonds. Symmetry expansion according to C_{3v} symmetry for the PR$_3$ rotors, and C_{2v} symmetry for the XMP$_3$ frame, yielded 1620 data points whose distribution in the three-dimensional parameter space was initially examined visually by means of stereoplots. Two distinct pathways were identified: the first maps a gearing motion of the three interlocked PR$_3$ fragments, while the second delineates a gearing motion on one side of the central PR$_3$ group, coupled to a gear slippage on the other. Subsequent principal component analysis revealed correlated distortions between the torsion angles and the

C−P−M bond angles that serve to increase the distance between neighboring PR_3 groups. Possible mechanisms for the transmission of structural information across metal complexes are discussed.

The second study [10] involved a more comprehensive description of the conformations of 62 $M(PPh_3)_2$ fragments from $[XYM(PPh_3)_2]$ complexes in terms of torsion angles about both the M−P and the P−C bonds (8 parameters), and the data were symmetry expanded to 2232 data points. A novel feature of this study is its treatment of the periodicity of torsion angles. As a consequence of this periodicity, points at, say, 0 and $2\pi, 4\pi, 6\pi$, etc. are identical, but commercially available statistical programs fail to take account of this symmetry. Instead of analyzing only one asymmetric unit of the parameter space, or a section large enough to incorporate all possible clusters (see Chapters 2 and 4), each of the parameter axes was bent into a circle, thereby constructing a torus with an eight-dimensional surface in a sixteen-dimensional space. An exhaustive cluster analysis of the data distribution on this surface, employing both single linkage- and Ward's clustering techniques (see Chapter 4), reveals three clusters that map a geared rotation of the PPh_3 groups about the M−P bonds, and the stepwise change in helicity of the PPh_3 propellers.

It is unlikely that we will ever know the details of the atomic motions taking place during fluxional behavior of these $[XM(PR_3)_3]$ and $[XYM(PR_3)_2]$ complexes, but it is important to stress that the method of structure correlation has at least suggested some geometrically feasible possibilities for these motions.

8.2.3 ML₅ Complexes

8.2.3.1 Associative Ligand Substitution Reactions and the Berry Rearrangement

Complexes of the ML_5 type, specifically those of copper and the d^8 metals of the platinum group, have been far more widely studied by the technique of structure correlation than any other type of transition metal compound. There are presumably many reasons for this. One of them is the significance of compounds of this stoichiometry as intermediates in substitution/elimination reactions of square-planar, tetrahedral, and octahedral complexes [11]. Figure 8.1 outlines the geometric deformations of five-coordinate complexes that have most commonly been invoked in mechanistic explanations of the chemical behavior of these compounds. In the context of a purely geometric perspective, step (**A**) maps the addition of a fifth donor atom to a square-planar complex, or the reverse reaction. The incipient stages of this path are characterized by pentavalent complexes whose metal atom lies close to the base of a square pyramid (SQP) − such conformers will be dubbed flattened square pyramids (fSQP). As the fifth ligand atom becomes more firmly bonded to the metal, the latter is drawn away from the basal plane of the fSQP, resulting in what is termed an elevated SQP (eSQP) [12a]. Step (**B**) maps the Berry intramolecular rearrangement [13] whereby the SQP deforms into a trigonal bipyramid (TBP) [12b], while step (**C**) maps the elimination of an axial ligand from the TBP, yielding a

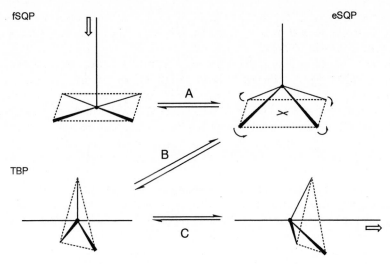

Fig. 8.1. Deformation and reaction pathways of five-coordinate complexes. fSQP = flattened square pyramid [12a], eSQP = elevated square pyramid, TBP = trigonal bipyramid

tetrahedral complex, or the reverse addition. In organic chemistry step (**C**) is involved in the S_N2 pathway (see Chapter 7). The two geometries of highest symmetry, usually chosen as reference geometries for ML_5 complexes, are the TBP (D_{3h} symmetry) and the SQP (C_{4v}); while the angular geometry of the TBP is fixed by its symmetry, this is not true for the SQP, which can adopt any value for the angles between *trans* ligands, so long as these angles are equal.

Because the ML_5 system has been so widely studied, the various papers offer interesting comparisons, illustrating how successively more complex methods of correlation analysis can yield significantly deeper insights into the stereochemistry of these complexes. Probably the earliest study of an ML_5 system, and also the one in which the structure correlation approach was explicitly outlined, is Bürgi's study of eleven five-coordinate cadmium complexes whose coordination polyhedra were considered as deformed TBP ([1] and Chapter 7, Section 7.2). The molecular geometry was described in terms of internal coordinates, and correlations were analyzed by means of scatterplots. Correlations between the displacement (Δz) of the central Cd atom out of the equatorial plane, and the distance changes between the metal and its two axial ligands ($\Delta x, \Delta y$), were interpreted in terms of an S_N2 distortion coordinate equivalent to step (**C**); these are shown in Figure 8.2. An analytic expression, which parallels that proposed by Pauling [14], was used to describe a geometric distortion coordinate along which changes in interatomic distance Δd are related to bond numbers n:

$$\Delta d = C \log n \qquad 0 < n < 1$$

On the basis of the analytic expression, Bürgi was able to show that the sum of the axial bond numbers equals unity at each stage of the reaction, implying that the bond between the incoming ligand and cadmium is strengthened at the expense of

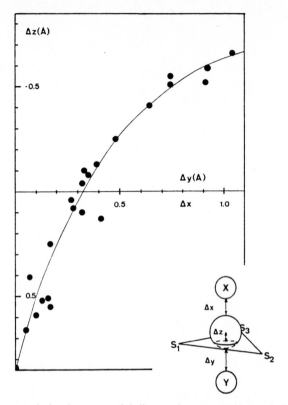

Fig. 8.2. Plot showing correlation between axial distance increments (Δx or Δy) and out-of-plane displacement (Δz) of the Cd atom. The distortion maps an $S_N 2$ coordinate equivalent to step (**C**) in Figure 8.1)

that between the departing group and the metal. Using an angular overlap molecular orbital approach, Burdett later modeled C_{3v} distortions of D_{3h} ML$_5$ fragments, analogous to the one discussed here (corresponding to step (**C**) above) as well as coupled distortions along paths (**A**) and (**B**) [15]. Despite discrepancies between his calculated curves and the experimental ones of Bürgi and others, Burdett reports a fairly good qualitative reproduction of the observed bond shortenings and elongations.

Although Bürgi's study revealed convincing correlations, it ignored certain deformations of the coordination polyhedron by assuming that the complexes all approached C_{3v} symmetry. A careful analysis of the data reveals, however, that some of the complexes are significantly distorted away from this symmetry, with the angle between axial ligands as low as 161°. An attempt by Muetterties and Guggenberger, rooted in internal coordinates, to explicitly describe the coordination geometry, involved the use of dihedral angles between normals to adjacent polytopal faces [6]. Here the distortion coordinates were expressed in terms of correlations between those dihedral angles undergoing the largest changes during a particular polyhedral rear-

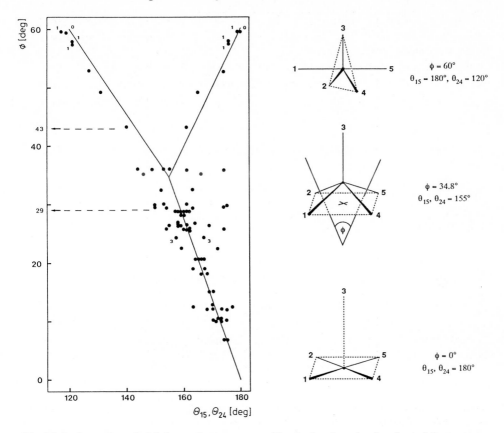

Fig. 8.3. Deformation of nickel complexes as mapped by the bond angles θ_{15}, θ_{24} and the parameter ϕ, as defined on the right of the diagram. Points on the lower right are fSQP, those in the center eSQP, and those at the top right and left are TBP conformers. The solid line maps the theoretical coordinate for the deformation described on the right of the scatterplot

rangement, and associated angles subtended at the central metal. Using this approach, Muetterties and Guggenberger arranged seven pentavalent complexes containing six different central atoms, into a series convincingly mimicking the Berry path (**B**) in Figure 8.1. The corresponding diagram yields such a good visual image of this structural pathway that it has found its way into an inorganic chemistry textbook [16].

Five-coordinate nickel complexes were later found to map the third of the three basic pathways, namely step (**A**) (Figure 8.1), as well as the Berry coordinate. Data on 78 five-coordinate nickel complexes were analyzed by the dihedral angle method [17], and the results are shown in Figure 8.3. Points on the lower right represent compounds mapping the incipient stages of step (**A**), i.e., flattened SQPs, those at the center represent elevated SQPs, while points in the upper right and left represent TBP complexes. Although this dihedral angle method sheds more light on the static deformations of pentavalent nickel complexes, no separation is apparent between the various conformers. In addition, it is not clear how to interpret the scatter about the theoretical coordinate: This may arise from structural variations that are not adequately modeled by the method, which includes only three geometric parameters – and may hence be biased by the choice of variables. Finally, this method of describing geometrical deformations is limited to angular distortions only; it yields no (direct) information on changes in bond lengths or non-bonded distances.

8.2.3.2 More Sophisticated Methods of Analysis

Section 8.2.3.1 outlined the three basic deformation paths of pentavalent complexes. In this section we shall describe how developments in methods of analysis have led to deeper insights into these distortion patterns.

An extension of Muetterties and Guggenberger's dihedral angle method, which incorporates all nine dihedral angles of the coordination polyhedron, was proposed by Holmes and Deiters [18]. Their method depends on determining the sum of the differences between the dihedral angles $\{\delta_{ij}(C)\}$ for each edge (ij) in the observed structures C, and the corresponding angles in the reference conformers $\{\delta_{ij}(SQP)$ and $\delta_{ij}(TBP)$, respectively$\}$. A plot of $217.9° - \sum_{ij} |\delta_{ij}(C) - \delta_{ij}(SQP)|$ versus $\sum_{ij} |\delta_{ij}(C) - \delta_{ij}(TBP)|$, where $217.9° = \sum_{ij} |\delta_{ij}(TBP) - \delta_{ij}(SQP)|$, then gives a good representation of the Berry pathway. Figure 8.4 illustrates the adherence to the Berry coordinate of the 78 nickel complexes mentioned earlier [17], as mapped by

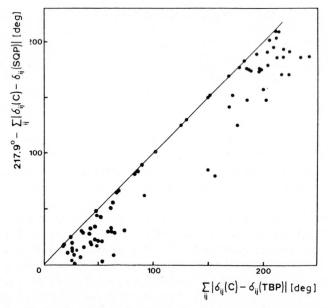

Fig. 8.4. Diagram showing the adherence of 78 nickel complexes to the Berry coordinate (solid line). The origin represents a TBP, the upper right end of the coordinate a SQP

the method of Holmes and Deiters. As can be seen, the data tend to cluster fairly evenly around both ends of the coordinate. This suggests a small energy difference between the TBP and SQP, since a large difference would presumably result in preferred clustering of observed geometries about the lower energy conformer, with a relative paucity of points around the higher energy one. However, there is fairly significant scattering away from the theoretical pathway linking the TBP and SQP. Whether this scatter arises randomly from the differing crystalline environments, or whether it reflects some underlying factors cannot be established from the scatterplot itself. Holmes has investigated subsets of the above dataset, dividing the compounds for which the spin state was known into low-spin and high-spin complexes (fourteen and nine compounds, respectively) [19]. His analysis showed that low-spin complexes have a preference for the fSQP conformation, high-spin ones for the eSQP, corresponding to the beginning and end of step (**A**) in Figure 8.1, respectively. The chemical implications of this (for spin-state transitions, for example) were not discussed.

Pentacoordinate zinc compounds have also been investigated by means of scatterplots and the dihedral angle method [20]; an analysis of 33 different fragments (represented by their internal coordinates) has enabled a mapping of steps (**B**) and (**C**) (Figure 8.1). Results are similar to those for the cadmium complexes and the 78 nickel complexes described earlier. A sample of 44 five-coordinate molybdenum complexes has been similarly treated, and coordinates for an S_N2 reaction and a Berry exchange were obtained [21]. A Berry coordinate has also been mapped from an analysis of nine structures of the type $[Fe(CO)_3P_2]$, where P_2 represents either a bidentate, or two monodentate phosphine-type ligands [22]. In the latter case, the fairly even distribution of observed geometries along the TBP→SQP trajectory again suggests a low energy barrier for this deformation; indeed, two of the compounds are fluxional even at temperatures as low as $-80\,°C$.

Although the dihedral angle method has helped to resolve the complexes more clearly into different conformers, there are nonetheless problems associated with it: for example, for intermediate geometries there is some arbitrariness in relating angles in observed structures to corresponding angles in the reference conformers. Moreover, the method requires an *a priori* definition of what constitutes a SQP, so that the $\delta_{ij}(SQP)$'s may be calculated.

Perhaps the most comprehensive analysis of the system discussed here is in a study involving a multivariate analysis of a dataset comprising 196 pentacoordinate d^8 complexes, containing the metals nickel, palladium, platinum, rhodium, and iridium [23]. The coordination geometries were expressed by means of a complete set of twelve symmetry deformation coordinates (Chapters 1 and 2) for both TBP and SQP conformations. Each ML_5 fragment was thus represented by a point in two different 12-dimensional parameter spaces: *T*-space, where each structure is related to a TBP, and *S*-space, where each is related to a SQP. For both spaces the data were symmetry expanded (D_{3h} and C_{4v}, see Chapters 2 and 4), leading to 2352 and 1568 data points in *T*- and *S*-space, respectively. The resultant twelve-dimensional data distributions were then analyzed by means of both hierarchical and non-hierarchical CA, and PCA (see Chapter 4). Four clusters are found in both spaces (see Figure 8.5): in *T*-space a central cluster (cluster **T3**) representing an ideal TBP is surrounded by

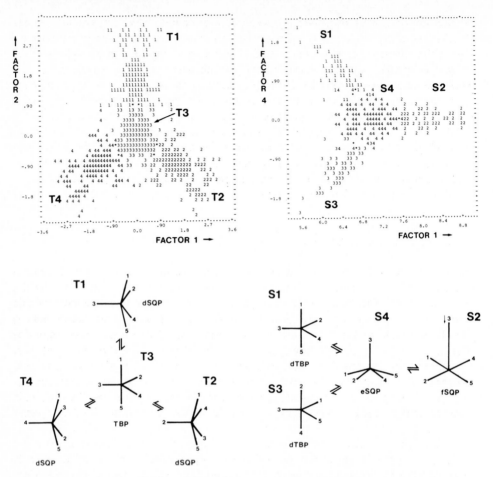

Fig. 8.5. Results of cluster analysis of d^8 metal complexes. The scatterplot on the left shows the clusters formed in *T*-space, that on the right those formed in *S*-space. Below each scatterplot is a qualitative interpretation of the cluster archetypes, and the deformation coordinates relating them

three identical clusters (**T1, T2, T4**) whose median conformation (archetype) is a distorted SQP, while in *S*-space the data cluster about a fSQP and an eSQP archetype (**S2** and **S4**), as well as two permutationally equivalent TBP archetypes (**S1, S3**).

PCA did not lead to a dramatic reduction in dimensionality — indicating the substantial simplifications made in the earlier studies [1, 17, 19–22], and attesting to the flexibility of these complexes — but it did reveal chemically meaningful results. For the TBP the two most important distortions are the S_N2 and the Berry deformations (steps **(C)** and **(B)**), accounting for 15 and 24 percent of the angular variance: this suggests that five-coordinate Rh and Ir compounds — which are significantly represented in the TBP cluster — may undergo dissociation via an S_N2

mechanism to form a (short-lived) tetrahedral intermediate, even though this may not be evident from the chemistry of their (square-planar) tetracoordinate complexes. The eSQP, meanwhile, was shown to distort along both the addition/elimination pathway (step (**A**), 14 percent) and the Berry coordinate (12 percent). In contrast, the fSQP maps only the incipient stages of the addition/elimination reaction (14 percent of variance), and was not seen to distort along the Berry path. This suggests that pentacoordinate Pt and Pd complexes – which are principally represented in the fSQP cluster – may not undergo intramolecular ligand exchange by the Berry mechanism, or, at least, have to undergo a considerable distortion of their coordination sphere to an eSQP, before this mechanism can operate. This observation may have a corollary in the lack of unambiguous experimental results that indicate Berry fluxionality in complexes of these metals [23 c].

8.2.3.3 Comparison of Structural Results with a Point-Charge Model

Kepert and co-workers have developed a simple repulsion model in which the ligand donor atoms – regarded as point charges – are assumed to be distributed on the surface of a hard sphere, with the metal at its center [24]. The main purpose of this model lies in classifying, rationalizing, and predicting the stereochemistries of $[ML_n]$ complexes, where L represents either unidentate, bidentate, or less commonly, tridentate ligands. For the stoichiometry $[M(unidentate)_5]$, for instance, the data – including both transition and non-transition metals – appear to fall into two categories: those that map the TBP→SQP transition, and those that do not, the latter to a large extent mapping the apical departure from a SQP instead [24, 25]. This finding was later corroborated in the analysis of 78 five-coordinate nickel complexes discussed above [17].

In some instances Kepert also compared the distribution of available data with the topography of the potential energy surface (PES) obtained from his model. In their study of $[M(unidentate)_5]$ complexes, Favas and Kepert define an axis in the TBP, such that the angles between this axis and each of the ligand atoms, A, B, and C, are equal; angles ϕ_D and ϕ_E are then measured between the axis and the bonds to D and E, respectively. Figure 8.6 depicts the PES for $[M(unidentate)_5]$ molecules with C_s symmetry, projected onto the ϕ_D/ϕ_E plane. The positions of the TBPs and SQPs are shown by T and S, respectively, and the PES is symmetrical about the line $\phi_E = 180° - \phi_D$. A scatterplot of the structural data for the 33 zinc complexes mentioned earlier [20], onto the corresponding ϕ_D/ϕ_E plane, strikingly mirrors the topography of the PES depicted in Figure 8.6. For complexes of the type $[MA_2B_3]$, where A and B are unidentate, the PES has similar general features to those shown in Figure 8.6, but the relative energies of stereochemically distinct TBPs and SQPs, i.e. of the potential minima (T_0, T_1, T_2, T_3) and the saddle points (S_1, S_2, S_3), depend on the relative lengths of the M−A and M−B bonds, and it is therefore not possible to obtain a unique representation of the PES in terms of this model. Nonetheless, a plot of the angular data for 51 mixed complexes of this kind [25] reproduces the broad outlines of the PES for these $[ML_5]$ complexes, as shown in Figure 8.7. An analysis of 16 $[M(tridentate)(unidentate)_2]$ compounds with a variety of transition

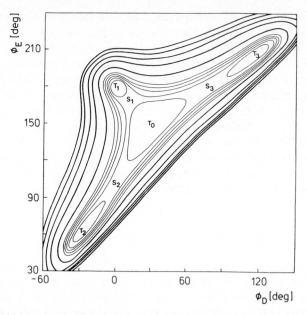

Fig. 8.6. PES for [M(unidentate)$_5$] complexes. T_0, T_1, T_2, and T_3 represent TBP conformers that may be interconverted through SQP intermediates, represented by S_1, S_2, and S_3. The angular parameters ϕ_D and ϕ_E are described in the text

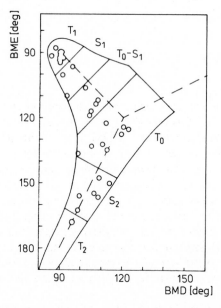

Fig. 8.7. Angular data for 51 mixed [M(unidentate A)$_2$ (unidentate B)$_3$] complexes [26b]. Note the similarity between the outlines of this distribution and those of the PES in Figure 8.6

and non-transition metals has provided an empirical mapping of a stereochemical interconversion of two distorted TBPs with a SQP intermediate, that maintains C_s symmetry [26a]. The process is not equivalent to the Berry mechanism and it has no readily recognizable counterpart in proposed [ML$_5$] reaction mechanisms. Nevertheless, the data partially lie along troughs in the PES for this system, as established using Kepert's model. Lastly, the amount of square-pyramidal distortion of eleven [Cu(planar tridentate)(unidentate)$_2$] complexes has been analyzed and quantified using the dihedral angle method of Muetterties and Guggenberger, and the data distribution compared to the PES for this system according to Kepert's model [27]. A comparison of results showed that the ordering of the complexes along the TBP→SQP coordinate, using the approach of Muetterties and Guggenberger, was similar to the ordering that emerged from Kepert's model.

8.2.3.4 Retrospective Comments

By and large the numerous studies on pentacoordinate complexes have served to substantiate earlier hypotheses concerning the likely geometric route that reaction pathways for ML$_5$ or (ML$_4$+L) systems might take. Even if those studies employing simpler two-dimensional methods of analysis may not have significantly deepened our understanding of the mechanisms, they nonetheless lend support to the structure correlation hypothesis. The multidimensional analysis, on the other hand, has revealed geometric details which suggest that our understanding of these reaction mechanisms is not yet complete and that they require more careful study, employing techniques that allow for shorter time-scales.

8.2.4 ML$_6$ Complexes

The first systematic attempt to analyze the structures of six-coordinate complexes with a view to obtaining information about their fluxional behaviour, was that of Muetterties and Guggenberger [6], who applied their dihedral angle method to a sample of 14 tris-chelate complexes of a variety of metals. As reference structures for the polytopal rearrangement they chose a regular octahedron and a trigonal prism (D_{3h} symmetry): these conformations may be interchanged by means of a trigonal twist − also known as the Bailar twist − about one of the C_3 axes of the octahedron. All of the structures in their study exhibited D_3 (or approximate D_3) symmetry, but they did not span the entire $O_h(D_{3d}) \to D_{3h}$ trajectory, tending to map the first half of this pathway only. At the time, the authors speculated that the lack of structures falling onto the second half of the $O_h \to D_{3h}$ pathway may be a reflection either of a lack of synthetic studies using appropriate ligands, or of the energetic nature of the process.

    As part of his general investigation of the [ML$_6$] system, Kepert has also modeled this trigonal twist mechanism for [M(bidentate)$_3$] polyhedra [24, 28]. The

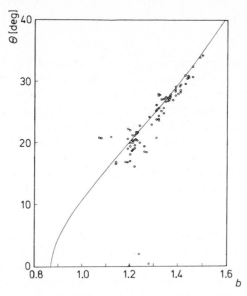

Fig. 8.8. Scatterplot of θ versus b for 158 complexes, and theoretical coordinate (solid line) for the Bailar mechanism joining the octahedron ($\theta = 30°$, $b = \sqrt{2}$) with the trigonal prism ($\theta = 0°$). The parameters θ and b are described in the text

geometric parameters that he employs to define the stereochemistries are the bite b of the chelate ligand, normalized with respect to bond distance, and a twist angle θ between two opposite triangular faces. A scatterplot [28] of the corresponding variables for a sample of 158 mainly transition metal complexes (Figure 8.8), reveals interesting results: the data mostly scatter about the octahedron ($\theta = 30°$, $b = \sqrt{2}$) and the first half of the coordinate joining this geometry to the trigonal prism ($\theta = 0°$), with a relative scarcity in the vicinity of the latter — in line with the earlier finding of Muetterties and Guggenberger. A comparison of the data clustering in Figure 8.8 with the various scatterplots illustrating the Berry mechanism for [ML$_5$] polyhedra suggests, firstly, that the energy barrier to the trigonal twist mechanism may well be higher than that for the Berry process, and, secondly, that the relative energy differences between the reference geometries for the [ML$_6$] case are larger than those for the [ML$_5$] system.

Neither Kepert, nor before him Muetterties and Guggenberger, considered alternative isomerization mechanisms of tris-chelate complexes, such as the Ray-Dutt mechanism, which involves a twist about a pseudo-C_3 axis of the complex, resulting in an intermediate trigonal prism with C_{2v} symmetry instead of D_{3h}. Rodger and Johnson, for instance, have argued that the Ray-Dutt twist will predominate when the ligand bite b is larger than the L−L hard-sphere distance (where L represents the coordinating atom), while the Bailar twist will only be operable when the converse holds [29].

8.2.5 ML_n complexes, $n > 6$

Increase in the coordination number brings greater complexity due to (i) the greater number of possible reference geometries [30], (ii) a consequent increase in the number of mechanistic options available to the ML_n moiety, and (iii) the greater number of geometric parameters necessary for the representation of the polyhedra. These problems, coupled to the significantly smaller number of available structures as n increases, have militated against these systems receiving much attention. Muetterties and Guggenberger, for instance, analyzed in detail the theoretical deformation pathways for $n = 7$ and $n = 8$ polyhedra, but had no more than half a dozen structures for either stoichiometry to map the pathways [6]. They fared only slightly better with their discussion of ML_9 structures: here they were able to use about a dozen complexes [31].

Seven-coordinate complexes have been examined also by Kepert [24, 32] and Drew [33]. In the case of seven-coordination, the problems outlined above are complicated by the absence of any regular polyhedron that could be used as a reference geometry against which to measure observed structures. As reference geometries, all three studies [6, 24, 32, 33] chose the pentagonal bipyramid (PB; D_{5h} symmetry), the capped octahedron (CO; C_{3v}), and the capped trigonal prism (CTP; C_{2v}), which are all related by means of simple edge stretches [6]. Both Kepert and Drew analyzed a fair number of seven-coordinate structures (of the order of 120 in Kepert's case) with a variety of transition and non-transition metals, and various combinations of uni- and bidentate ligands. Some of the complexes were discussed in the context of their positions along the various deformation pathways connecting the three reference polyhedra, but no correlations between the structures were extracted. Muetterties and Guggenberger – using their dihedral angle method – similarly commented briefly on the distribution of data about the reference structures, without much discussion of the mappings of distortion pathways produced by them. Nonetheless, in all three studies few complexes were found to cluster about the CO, most lying about the PB and the CTP. By contrast, the CTP does not appear in the energetically feasible deformation pathways proposed from theoretical considerations by Rodger and Johnson for seven-vertex clusters, while the PB and the CO do [34].

Eight-coordinate complexes have been studied by Kepert [24, 35], Drew [36], and Muetterties and Guggenberger [6]. The most commonly employed reference geometries were the dodecahedron (D; D_{2d}), the square antiprism (SA; D_{4d}), and the bicapped trigonal prism (BTP; C_{2v}), which again are interrelated through edge-stretch mechanisms. The studies agree that most of the compounds cluster about the D and SA, along the coordinate joining them, and along the first part of the pathway connecting each of them to the BTP; indeed, Drew comments that "the vast majority of eight-coordinate complexes can be described in terms of one or both of these geometries" – meaning the D and SA [36]. Other (minor) observed geometries include the cube (O_h) and the hexagonal bipyramid (D_{6h}). Two energetically feasible rearrangement mechanisms for eight-vertex clusters have been proposed from theoretical considerations: both are based on deformations of the D, and neither has SA or BTP intermediates/components [34].

Results of hard-sphere repulsion calculations on ML_9 and M_9 systems were first compared with structural data by Muetterties and Guggenberger with a view to mapping deformations paths [31]. They report that all calculations on hand yielded the tricapped trigonal prism (TTP; D_{3h}) as the minimum energy structure, with the alternative polytope, the monocapped square antiprism (MSA; C_{4v}) as an intermediate in a TTP→TTP rearrangement that maintains C_{2v} symmetry. The structures of seven ML_9 complexes, where L stands for equivalent monodentate ligands, and three M_9 clusters accordingly agglomerated around the TTP reference, with very little displacement in the direction of the MSA. Favas and Kepert's later calculations indicate a much smaller energy difference between the two polytopes than is suggested by the clustering observed by Muetterties and Guggenberger, and they even yield a lower-energy MSA for [M(bidentate)$_4$(unidentate)] complexes, with some structures falling into this pattern [24, 37]. Drew found no evidence that polyhedra other than the TTP and MSA form any minimum on the PES, and he furthermore maps the TTP→MSA→TTP pathway from values of the dihedral angles in the structures of several intermediate compounds – largely non-transition, mono- and polymeric in nature [36]. Theoretical considerations have yielded essentially similar mechanisms for intramolecular rearrangements in M_9 clusters [34].

Drew and also Favas and Kepert have analyzed possible isomerization mechanisms for values of n greater than nine, but the structural information for complexes of these coordination numbers is so scanty that nothing can be said about their deformation paths [36, 24, 37]. Gross simplifications are involved in the studies discussed particularly in the last part of this section, since systems containing upwards of eight ligand atoms have generally been represented by two or three geometric parameters, while they in fact have $2n$-3 independent LML-angles – obviously much structural information must have been lost in this approach.

8.3 Discriminating Between Reaction Mechanisms: Metal Cluster Rearrangements

Metal clusters containing trigonal bipyramidal M_2Ru_3 fragments (M=Cu, Ag, Au) have been shown to exhibit dynamic behavior in solution, involving an intramolecular site exchange of the M atoms in the TBP [38]. In the absence of any geometric information about the interconversion pathways, Rodger and Johnson have put forward three hypotheses concerning possible mechanisms for such polyhedral rearrangements [34]. They postulate that there are only two possible mechanisms [39] for site exchange in five-atom clusters (Figure 8.9a): the first involves the stepwise cleavage and re-formation of a polyhedral edge, with an edge-bridged tetrahedral intermediate, while the second involves a concerted process and a SQP intermediate, akin to the Berry process. An examination of the structures of

19 Au_2Ru_3 fragments contained in clusters with a variety of carbonyl, phosphine and μ_3-ligands (e.g. S, COMe, etc.), has revealed [40] that site exchange most likely occurs by means of the Berry process. It proved possible to rule out tetrahedral edge-bridged intermediates and the alternative turnstile mechanism [41], on the basis of the structural evidence. The mechanism involves an exchange of the two Au atoms and two of the Ru atoms in the axial and equatorial positions of the TBP, while the integrity of the Ru_3 triangle is maintained, as illustrated in Figure 8.9. The data, composed of interatomic distances and selected bond and dihedral angles, were symmetry expanded according to the C_s symmetry group common to the idealized TBP and SQP clusters, and the analysis included multivariate techniques and scatterplots. PCA revealed one principal component (PC) accounting for 92% of the variance in the ten metal-metal distances chosen to represent the fragments (see Figure 8.9a); by far the major components of this PC are the symmetry-equivalent distances $Au(1)-Ru(2)$ and $Au(2)-Ru(1)$. The proposed mechanism is fully consistent with

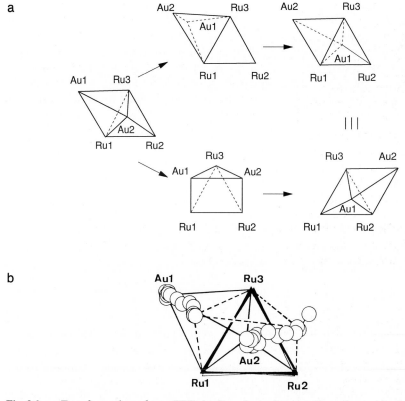

Fig. 8.9. a: Transformation of one TBP Au_2Ru_3 cluster into another, via an edge-bridged (upper line) or SQP intermediate. Note that the Ru_3 triangle remains unchanged throughout. The fragments in the study were defined by the nine distances indicated in the initial TBP, plus that between $Au(1)$ and $Ru(2)$

b: Trajectory mapped by the Au atoms when the Ru_3 triangles of the structures are overlapped

the results of NMR studies on the dynamic behavior of these fragments in solution [38].

8.4 Reactions of Organometallic Compounds

8.4.1 Ring-Whizzing Reactions

In organic chemistry, walk rearrangements (or circumambulations) involve the migration of a CH_2 (or CR_2) group around the periphery of a polyene ring, by a series of sigmatropic reactions [42]. A similar process has been observed in $[(C_n)ML_2]$ complexes (C_n = a cyclic polyene or a cyclopropene, and L = ligands such as PPh_3 or CO), wherein the ML_2 fragment moves around the perimeter of the cyclic ene, from one η^2 position to another. The structures of four cyclopropenium complexes, with M = Ni, Pd, and Pt, have been shown to chart the reaction path for this site exchange, as determined by molecular orbital calculations of the extended Hückel type [43]. The migration of the metal atom is accompanied by a simultaneous rotation of the entire ML_2 fragment with respect to the cyclic ene, as shown in Figure 8.10. In a subsequent publication further calculations on polyene complexes are reported, and a brief comparison with published structures is made [44].

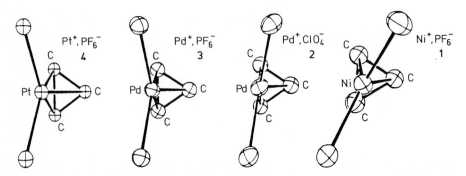

Fig. 8.10. Migration of the ML_2 fragment as mapped by four structures. The structure on the left closely resembles the η^2 ground state, the one on the right the transition state, as determined by extended Hückel calculations on $H_3C_3Ni(PH_3)_2$. Note that the ML_2 moiety rotates w.r.t. the cyclic ene, as M migrates from one C−C bond to the next

8.4.2 Agostic Interactions as Precursors to H-Transfer Reactions

Recently it has been shown [45] that $C-H$ bonds of (terminal) methyl groups in alkyl or aryl ligands can be oxidatively added to coordinatively unsaturated, complexed metal ions, as in

$$C-H + M \rightarrow C-H \ldots M \rightarrow C-M-H .$$

In the solid state, complexes of such ligands often exhibit agostic $C-H \ldots M$ interactions between a hydrogen atom in the ligand and the metal atom. Although cyclometallation in these complexes is rare, relative to activation of an external $C-H$ bond [46], such complexes may be viewed as precursors in oxidative addition reactions, and an analysis of 17 structures containing a $C-H \ldots M$ bridge (with a variety of metals M) has revealed the reaction trajectory outlined in Figure 8.11 [46].

The initial approach angle ($C-H-M$) of about $130°$ was rationalized in terms of a proposed interaction between the $C-H$ σ bond and an empty metal d orbital, and back-donation from a filled d orbital to a σ* orbital of the $C-H$ system, but this angle may also result from geometrical factors. The role of steric and conformational effects in the balance between cyclometallation and attack on an external substrate (such as an alkane), was discussed in the context of the above reaction path. The conclusion was that sterically uncongested metal systems are more likely to activate alkanes, than to undergo cyclometallation.

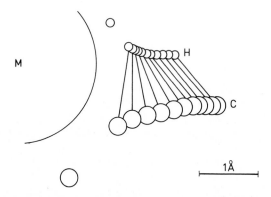

Fig. 8.11. Trajectory of approach of the $C-H$ bond to the metal atom M in an oxidative addition. The radius of M is proportional to the metal covalent radius [46], but the radii for C and H are arbitrary

8.4.3 Carbonyl Transfer Reactions

Some years ago, Cotton [47] proposed that semi-bridging CO groups in carbonyl complexes (as in types (2) and (4)) may be thought of as intermediates between terminal (1) and symmetrical bridging (3) arrangements.

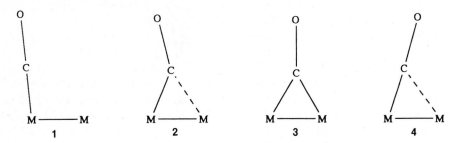

Indeed, Crabtree and Lavin found a fairly smooth continuum of carbonyl structures between the extremes of (1) and (3) in an examination of 111 binuclear metal-carbonyl fragments [48]. The data comprised mainly iron complexes, though other metals were used when "suitable data were not available for iron". In such cases, observed interatomic distances were adjusted for differences in covalent radii. Figure 8.12 depicts the reaction path for transfer of CO from one metal atom to the other, as mapped by the binuclear complexes studied. Terminal (1) and symmetrical bridging (3) CO's are more prevalent than bent semi-bridging (2) CO's, but amongst the latter no specific geometry predominates. This suggests a broad low-energy barrier between the more stable structures (1) and (3). Nonetheless, an empirical differentiation between bent (2) and linear semi-bridging (4) CO arrangements was made, and the latter (46 fragments) were, in fact, not included in the mapping of the reaction trajectory. It was argued that the group of linear semi-bridging CO's (with $M-C-O$ bonds $165\,°-180\,°$) could be subdivided into three major types, whose chemical and structural features were discussed at length.

Finally, trinuclear carbonyls of the type $[Fe_3(CO)_{12-n}L_n]$ ($n = 0-3$; $L = PPh_3$, $CNBu^t$, etc.) are known to exhibit fluxional behavior that exchanges CO's between

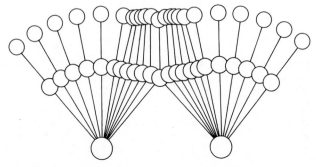

Fig. 8.12. Carbonyl transfer trajectory for binuclear iron complexes

the metal atoms. It has been suggested [49] that the X-ray structures of five such complexes can be viewed as mapping incipient stages in a CO exchange involving a concerted double-bridge formation between two of the iron atoms, and a concomitant opening of an already existing bridge. Expressed differently, this mechanism would appear to circulate the double-CO-bridges (between two metal atoms) around the Fe_3 ring.

8.4.4 Substitution Reactions at Sn(IV)

Britton and Dunitz have examined in great detail the structures of 186 four-, five-, and six-coordinate Sn(IV) complexes, classifying them into various subsets according to the kinds of atom bonded to the tin ([50] and Chapter 7, Section 7.3). The structures were all represented by their internal coordinates, and the correlation analysis involved only the use of scatterplots. Using the approach devised earlier by Bürgi in his study of five-coordinate Cd complexes [1], the penta-coordinated organotin structures ($YSnR_3X$; R = organic ligand) were shown to map an S_N2 pathway with inversion of configuration:

$$Y + SnR_3X \rightleftharpoons [Y \ldots SnR_3 \ldots X] \rightleftharpoons YSnR_3 + X$$

A very small number (seven) of the penta-coordinate complexes have an arrangement of the five ligand atoms that suggests a mapping of a similar bimolecular pathway, but with retention of configuration. The distribution of these structures also suggests that the five-coordinate intermediate is very unlikely to undergo isomerization by means of the turnstile rotation [41]; possible isomerization through the alternative Berry rearrangement was not investigated (see also Chapter 7, Section 7.4.6).

Correlations found for the six-coordinate complexes ($SnR_2X_2Y_2$) reveal a possible mechanism for an (unprecedented) termolecular reaction involving a concerted double addition and elimination process in which the R ligands adopt transoid positions, and the X and Y groups are mutually *cis* in the intermediate (or transitional) octahedral conformer; Britton and Dunitz dubbed this an S_N3 reaction:

$$2Y + SnR_2X_2 \rightleftharpoons [Y_2 \ldots SnR_2 \ldots X_2] \rightleftharpoons Y_2SnR_2 + 2X \ .$$

At the time the authors knew of no experimental observations that indicated such an entropically disfavored pathway, but they speculated that it "may be important for enzymatic reactions". Interestingly, a very similar S_N3 pathway may be operable in the base-catalyzed inversion of Λ-*cis*-(+)-$[Co(en)_2Cl_2]^+$ to Δ-*cis*-(−)-$[Co(en)_2(OH)_2]^+$, which Jackson and Begbie found to take place via a route that involves the simultaneous loss of the two Cl^- ions in a single observable step, and to be promoted by OH^- in "some unprecedented way" [51a]. The authors propose a complicated modification of the classical S_N1 conjugate base mechanism in order to avoid coming to the conclusion that the mechanism is indeed a termolecular one − and then concede that their account is "not especially compelling" [51b]. It remains to be seen whether further studies will reveal other systems in which such termolecular reactions appear to take place; but in any event, structure correlation has suggested a geometrically feasible pathway whereby this might proceed.

8.5 Beyond Geometry: Empirical Potential Energy Surfaces?

The hypothesis of structure correlation begins by assuming that crystal structures can be represented by points lying in regions of low potential energy – along minimum energy paths – on the corresponding PES; correlations between structural parameters describing the conformations of the fragments lying along these paths yield information about the geometric changes occurring along the particular reaction pathway. The studies discussed in the previous sections of this chapter typify this strictly geometric approach to reaction pathways. But at least on a theoretical level there would also appear to be a relationship between the position of a representative point on the PES, and the energetics of the process. This raises the alluring thought that the results of a structure correlation analysis might possibly also yield insight into the energetics of a given reaction.

In this section we shall first review some studies of copper(II) complexes in which explicit attempts were made to couch the geometric reaction paths in an energetic context, and shall then go on to discuss and contrast two divergent approaches to obtaining empirical PESs from structural data. One approach, discussed in Section 8.5.2, below, is to assume that the distribution of representative points on the PES can be approximated by a Boltzmann-type function, which is then used to model the distribution density of observed points in the space spanned by the geometric parameters. An alternative approach, discussed in Section 8.5.3, begins by assuming that each structure is characterized by a unique PES, and then proceeds to relate experimentally determined activation energies for each individual compound to small changes in molecular structure along the reaction path.

8.5.1 The Case of Cu(II) Complexes: Problems and Possibilities

Complexes of copper(II) exhibit a wide variety of regular and non-regular stereochemistries, partially as a result of the lack of spherical symmetry of this d^9 ion. In particular, it is often impossible to distinguish between penta- and hexa-coordination. This feature has led to attempts to classify these complexes as $(4+2)$ when there are four short and two long bonds, or as $(4+1+1^*)$ in which case there are four short bonds, one long one and one (1*) of intermediate length [52]. These configurations could alternatively also be considered as a perturbed ML_4 geometry (4+2; or 4+1+1*), perturbed ML_5 (4+1+1*), or distorted ML_6 (4+2; or 4+1+1*). Such a structural spectrum inevitably invites attempts to map deformation pathways. But it might also frustrate those attempts, since the broad range of conformations could be indicative of a flat PES, in which case correlations may not be easily identifiable.

8.5.1.1 Cu(II)L$_5$ Complexes

The geometry of the $[Cu(bipy)_2Cl]^+$ cation (bipy = 2,2' bipyridyl) has been shown to be closer to TBP, but distorted towards SQP, in a series of five complexes with various anions [52−56]. Correlations among some of the internal parameters of the CuN_4Cl chromophore led Hathaway and colleagues to suggest that these structures map out a TBP→SQP transformation. In terms of the Berry mechanism there are three possible deformations of a trigonal bipyramidal CuN_4Cl polyhedron, of which two are equivalent (see Figure 8.13). The data for the five $[Cu(bipy)_2Cl]^+$ complexes seem to map pathway (**E**) (or the equivalent path **F**), not path (**D**) (as defined in Figure 8.13). Since the distortion along path (**E**) does not maintain C_2 or C_{2v} symmetry, Hathaway preferred to rationalize the deformation in terms of a linear combination of the normal modes of vibration of a trigonal bipyramidal $[CuN_4Cl]^+$, rather than refer to it as a Berry mechanism. Path (**D**), however, retains C_2 symmetry, and can hence "be considered as a Berry Twist-type mechanistic pathway"; the latter path is mapped by three out of four $[Cu(bipy)_2(OH_2)]^{2+}$ complexes [54].

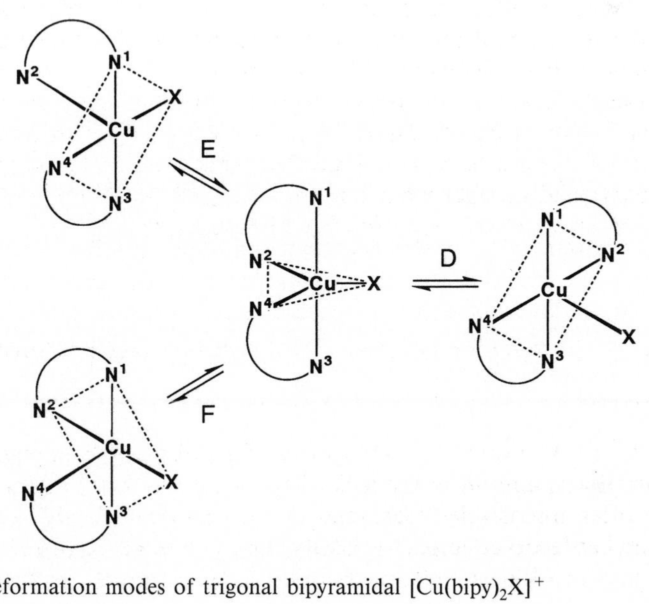

Fig. 8.13. Deformation modes of trigonal bipyramidal $[Cu(bipy)_2X]^+$

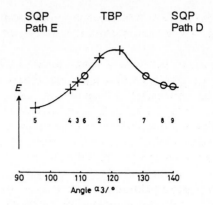

Fig. 8.14. The curve depicts the extended Hückel energy profile for the model complex $[Cu(NH_3)_4Cl]^+$, as a function of one of the $N-Cu-N$ angles, defined as α_3 ($N(2)-Cu-N(4)$) in Figure 8.13. The crosses represent the observed values of α_3 for complexes of the type $[Cu(bipy)_2X]^+$ with X = Cl, the circles those with X = OH_2; these values of α_3 have simply been plotted onto the curve obtained from the Hückel calculations

Extended Hückel molecular orbital calculations [52, 53] were then used to estimate the relative total energies for a $[Cu(NH_3)_4Cl]^+$ cation, as a function of one of the $N-Cu-N$ angles, defined as α_3 (equivalent to the angle $N(2)-Cu-N(4)$ in Figure 8.13). The resulting energy profile is shown by the curve in Figure 8.14. Plotted onto the profile are the values of α_3 for the chloro complexes (crosses) and the aquo complexes (circles). The diagram should not be misconstrued as suggesting a correlation between the observed values of α_3 and the energy of the complex, but it does serve to illustrate the idea behind modeling the data distribution density: if a large enough sample pertains, then it might be possible to model the topography of the energy well (or minimum energy path), and thereby perhaps obtain information concerning the shape of the PES. However, in this case the calculations tellingly revealed the TBP to be of a slightly higher energy than the SQP, even though most of the structures approximate more closely to the TBP than the SQP; perhaps this illustrates the point that the PESs for the various observed structures and the model complex are significantly different, even though the complexes have similar stoichiometries. In a subsequent paper the results of variable temperature experiments on the $[Cu(bipy)_2X]^+$ cations (X = Cl or I) are reported, and it is shown that they exhibit no fluxionality [57]; in view of this, it is not clear how to interpret the structural correlations obtained for these complexes in the earlier reports. In another study, Hathaway and colleagues have proposed that three related complexes with the CuN_5 chromophore may be viewed as stationary points along a Berry pathway, but no further reports have emerged for this system [58].

8.5.1.2 Cu(II)L$_6$ Complexes

In early communications the geometry of the chlorocuprate fragment $[CuCl_6]^{4-}$ in 13 different crystalline environments was cursorily discussed [7, 8]. The structures were all found to be of the (4+2) type, and a hyperbolic pathway for their dissociation to $[CuCl_4]^{2-}$ was obtained in terms of the average axial and equatorial bond lengths [8].

Much attention has since been focussed on a series of 18 complexes of the type $[Cu(LL)_2X]^+$, where LL is either 2,2'-bipyridyl (bipy) or 1,10-phenanthroline (phen), and X represents potentially bidentate ions that coordinate through two oxygen atoms, such as ONO^-, CH_3COO^-, or $HCOO^-$ [52, 59, 60]. The stereochemistry of the CuN_4O_2 chromophore in many of these complexes is significantly dependent on temperature, which has led to them being termed "fluxional". Moreover, for some of the complexes the bidentate ligand X — which necessarily occupies *cis* positions — is coordinated far more strongly through one O atom than the other, so that these coordination polyhedra are considerably distorted from approximate O_h or tetragonal symmetry. The extremal distortions are of the (4+1+1*) kind [59b]. The geometric distortions of the N_4O_2 coordination polyhedron were mapped principally by the parameters $\Delta(Cu-N_{eq})$ and $\Delta(Cu-O_{eq})$, defined as the respective differences between the two Cu−N and Cu−O bonds in the equatorial plane of the (presumed) octahedron [60]. A linear relationship (Figure 8.15a) that describes a distortion of the CuN_4O_2 chromophore away from a "*cis*-distorted octahedron" with C_2 symmetry toward the extremal, low energy conformations, (Figure 8.15b, bottom) was found. No attempt to interpret this "structural pathway" in terms of the chemical reactivity of the $[Cu(LL)_2X]^+$ fragment was made [61]. The relationship between the structural variations mapped by some of the complexes and their electronic spectra was also discussed, but no empirical relationships between geometry and energy were reported.

An empirical PES for the deformation of $[Cu(bipy)_2(ONO)](NO_3)$ according to the scheme in Figure 8.15b, bottom, was obtained [60] by assuming an adiabatic PES with ^2A ground and ^2B excited electronic states, and two non-equivalent ground-state conformers, as shown in Figure 8.15b, top. From a series of crystal structures of the complex at temperatures of 20, 100, 165, and 296 K, the temperature variation of the coordination polyhedron was determined, and by applying Boltzmann statistics to this, an energy difference $\Delta E = 220$ cal mol^{-1} between the two ground-state conformers could be estimated. It was pointed out that this value is comparable with those obtained from variable temperature electron spin resonance experiments. Importantly, the approach to the empirical PES is in this case premised on the application of Boltzmann statistics to the variable temperature data for a single molecule, not for an assemblage of different ones (Chapter 5, Section 5.4).

8.5.2 "Pseudorotation" in Co(ethylenediamine) Chelate Rings

Pseudorotation — the process whereby envelope (E) and twisted (T) conformers (with C_s and C_2 symmetry, respectively) of cyclopentane are interconverted — was first proposed on the basis of a study of the dynamic puckering in this molecule [62]. The process may be conveniently mapped by two parameters: the puckering ampli-

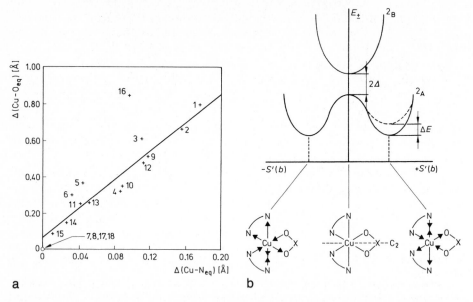

Fig. 8.15. a: Least-squares plot of $\Delta(Cu-N_{eq})$ versus $\Delta(Cu-O_{eq})$ for CuN_4O_2 chromophore
b: Lower part: distortions of the CuN_4O_2 chromophore mapped by the coordinate in a; upper part: proposed adiabatic PES model for the pseudodegenerate ^2A ground and ^2B excited electronic states of the CuN_4O_2 chromophore

tude q, measuring the root-mean-square displacement of the ring atoms from the mean plane, and a phase angle (ϕ or P) that expresses how the puckering varies around the ring. Alternative mathematical treatments have since been offered [63].

Pavelcik and Luptakova have applied this concept to a conformational analysis of a symmetry-expanded sample of 2972 Co(en) chelate rings (en = ethylenediamine) from 743 independent fragments [64]. While the data were expanded according to an idealized C_{2v} planar ring conformation (not C_2, as stated in the paper), the analysis revealed the mean geometry of the rings to be T, approximating C_2 symmetry, with the C_2 axis passing through the Co atom and the C-C bond. The density distribution of the data points in the two-dimensional conformation space (spanned by q and P) was modeled by a Boltzmann-type equation in order to establish an empirical PES, which was then compared to one obtained by means of molecular mechanics (MM) calculations (see Figure 8.16). The MM analysis yielded an E conformer transition state with Co out of the plane of the en ligand and an energy of 27.29 kJ mol^{-1} (at $P = 18°$), two enantiomeric T conformer ground states with energies of 2.34 kJ mol^{-1} (at $P = -72°$ and 108°), and an energy of 38.97 kJ mol^{-1} for the planar conformer (at the origin). Since the data do not fully span the semi-circular T→T itinerary, the authors conclude that "actual pseudorota-

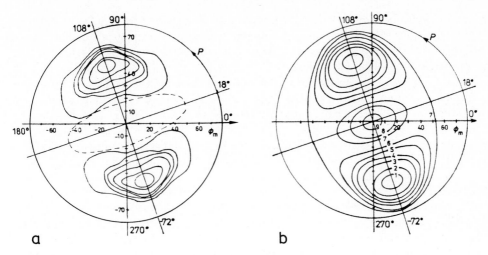

Fig. 8.16. a: Empirical PES for conformational changes of the Co(en) ring (contours are on an arbitrary scale)
b: Molecular-mechanics PES (energy values in kJ mol^{-1}). The E conformer transition state lies at $P = 18°$, two enantiomeric ground state T conformers at $P = -72°$ and $108°$

tion was not found"; instead, the data essentially map out a libration of the C–C bond of en about the C_2 axis passing through it and the Co atom. The structural results obtained from this study are not unexpected: the notion that the Co(en) chelate ring might exhibit pseudorotation ignores crucial (and basic) distinctions between the non-equilateral chelate ring, containing three different kinds of atoms, and the equilateral cyclopentane pentagon.

However, the main criticism that might be levelled at the paper is that it assumes exactly the same topography of the PES for each observed structure, an untenable assumption, given the widely divergent compounds involved. Indeed, Bürgi and Dunitz have criticized this approach in detail [65], making the point that the ensemble of different crystal structures does "not even remotely resemble a closed system at thermal equilibrium", the condition for Boltzmann-like statistics [65]; absolute deformation energies cannot therefore be obtained from such statistical analyses "without introducing arbitrary and unwarranted assumptions" (see also Chapter 5, Section 5.4).

8.5.3 Ring Inversion in Metallacyclopentenes

In part, Simmons's attempt [60] to detail an empirical PES (see Section 8.5.1 and Figure 8.15) for [Cu(bipy)$_2$(ONO)](NO$_3$) from variable-temperature crystallographic data on *one structure* reflects the philosophy of the alternate approach to the one discussed above. A study of metallacyclopentene ring inversion in (*s-cis-η^4*-butadiene)metallocene complexes (M = hafnium or zirconium) takes this approach further, by correlating experimentally obtained activation energies to displacements along the reaction path for *each observed structure* [66]. The data set comprised 29 fragments after symmetry expansion (C_s), from 15 original complexes [67]. A geometrical reaction coordinate was established from a PCA of both internal and inertial coordinates of the metallacyclic rings. The PCA revealed that two PCs accounted for 99.3% of variance, the first PC corresponding to the ring inversion coordinate, the second to one along which the bite size of the ligand increases and decreases; the deformations described by the PCs are illustrated in Figure 8.17a. The geometric reaction coordinate could be correlated with activation energies of the compounds in the data set, the latter determined from dynamic NMR experiments. Two energy profiles were drawn up for the reaction, both as functions of the displacement (x_0) of the observed structures along the reaction coordinate. The first profile assumed a quadratic dependence of $\Delta G^{\ne}$ on x_0, the second a fourth-order one; these are depicted as solid lines on the right and left sides of Figure 8.17b, respectively. Barrier heights (ΔE) to automerization were also modeled for each of the structures, and the dependence of ΔE on x, i.e. the reaction profile for one in the series of molecules, is illustrated by the broken lines in Figure 8.17b. The analysis showed that the free energy of activation for ring inversion decreases as the bite size of the ligand increases and the ring becomes progressively more planar.

8.6 Concluding Remarks

We have seen that the method of structure correlation has found some interesting applications in transition metal chemistry [68]. It has suggested solutions to some as yet unanswered questions — such as e.g. the geometric mechanism of M_5 cluster rearrangements, or of fluxional behavior in metal phosphine complexes. It has also raised new questions: the speculations about isomerization mechanisms of penta-coordinate Pt^{II} and Pd^{II} complexes, the dissociation pathway of five-coordinate Rh^I and Ir^I complexes, as well as the proposed S_N3 mechanism, are cases in point. Other possibilities concern the isomerization mechanisms of clusters and ligand polyhedra with seven or more vertices — the study of clusters, in particular, would seem to offer fruitful ground, considering recent increases in the number of solved structures, developments in the use of multivariate statistical techniques, and the lack of detail regarding geometric deformation paths. Future work might focus more on obtaining energetic information along with geometric detail, by combining kinetic and structural data to construct individual reaction-path — energy profiles.

However, a more fundamental question raises its head: Why is it that whole series of $[ML_n]$ complexes exhibit such interesting and consistent correlations, when sometimes all they have in common is a metal atom — often not even this is constant — surrounded by *n* coordinated ligand atoms?

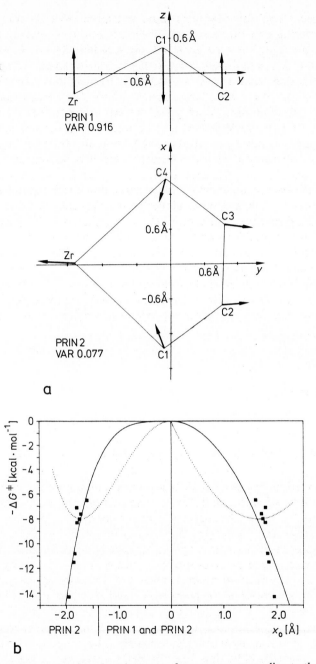

Fig. 8.17. a: Deformations of a zirconiacyclopentene fragment corresponding to the first two PCs; VAR is the amount of variance accounted for

b: Dependence of activation energy $\Delta G^{\ddagger}$ on ground-state structure x_0 (solid lines), and energy E on structural coordinate x for one particular molecule (broken lines), for a quadratic model (right) and a quartic model (left)

Although the number of transition metal systems that have been studied thus far is limited, the results are encouraging and should go some way toward reassuring inorganic chemists that the technique of structure correlation has great potential use, the relative complexities of inorganic systems notwithstanding. The advent of more sophisticated statistical methods has brought with it tools of considerably more power than the routine examination of scatterplots could ever hope to offer, even though the latter should always represent a point of departure in any correlation analysis. Moreover, as computational modeling of inorganic systems becomes more feasible, the lack of rigorous categorization of inorganic reaction mechanisms − by comparison with organic systems − will begin to fade away, and more comparable data should become available. Indeed, one could make the argument that it is precisely the relative lack of information on inorganic reaction pathways that should spur further work in the area.

References

[1]   Bürgi, H. B., *Inorg. Chem.* **1973**, *12*, 2321−2325
[2]   Bürgi, H. B., *Angew. Chem. Int. Ed. Engl.* **1975**, *14*, 460−473; Bürgi, H. B., Dunitz, J. D., *Acc. Chem. Res.* **1983**, *16*, 153−161
[3]   Allen, F. H., Kennard, O., Taylor, R., *Acc. Chem. Res.* **1983**, *16*, 146−153
[4]   Barrow, M., Bürgi, H. B., Johnson, D. K., Venanzi, L. M., *J. Am. Chem. Soc.* **1976**, *98*, 2356−2357
[5]   Bürgi, H. B., Murray-Rust, J., Camalli, M., Caruso, F., Venanzi, L. M., *Helv. Chim. Acta* **1989**, *72*, 1293−1300
[6]   Muetterties, E. L., Guggenberger, L. J., *J. Am. Chem. Soc.* **1974**, *96*, 1748−1756
[7]   Murray-Rust, P., Murray-Rust, J., *Acta Cryst.* **1975**, *A 31*, S 64
[8]   Murray-Rust, P., *Mol. Struct. Diff. Meth. Spec. Rep.* **1977**, *5*, 350−371
[9]   Chandrasekhar, K., Bürgi, H. B., *J. Am. Chem. Soc.* **1983**, *105*, 7081−7093
[10]   Nørskov-Lauritsen, L., Bürgi, H. B., *J. Comp. Chem.* **1984**, *6*, 216−228
[11 a] Basolo, F., Pearson, R. G., *Mechanisms of Inorganic Reactions*, 2nd edn. Wiley, New York, **1967**
[11 b] Anderson, G. K., Cross, R. J., *Chem. Soc. Rev.* **1980**, *9*, 185−215
[11 c] Twigg, M. V., *Mechanisms of Inorganic and Organometallic Reactions*, Plenum, New York, **1984**, Vol. 2
[11 d] Cross, R. J., *Adv. Inorg. Chem.* **1989**, *34*, 219−292
[12 a] Clearly, when the ligands are not all identical in the tetra-coordinate complex the symmetry of the SQP will be lower (than C_{4v}). The term SQP then implies that all apical equatorial angles are only approximately equal
[12 b] The SQP→TBP rearrangement implies that pairs of opposite apical-equatorial angles remain approximately equal during the process
[13]   Berry, R. S., *J. Chem. Phys.* **1960**, *32*, 933−938
[14]   Pauling, L., *J. Am. Chem. Soc.* **1947**, *69*, 542−553
[15]   Burdett, J. K., *Inorg. Chem.* **1979**, *18*, 1024−1030
[16]   Huheey, J. E., *Inorganic Chemistry*, 2nd edn., Harper and Row, New York, **1979**, 196
[17]   Auf der Heyde, T. P. E., Nassimbeni, L. R., *Inorg. Chem.* **1984**, *23*, 4525−4532
[18]   Holmes, R. R., Deiters, J. A., *J. Am. Chem. Soc.* **1977**, *99*, 3318−3326
[19 a] Holmes, R. R., *J. Am. Chem. Soc.* **1984**, *106*, 3745−3750

[19b] Holmes, R. R., *Prog. Inorg. Chem.* **1984**, *32*, 119–235
[20]   Auf der Heyde, T. P. E., Nassimbeni, L. R., *Acta Cryst.* **1984**, *B40*, 582–590
[21]   Liao, D., Fu, H., Tang, Y., *Wuli Huaxue Xuebao* **1987**, *3*, 449–452; *CAS* **1988**, *108*, entry 119883
[22]   Casey, C. P., Whiteker, G. T., Campana, C. F., Powell, D. R., *Inorg. Chem.* **1990**, *29*, 3376–3381
[23a] Auf der Heyde, T. P. E., Bürgi, H. B., *Inorg. Chem.* **1989**, *28*, 3960–3969
[23b] Auf der Heyde, T. P. E., Bürgi, H. B., *Inorg. Chem.* **1989**, *28*, 3970–3981
[23c] Auf der Heyde, T. P. E., Bürgi, H. B., *Inorg. Chem.* **1989**, *28*, 3982–3989
[24]   Kepert, D. L., *Inorganic Stereochemistry*, Springer, Heidelberg, Berlin **1982** (and references therein)
[25]   Favas, M. C., Kepert, D. L., *Prog. Inorg. Chem.* **1980**, *27*, 325–463
[26a] Kepert, D. L., Kucharski, E. S., White, A. H., *J. Chem. Soc. Dalton* **1980**, 1932–1938
[26b] The angles BME and BMD are defined within the same plane in the structures as the angles ϕ_D and ϕ_E; the two pairs of angular parameters are therefore geometrically related
[27]   Mesa, J. L., Arriortua, M. I., Lezama, L., Pizarro, J. L., Rojo, T., Beltran, D., *Polyhedron* **1988**, *7*, 1383–1388
[28]   Kepert, D. L., *Prog. Inorg. Chem.* **1977**, *23*, 1–65
[29]   Rodger, A., Johnson, B. F. G., *Inorg. Chem.* **1988**, *27*, 3061–3062
[30]   Britton and Dunitz, for instance, have catalogued the 257(!) graph-theoretically allowed non-isomorphic polyhedra for ML_8; see: Britton, D., Dunitz, J. D., *Acta Cryst.* **1973**, *A29*, 362–371
[31]   Guggenberger, L. J., Muetterties, E. L., *J. Am. Chem. Soc.* **1976**, *98*, 7221–7225
[32]   Kepert, D. L., *Prog. Inorg. Chem.* **1979**, *25*, 41–144
[33]   Drew, M. G. B., *Prog. Inorg. Chem.* **1977**, *23*, 67–210
[34]   Rodger, A., Johnson, B. F. G., *Polyhedron* **1988**, *7*, 1107–1120
[35]   Kepert, D. L., *Prog. Inorg. Chem.* **1978**, *24*, 179–249
[36]   Drew, M. G. B., *Coord. Chem. Rev.* **1977**, *24*, 179–275
[37]   Favas, M. C., Kepert, D. L., *Prog. Inorg. Chem.* **1987**, *28*, 309–367
[38a] Blaxill, C. P., Brown, S. S. D., Frankland, J. C., Salter, I. D., Sik, V., *J. Chem. Soc. Dalton* **1989**, 2039–2047 (and references therein)
[38b] Brown, S. S. D., Salter, I. D., Sik, V., Colquhoun, I. J., McFarlane, W., Bates, P. A., Hursthouse, M. B., Murray, M., *J. Chem. Soc. Dalton* **1988**, 2177–2185
[38c] Farrugia, L. J., Freeman, M. J., Green, M., Orpen, A. G., Stone, G. A., Salter, I. D., *J. Organomet. Chem.* **1983**, *249*, 273–288
[39]   Britton, D., Dunitz, J. D., *J. Am. Chem. Soc.* **1975**, *97*, 3836
[40]   Orpen, A. G., Salter, I. D., *Organometallics* **1991**, *10*, 111–117
[41]   Ugi, I., Marquarding, D., Klusacek, H., Gillespie, P., Ramirez, F., *Acc. Chem. Res.* **1971**, *4*, 288–296
[42a] Woodward, R. B., Hoffmann, R., *The Conservation of Orbital Symmetry*, Verlag Chemie, Weinheim, **1970**
[42b] Klärner, F. G., *Top. in Stereochem.* **1984**, *15*, 1–42
[43]   Mealli, C., Midollini, S., Moneti, S., Sacconi, L., Silvestre, J., Albright, T., *J. Am. Chem. Soc.* **1982**, *104*, 95–107
[44]   Silvestre, J., Albright, T. A., *Nouv. J. Chim.* **1985**, *9*, 659–668
[45a] Schwarz, H., *Acc. Chem. Res.* **1989**, *22*, 282–287
[45b] Crabtree, R. H., Hamilton, D. G., *Adv. Organomet. Chem.* **1988**, *28*, 299–338
[46]   Crabtree, R. H., Holt, E. M., Lavin, M., Morehouse, S. M., *Inorg. Chem.* **1985**, *24*, 1986–1992
[47]   Cotton, F. A., *Prog. Inorg. Chem.* **1976**, *21*, 1–28
[48]   Crabtree, R. H., Lavin, M., *Inorg. Chem.* **1986**, *25*, 805–812
[49]   Adams, H., Bailey, N. A., Bentley, G. W., Mann, B. E., *J. Chem. Soc. Dalton* **1989**, 1831–1844
[50]   Britton, D., Dunitz, J. D., *J. Am. Chem. Soc.* **1981**, *103*, 2971–2979
[51a] Jackson, W. G., in: *Stereochemistry of Organometallic and Inorganic Compounds*, Bernal, I. (ed.), Elsevier, Amsterdam **1986**, Vol. 1

[51 b] Jackson, W.G., Begbie, C.M., *Inorg. Chem.* **1983**, *22*, 1190–1197

[52]   Hathaway, B.J., *Struct. Bond.* **1984**, *57*, 55–118

[53]   Harrison, W.D., Kennedy, D.M., Hathaway, B.J., *Inorg. Nucl. Chem. Lett.* **1981**, *17*, 87–90

[54]   Harrison, W.D., Kennedy, D.M., Power, M., Sheahan, R., Hathaway, B.J., *J. Chem. Soc. Dalton* **1981**, 1556–1564

[55]   Tyagi, S., Hathaway, B.J., *J. Chem. Soc. Dalton* **1981**, 2029–2033

[56]   Hathaway, B.J., *Coord. Chem. Rev.* **1982**, *41*, 423–487

[57]   Tyagi, S., Hathaway, B.J., Kremer, S., Stratemeier, H., Reinen, D., *J. Chem. Soc. Dalton* **1984**, 2087–2091

[58]   Ray, N.J., Hulett, L., Sheahan, R., Hathaway, B.J., *J. Chem. Soc. Dalton* **1981**, 1463–1469

[59 a] Fereday, R.J., Hodgson, P., Tyagi, S., Hathaway, B.J., *J. Chem. Soc. Dalton* **1981**, 2070–2077

[59 b] Fitzgerald, W., Murphy, B., Tyagi, S., Walsh, B., Walsh, A., Hathaway, B., *J. Chem. Soc. Dalton* **1981**, 2271–2279

[59 c] Fitzgerald, W., Hathaway, B., Simmons, C.J., *J. Chem. Soc. Dalton* **1985**, 141–149

[60]   Simmons, C.J., Hathaway, B.J., Amornjarusiri, K., Santarsiero, B.D., Clearfield, A., *J. Am. Chem. Soc.* **1987**, *109*, 1947–1958

[61]   Instead of viewing the distortions of the complexes in terms of "structural pathways", Stebler and Bürgi have chosen to regard the structures as disordered as a consequence of pseudo-Jahn-Teller distortion; see Stebler, M., Bürgi, H.B., *J. Am. Chem. Soc.* **1987**, *109*, 1395–1401

[62]   Kilpatrick, J.E., Pitzer, K.S., Spitzer, R., *J. Am. Chem. Soc.* **1947**, *69*, 2483–2488

[63 a] Altona, C., Geise, H.J., Romers, C., *Tetrahedron* **1968**, *24*, 13–32

[63 b] Altona, C., Sundaralingam, M., *J. Am. Chem. Soc.* **1972**, *94*, 8205–8212

[63 c] Cremer, D., Pople, J.A., *J. Am. Chem. Soc.* **1975**, *97*, 1354–1358

[64]   Pavelcik, F., Luptakova, E., *Coll. Czech. Chem. Comm.* **1990**, *55*, 1427–1434

[65]   Bürgi, H.B., Dunitz, J.D., *Acta Cryst.* **1988**, *B44*, 445–448

[66]   Bürgi, H.B., Dubler-Steudle, K.C., *J. Am. Chem. Soc.* **1988**, *110*, 4953–4957

[67]   One of the structures is symmetrical, and lies on a mirror plane in the parameter space – it is therefore transformed into itself by the symmetry expansion operation

[68]   A potentially important application that has not been discussed here, concerns the optimization of molecular modeling software using averaged geometries and deformation paths obtained from structure correlation studies; see e.g. Vedani, A., Huhta, D.W., *J. Am. Chem. Soc.* **1990**, *112*, 4759–4767

9 Conformational Analysis

W. Bernd Schweizer

9.1 Conformational Analysis

9.1.1 Introduction

The conformation of a molecule is defined as an arrangement of atoms of given configuration corresponding to a potential energy minimum. Normally, different conformations are obtained by rotation about bonds. At a certain temperature, one or more of the possible conformations can be observed, depending on their relative energies and the energy barriers that have to be overcome. The potential energy and energy barriers are influenced by intra- and intermolecular forces. In vacuum – the condition to which most of the theoretical calculations refer – intermolecular forces do not exist. In the gas phase these forces are normally negligible, but in solution, and certainly in the solid state, distortions of the stable conformations of the free molecule can be expected.

However, if we exclude hydrogen bonding, the intermolecular forces in the crystal will generally be small. Therefore, X-ray analysis provides useful information about preferred conformations, because an atomic arrangement of a molecule observed in a crystal can still be expected to be rather close to an equilibrium conformation of the isolated molecule. The perturbation produced by the environment of the molecule in any individual crystal structure may cause a deviation from the minimum energy conformation of the isolated molecule, but the relative frequency of similar conformations of the same molecule or fragment in different crystals can give us an idea about their relative stabilities. It can be expected that conformations with lower potential energy will show up more often than those with higher energy and that evenly populated clusters are similar in energy.

Depending on the available experimental structural data, different approaches can or have to be used to extract the information. Whereas careful inspection of the conformation of single structures can give the spark to explain unusual chemical behavior, analyses of the variation in molecular geometry from a large number of chemical fragments need to be done by statistical and graphical methods.

In the following sections a number of examples are summarized with an emphasis on the chemical information that can be obtained from them. Some have already been mentioned in previous chapters to illustrate specific methodological aspects of structure-structure and of structure-energy correlation. Others will merely be mentioned to provide a reference to the primary literature for the interested reader.

9.1.2 Description of Local Conformations

In the various atomic arrangements that can be adopted by a given molecule, bond distances and angles do not change much. The main changes are in the torsion angles. These are best specified by their numerical values (see Chapter 1 for how these are calculated) and best visualized in terms of a projection along the central bond (Newman projection, Figure 9.1), but various qualitative descriptions have

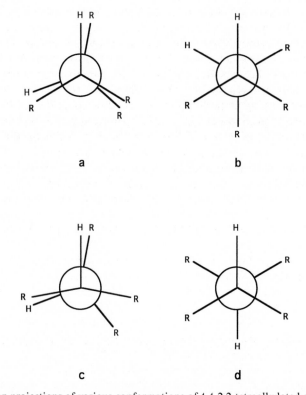

Fig. 9.1. Newman projections of various conformations of 1,1,2,2-tetraalkylated ethane; qualitative terms for the description of the conformations are:
a: *syn*, +*syn*-periplanar, eclipsed; b: *gauche* (H, H), skew (H, H), −*syn*-clinal (H, H), staggered;
c: *gauche* (H, H), −*anti*-clinal (H, H), eclipsed; d: *anti* (H, H), *anti*-planar (H, H), staggered

come into popular use. Some authors use the terms *eclipsed* or *syn* (for a torsion angle close to zero), *anti* (for one close to 180°), and *skew* or *gauche* (for one which is neither of these, say between 45° and 135°). Another set of terms, introduced by Klyne and Prelog [1], is somewhat more explicit: ±*syn-(peri)planar* or ±*sp* (for torsion angles of 0°±30°), +*syn-clinal* or +*sc* (+60°±30°), +*anti-clinal* or +*ac* (+120°±30°), ±*anti(peri)planar* or ±*ap* (180°±30°), −*anti-clinal* or −*ac* (−120°±30°), −*syn-clinal* or −*sc* (−60°±30°), where +, − and ± refer to the sign of the torsion angle and hence to the sense of helicity of the four-atom fragment involved, positive for a right-hand screw.

In addition, terms such as *eclipsed* and *staggered* have been used to describe the relationship between the projected directions of the three bonds at the two opposite sides of the Newman projection (Figure 9.1). As long as the three outer bond (or lone-pair) angles at each central atom are approximately equal, the projected angles are all about 120° and the projection has approximate threefold symmetry. Under these circumstances, the terms *eclipsed* and *staggered* can be used without too much danger of misunderstanding. But for large deviations from threefold symmetry these terms can become ambiguous and even misleading.

For example, in the Newman projection for the calculated stable conformation of 1,1,2,2-tetra-*t*-butylethane [2] (Figure 9.2b, **1**), each of the two vicinal hydrogens is nearly eclipsed with a bulky substituent, while the torsion angle between the two remaining substituents is 69°. Mislow [3] has introduced an *ad hoc* nomenclature to deal with this kind of situation. In the Newman projection, ligands in the front are labeled F and those in the back B. A pair of eclipsed ligands are labeled E. By reading the ligand labels in a clockwise order a descriptor is obtained. For example, a system with an alternating conformation has the descriptor FBFBFB. To define an unambiguous starting point, rank digits 2, 1, 0 for the labels F, B, E respectively are assigned. The largest number out of those obtained by replacing the labels by the corresponding rank digits serves as a unique descriptor.

We take again tetra-*t*-butylethane (**1**) as an example, assuming exactly eclipsed hydrogens (Figure 9.2b). Starting with the *tert*-butyl substituent pointing to the right the following descriptors (and corresponding numbers) are obtained:

EBFE (0120), BFEE (1200), FEEB (2001), EEBF (0012) .

The descriptor for the idealized conformation of **1** would therefore be FEEB as 2001 is the largest number.

9.2 Conformational Analysis of Single Molecules

In this section we illustrate by way of a few examples some of the chemical questions that can be addressed with the help of conformational data on a particular molecule or a few related ones.

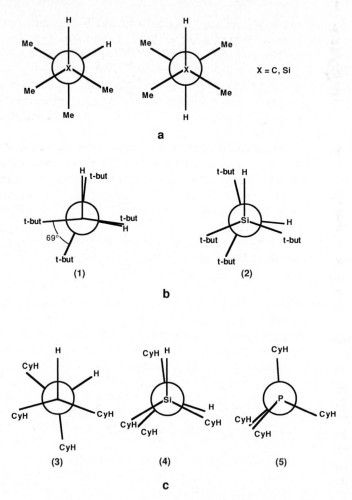

Fig. 9.2. Newman projections of major calculated conformations (X-ray data for **5**).
a: 1,1,2,2-tetramethyl ethane and disilane; relation *sc* to *ap* = 2:1; b: 1,1,2,2-tetra-*tert*-butyl ethane and disilane; c: 1,1,2,2-tetra-cyclohexyl ethane, disilane and diphosphine

9.2.1 Is the Conformation of Tetraalkyldiphosphines Caused by Stereoelectronic Influence? [3]

According to microwave spectroscopy [4], the stable conformation of diphosphine has a torsion angle of 19° between the nearest pair of hydrogens (74° rotation from the *syn* or eclipsed conformation of the molecule). Although the P−P distance in this molecule is relatively long, 2.22 Å, the observed conformation was interpreted as an example of the "*gauche* effect", the tendency of molecules "to adopt that structure which has the maximum number of *gauche* interactions between the adjacent electron pairs and/or polar bonds" [5]. To evaluate the importance of the *gauche* effect, Mislow and co-workers compared tetraalkyldisilanes with tetraalkyldiphosphines and tetraalkylethanes [6], using both observed structures and those derived by empirical force-field (EFF) calculations. The ethanes and disilanes are similar in their electronic structures – in particular, neither have lone pairs on the central atoms that could provoke a *gauche* effect – while the diphosphines and disilanes are similar in their steric interactions.

In Figure 9.2 the major calculated (EFF) conformations are depicted for the tetramethyl, tetracyclohexyl and tetra-*t*-butyl substituted disilanes and ethanes. For the cyclohexyl derivatives the choice of the stable conformation was not clear cut from the calculations. However, crystal structure determinations gave conformations close to the ones shown in the figure. The greater length of the P–P and Si–Si bonds (about 2.33 Å) compared with the central C–C bond of the ethanes (1.55–1.58 Å, depending on the substituents) must clearly lead to a reduction in the steric repulsion between the bulky substituents at opposite ends of the molecules. This is obviously less important for the methyl group, the least bulky of the three substituents.

According to the calculations, in 1,1,2,2-tetra-*t*-butylethane (**1**), the most overcrowded molecule in the series, the steric repulsion is alleviated by the extreme deformation mentioned in the previous section and shown in Figure 9.2b. The authors refer to this highly distorted conformation as a non-alternating FFBFBB structure (the ideally eclipsed structure would be called FEEB, as has been shown before). In contrast, 1,1,2,2-tetra-*t*-butyldisilane (**2**) adopts an alternating (FBFBFB), nearly eclipsed conformation as shown in Figure 9.2b. The staggered conformation with antiperiplanar relationships between the opposite groupings is calculated to be $11.9 \, \text{kcal mol}^{-1}$ higher; NMR experiments failed to indicate the presence of this form.

For 1,1,2,2-tetracyclohexylethane (**3**) the calculated energies of 30 starting conformations were minimized. The global minimum was obtained for the conformation shown in Figure 9.3 top with other conformations only about $1 \, \text{kcal mol}^{-1}$ higher in energy. Similar calculations for 1,1,2,2-tetracyclohexyldisilane (**4**) [7] gave an even less clear separation and depended on the choice of force field. The X-ray structures of tetracyclohexyldiphosphine (**5**) [8] and disilane (**4**) are remarkably similar as far as the molecular conformations are concerned (Figure 9.3 middle and bottom). Both conformations have approximate C_2 symmetry. The length of the P–P bond is 2.22 Å, that of the Si–Si bond is 2.33 Å. If there were a substantial electronic effect on the conformation we would expect to see a significant difference between the electronically unlike disilanes and diphosphines. But the disilane structures have more in common with the diphosphines than with the ethanes, pointing to the general importance of steric interactions in this series of molecules.

9.2.2 Conformation and Chemical Reactivity

Knowledge of the preferred conformation of a single molecule can sometimes explain or predict the course of a chemical reaction. An early example where the course

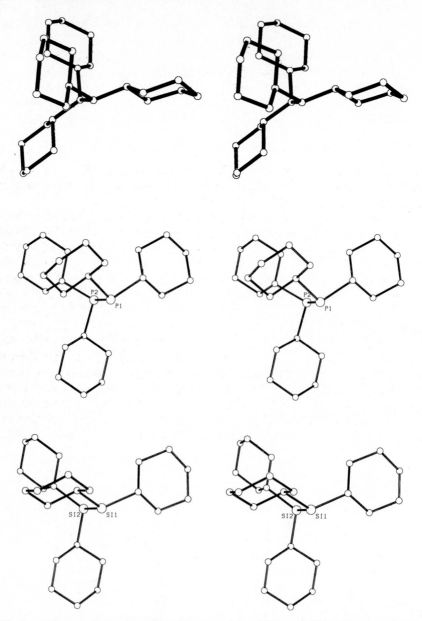

Fig. 9.3. Top: Stereoscopic drawing of 1,1,2,2-tetracyclohexylethane; Middle: Stereoscopic drawing of 1,1,2,2-tetracyclohexyldiphosphine; Bottom: Stereoscopic drawing of 1,1,2,2-tetracyclohexyl-disilane

of an unexpected reaction was explained in terms of the conformation found by X-ray analysis was given by Prelog, Küng and Tomljenovic for the acetolysis of cyclo-decyltoluene-*p*-sulfonate (**6**) [9].

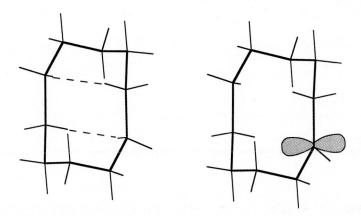

The cyclodecyl cation formed in the first step eliminates a proton to give a cyclo-olefin. The resulting product mixture contains only 20% of the thermodynamically more stable *cis* isomer but 80% of the less stable *trans* isomer.

X-ray analysis of several crystalline cyclodecane derivatives had shown that the preferred cyclodecane ring conformation shows energetically unfavorable short non-bonded H . . . H contacts and also CCC angles as wide as 120° (Figure 9.4, left). If the cyclodecyl cation is assumed to have the related conformation shown in Figure 9.4 right, then either of the two methylene groups next to the positively charged carbon atom can eliminate a proton to given the olefin, one leading to the *trans* and one to the *cis* isomer. A favorable transition state would have the empty p orbital parallel to the C−H bond of the leaving proton. The angle between the p orbital and the C−H bond for the formation of the *trans* isomer is only about 5° but it is around 25°−30° for the *cis* isomer. The conformational change for the transition state leading to the *cis* isomer is thus larger and presumably associated with a higher ac-tivation energy, which can explain the preferred formation of the *trans* isomer in a kinetically controlled process. The carbon atom most likely to eliminate a proton is thus the one with an intra-annular H atom involved in a short H . . . H distance, as shown by the dashed lines in the cyclodecane structure [10] in Figure 9.4 left.

Fig. 9.4. Left: Structure of cyclodecane. The dotted lines show the short H . . . H interactions. Right: The cyclodecyl cation intermediate after removal of the tosylate leaving group (see **6**)

9.2.3 Chair-Boat Interconversion of Six-membered Rings in the Solid

The idea that molecules in crystals are static and undeformable objects has been abandoned for a long time now. Many experiments have shown that substantial atomic displacements occur in the solid state. A nice example is the interconversion from the chair to the boat form of six-membered rings, as studied by Sim [11]. The chair-boat interconversion of cyclohexane is a classical problem in organic chemistry and has been studied extensively. It is well known that the chair form is the more stable conformer and that the boat form, which is 5 kcal mol^{-1} less stable [12], is hardly present at normal temperatures. There is therefore no hope that in a more or less unstrained cyclohexane ring a mixture of chair and boat forms can be observed in a crystal. In fused ring systems, however, the chance to find the boat form is much higher, even excluding those systems where the ring skeleton is forced by geometric constraints into a boat (e.g. bicyclo[2.2.1]heptane). For bicyclo[3.3.1]nonane, energy calculations give a difference of about 3 kcal mol^{-1} between the more stable chair-chair (*cc*) (**7a**) and the chair-boat (*cb*) conformation (**7b**) [13]. In rough agreement, 25% of the *cb* form has been observed at 673 K in an electron diffraction study [14]. The energy difference is still too high for both forms to be expected in a crystal. For steric reasons the energy difference between the chair and boat form in cyclohexanone is about 1.5 kcal mol^{-1} less than in cyclohexane. The energy difference between the *cc* and *cb* forms of bicyclo[3.3.1]nonan-9-one (**8**) should therefore be considerably less than in the parent hydrocarbon. Indeed, 22% of the *cb* conformer can

(7a)   (7b)   (8)

9 R=CH$_3$

10 R=CH$_3$-O

be observed at room temperature by NMR [15]. Derivatives of **8** seemed therefore promising candidates for observing both the *cc* and *cb* conformations in the crystal.

The *p*-toluenesulfonylhydrazone (**9**) and *p*-methoxybenzenesulfonylhydrazone (**10**) derivatives of **8** were chosen for X-ray analysis at several temperatures. At 130 K for **9** and 145 K for **10** the crystal structures are ordered, and both molecules show the energetically lower *cc* conformation. At higher temperature, the structures become more and more disordered. In both compounds only one six-membered ring shows the chair-boat interconversion. At room temperature the *cb* population rises to 31% for **9** and 20% for **10**. These changes are reversible. After a heating-cooling cycle, the same cell dimensions and reflection intensities were obtained. This means that the *cc-cb* interconversion is a dynamic process in the crystal. Sim used the occupation parameters obtained at the different temperatures to derive equilibrium constants and hence estimates of the difference in enthalpy for the two conformations in their crystal environments, obtaining ΔH values of about 1.5 kcal mol^{-1} for **9** and 3 kcal mol^{-1} for **10**.

9.3 Conformational Analysis on Multiple Molecular Fragments

The above examples show the use of individual structures of particular molecules to derive conclusions about chemical behavior. For many chemical questions, however, the method of choice is the analysis of the same fragment in a large number of molecules with different intra- and intermolecular environments. With the increasing number of available structures, this method is finding more and more applications. Essential tools for this kind of work are the crystallographic databases like the Cambridge Structural Database (CSD) [16] or the Inorganic Crystal Structure Database (ICSD, [17]; see also Chapter 3).

9.3.1 Low Energy Conformations of Macrocyclic Ring Systems

The calculation of molecular geometry by empirical force-field methods has become a widely used method nowadays. As far as bond distances and angles are concerned, sets of standard values are available and they are usually good enough for a first approximation. The straightforward way of finding the low energy conformations of ring systems by molecular mechanics is to examine all combinations of a set of torsion angles for each ring bond, checking for ring closure and calculating the energy. Because the number of possible conformations to be checked increases exponentially with the number of bonds, this method is restricted to systems of relatively small ring size. Many ways of coping with this problem have been attempted. One is the stochastic search method, which has been applied to find *all* conformers of cycloalkanes with 9 to 12 members [18].

For more complex molecules, it is advisable to combine computations with a knowledge of the various attainable conformations, which can be easily obtained by a systematic search through the known crystal structures, as demonstrated by the following examples. Gerber et al. [19a] have developed an algorithm for the systematic generation of conformations of macrocyclic systems by using a generalized pseudorotation concept, where a ring is described by phases and amplitudes [19b] (see 9.3.2 and Chapter 2, Sections 2.4.4 and 2.4.5), which leads to a complete two-dimensional Fourier representation of the ring shape. Generic shapes of a ring system are generated by using and modifying only two coefficients (radial and axial). To decide how many start structures have to be generated and what changes in the coefficients are important, extensive tests had to be done with numerous known macrocyclic structures found in the CSD.

Böhm et al. [20] tested different approaches for the conformational analysis of 9-ring lactams. A comparison was made of all the local minima obtained (1) by molecular dynamics, (2) by the above mentioned generic structure generation, (3) from a systematic search (with restrictive conditions for torsional change), (4) by the Monte-Carlo method (energy optimized with different force fields) and (5) from the 9-ring fragments found from X-ray structures in the CSD. With the exception of the systematic search, where the constraints for a manageable number of start structures turned out to be too narrow, all methods yielded the same global minimum. Many of the other, energetically higher minima were also found, but a disappointing feature was that the sequence and the energy differences showed no similarity between the optimization with the different force fields. The crystallographic data by themselves do not allow a quantitative conclusion on the relative energy levels of the conformations (see Chapter 5, Section 5.4).

9.3.2 Conformational Analysis of Carboxylic Esters and Amides

One motive for a conformational analysis of carboxylic esters [21] was to get standard geometrical parameters for numerous modifications of the ester group. The analysis of over 1 000 acyclic ester fragments showed one outstanding feature. There was not a single example of a (Z)-ester, i.e. no $C-C(O)-O-C$ torsion angle close to 0° could be found. For this Chapter, the search was repeated with the July 1991 release of the CSD. Disordered structures and molecules with a standard deviation of the $C-C$ bond length $\sigma > 0.01$ Å were excluded. Figure 9.5 shows the histograms of the conformations of the alcohol substituent of acyclic ester fragments. It is remarkable that among the 3952 fragments included there is still no example of an acyclic (Z)-ester. The largest deviation from the *s-trans* arrangement is 25°, found in **11** [22]. It is almost certainly caused by intermolecular packing forces, since there

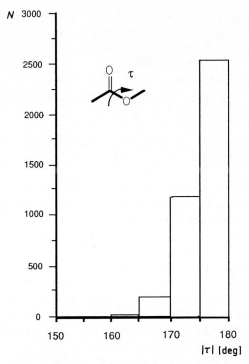

Fig. 9.5. Histogram of the absolute value of the C−C(O)−O−C torsion angle τ of 3950 acyclic esters found in the CSD release of July 1991. Only structures without disorder, no error flag, reported mean error in C−C bond length $\sigma \leq 0.01$ Å and agreement factor $R \leq 0.08$ were included

are two independent molecules in the crystal and the ester group of the second one shows a deviation of only 8° from the *s-trans* arrangement.

(11)

For a more detailed analysis, the sample was split into several sub-populations, depending on the nature of the substituents. Figure 9.6 shows the scattergrams of the absolute value of the torsion angle τ about the alcoholic C−O bond versus the bond angle α at the C substituent for esters of primary, secondary and tertiary alcohols.

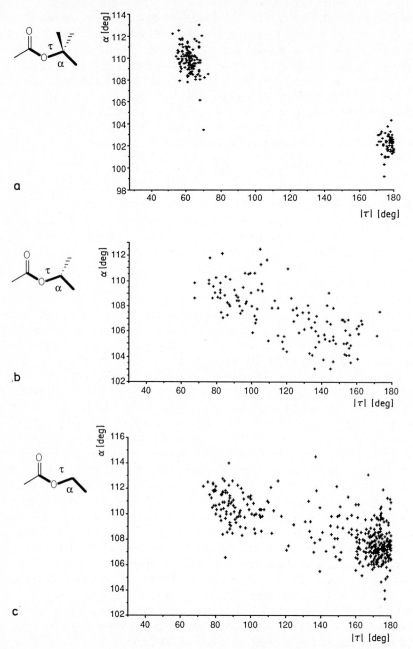

Fig. 9.6. Scattergrams of the absolute value of $C(O)-O-C-C$ torsion angle τ vs. $O-C-C$ angle α in acyclic esters found in the CSD release of July 1991. Exclusion criteria as in Figure 9.5 except that only structures with a reported mean error in $C-C$ bond length $\sigma \leq 0.005$ Å were included. Top: 74 esters of tertiary alcohols. Every fragment contributes three points to the scattergram. Middle: 31 esters of secondary alcohols. Every fragment contributes two points to the scattergram. Bottom: 395 esters of primary alcohols

A general feature is the increase of the $O-C-C$ angle α with increasing deviation of the $C(O)-O-C-C$ torsion angle from 180°, the antiperiplanar position to the ester $C(O)-O$ bond. The biggest distortions of the bond angles are observed in esters of tertiary alcohols. The two well defined clusters in Figure 9.6 (top) at 180° and 60° show that there is always a $C-C$ bond in an antiperiplanar position. As a consequence, the two other substituents have torsion angles of ±60°. For tertiary alcohols each ester group therefore contributes three points to the scattergram, one near 180° and two near 60°. The difference in the bond angle α between the centers of the two clusters is about 8°. That means that the local threefold axis relating the three substituents on the tertiary C is tilted from the alcoholic $O-C$ direction away from the carbonyl oxygen by about 5°. The pattern for esters of secondary alcohols (Figure 9.6 middle) is strikingly different in that it shows a rather even distribution of conformations over the whole range between 80° and 160° for the $C(O)-O-C-C$ torsion angle; it avoids the exact synclinal and antiperiplanar positions. The middle of the distribution is around 120°, corresponding to a synperiplanar orientation of the hydrogen atom to the $C(O)-O$ bond, a preference that was already discerned by Mathieson [23] in 1965. Here the mean difference in the bond angle α is about 5° between the two ends of the conformational range. The smaller value compared with the tertiary alcohols may be due simply to the smaller range of τ; the slopes are very similar. (Because of the periodicity of the geometrical changes associated with rotation around a bond, the angle deformation cannot be strictly linear with the change in τ; for symmetry reasons there have to be turning points at $\tau = 0$ and 180°.) This does not allow the expansion of the range from the secondary alcohols linearly to that of the tertiary alcohol esters to compare the two differences.

Most of the esters of the primary alcohols have the $C(O)-O-C-C$ torsion angle close to 180°, but there is a second prominent cluster around 85° (Figure 9.6 bottom). The change in the bond angle α is the smallest of the three groups, the difference between the midpoints of the two clusters being only about 3.5°. In a similar analysis by Murray-Rust [24] the conformational distribution of ethyl esters was compared with a potential energy curve for rotation about the O-ethyl bond obtained by Wilson [25] from a microwave study on ethyl formate. The potential energy curve (Figure 9.7 top) shows the global minimum for a conformer at 180° and a second local minimum, about $0.19\,\text{kcal}\,\text{mol}^{-1}$ higher, at 85°. There is a convincing correlation between the type of frequency distribution expected from such a curve and the actual histogram of the torsion angle τ (Figure 9.7 bottom, right). The two energy minima correspond nicely with the maxima in the distribution which correspond to fragments in the *s-trans* and *gauche* forms. The same effect is seen for all primary alcohols (Figure 9.7 bottom, left). Because of the huge number of relevant fragments, only those with $\sigma(C-C) < 0.005$ Å were included. Murray-Rust argued that the observed distribution could be interpreted as a Boltzmann distribution, but there are problems in defining the appropriate "temperature" [26], as discussed in Chapter 5.

The $C(O)-O$ bond length in esters is significantly shorter than the $C(sp^2)-O$ bond of aryl or vinyl ethers. In a study of the geometry of thioesters in crystal structures by Zacharias et al. [27] it was found that the bond length does not differ from that of the corresponding bond in aryl and vinyl monosulfides (O: 1.336 [21] vs. 1.367 Å; S: 1.767 vs. 1.763 Å [27]). This suggests that there is more electron delocal-

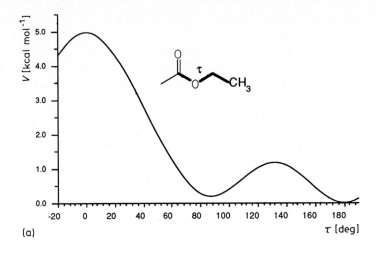

(a)

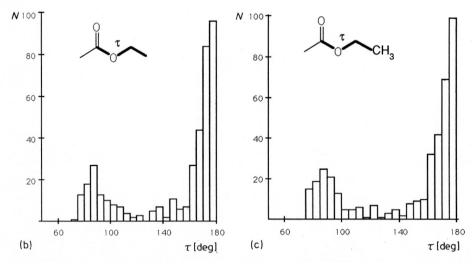

(b)   (c)

Fig. 9.7. Top: Potential energy curve for the rotation about the O-ethyl bond in ethyl formate according to a microwave study by Wilson [25]. Bottom: Histograms of $C(O)-O-C-C$ torsion angles of 395 esters of primary alcohols with $\sigma(C-C) \leq 0.005$ Å (left) and 397 ethyl esters with $\sigma \leq 0.01$ Å (right) found in the CSD release of July 1991

tion from oxygen into the carbonyl group than from sulfur. Acyclic thioesters all show the (E) form with a maximum deviation of about $10°$ from the antiperiplanar arrangement (CSD search, January 1992 release, 41 fragments).

A study similar to that of the carboxylic esters was done for carboxylic amides by Chakrabarti and Dunitz [28]. A general conclusion for the amides from primary, secondary and tertiary $C(\alpha)$ amines is that the $C(\alpha)-C(\beta)$ bond of an alkyl substituent avoids the synperiplanar arrangement to the $C(O)-N$ bond. For the *tert*-alkyl substituents one of the $C-C$ bonds is therefore always in an antiperiplanar position

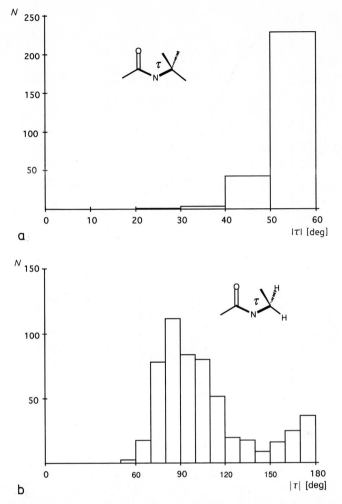

Fig. 9.8. Histograms of $C(O)-N-C(\alpha)-C(\beta)$ torsion angles τ (absolute values) in N-alkyl amides derived from (a) tertiary $C(\alpha)$ and (b) primary $C(a)$ amines. Of the three values for the tertiary case, the one closest to 60° was chosen

(Figure 9.8a). For the cases with primary substituted $C(\alpha)$ there seem to be two preferred conformations as can be seen from the histogram in Figure 9.8b. The dominant form is the *s-gauche* conformation and a second peak in the population can be observed for the *s-trans* form. It is interesting to compare the conformational preferences of the ethyl esters (Figure 9.7) with those of the amides. Both fragments favor the same torsion angles around 180° and 85°. Only the frequency distribution is opposite and the spread of the angles for the amides is somewhat wider. This agrees with the observation that the conformational preferences for the amides are less pronounced than for the esters.

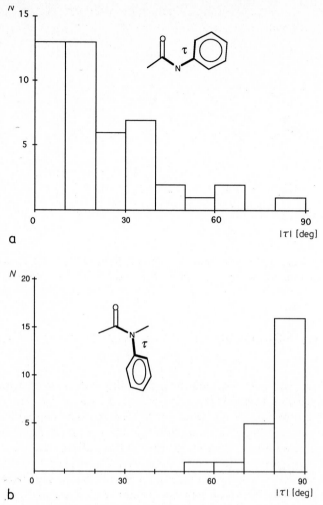

Fig. 9.9. Histograms of $C(O)-N-C-C$ torsion angles τ (absolute values) for N-aryl amides (top: secondary, bottom: tertiary)

A distinct difference in the conformation can be seen for secondary and tertiary N-aryl amides (Figure 9.9). The aromatic system of the secondary amides tends to lie close to the plane of the amide group (*ortho* substituted aryl systems were excluded). On the other hand, the aromatic substituents in the tertiary amides are nearly perpendicular to the amide plane.

The conformation about the $C-N$ bond has been studied in cyclic amides (O) [29a] and enamines (D) [29b]. Gilli et al. did a similar analysis and expanded the selection of fragments with thioamides (S), amidines (N), and anilines (A) [29c]. The 90 molecular structures included in the study map a pathway whereby the rotation about C−N is correlated with pyramidalization at N, the latter being largest when the nitrogen lone pair is in the X=C−N plane.

A study by Kaftory and Agmon describes a pathway for the simultaneous inversion of the two neighboring nitrogen atoms in the 1,2,4-triazolidinedione ring [30].

9.3.3 Conformational Studies of the Methoxyphenyl Group

A detailed study of the conformational flexibility of a methoxy group on phenyl rings was done by Hummel, Huml and Bürgi [31 a]. An earlier analysis of *ortho* unsubstituted methoxyphenyls by Nyburg and Faerman [31 b] showed that the coplanar conformation of the ring and the methoxy group was preferred. Substituents in the *ortho* position can be expected to force the methoxy group out of the ring plane. In Figure 9.10 the distribution of the torsion angle τ of the methoxy group to the phenyl ring is shown. The dominant conformation is still the coplanar arrangement (the shaded area shows the *ortho*-unsubstituted rings) with a second, more diffuse maximum at $\tau = 90°$. The rotation of the methoxy group is coupled with a change of the bond lengths and bond angles. As the methoxy group turns out of the ring plane a decrease of conjugation with the ring electrons can be expected. There is a very slight increase in the mean values of the $CH_3−O$ and $O−Ph$ distances: ($\tau \approx 0°$: 1.423(9), 1.368(8) Å; $\tau \approx 90°$: 1.427(11), 1.376(7) Å). The scatter of bond lengths caused by different experimental conditions and crystal quality are of the same order of magnitude as the bond length differences, which makes it difficult to be sure of the significance of the correlated change of distances and torsion angle.

A systematic increase of the bond angles α and β (Figure 9.11) can be seen when the $CH_3−O$ group turns into the plane of the phenyl ring. This avoids too close contact of the methyl group with the atom in the *ortho*-position. The change of β versus τ can be approximated very nicely by a cosine function [31 a]:

$$\beta(°) = 120 + 4.63 \cos(\tau)$$

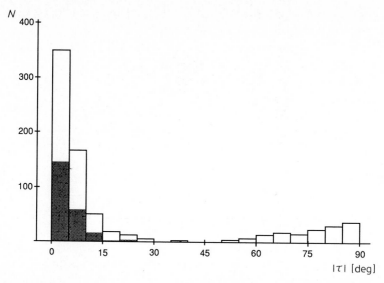

Fig. 9.10. Histogram of the absolute value of the twist angle τ of 768 methoxyphenyl rings found in the January 1992 release of the CSD. The shaded area represents the compounds with hydrogen atoms in *ortho*-positions

whereas in the case of α versus τ a regression through the points around $\tau \approx 90°$ extrapolates to $\alpha \approx 123°$ for $\tau \approx 0°$ instead of the actual value of ca. 118°. This reflects the difference in the nature of the *ortho*-substituent. The points at $\tau \approx 0°$ belong to non-substituted rings which allow smaller non-bonded distances to the methyl group and hence less deformation of the bond angle α compared with the substituted rings that contribute to the points around $\tau \approx 90°$.

A similar study for the conformational flexibility of the acetoxyphenyl group was done by Hummel, Roszak and Bürgi [32]. In these molecules the steric repulsion does not allow a coplanar arrangement of the ring and the ester group. The same tendency of increasing bond angle β with smaller torsion angle τ was observed.

9.3.4 Conformation and Pseudorotation of Five-membered Rings

The conformational freedom in cyclic molecules is reduced by ring closure conditions. This makes them interesting for a systematic analysis because, especially for small and medium rings, the number of known fragments is high and the overdetermination of the structure by the geometrical parameters leads necessarily to correlations among them. The conformation of a five-membered ring can be described by two parameters, a puckering amplitude q which expresses the overall distortion of the ring from planarity and a phase angle ϕ which describes the nature of the ring

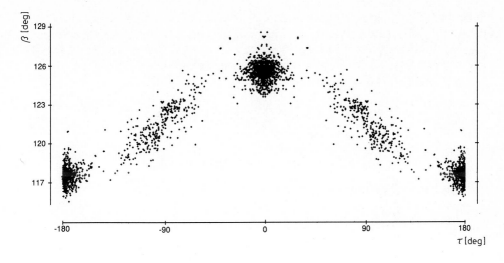

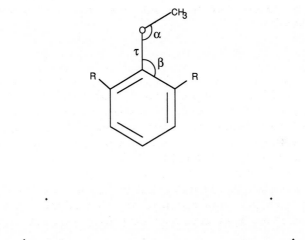

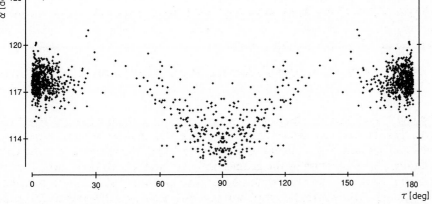

Fig. 9.11. Scattergrams of twist angle τ versus bond angles α and β in methoxyphenyl groups

puckering. These parameters can be estimated from the torsion angles in the ring [33] or from the displacements z_j of the individual atoms from the mean plane through the ring. From the general formulae for ring puckering of Cremer and Pople [34] the parameters for a five-membered ring are calculated as follows:

$$\tan \phi = - \frac{\sum\limits_{j=1}^{5} z_j \sin [4\pi (j-1)/5]}{\sum\limits_{j=1}^{5} z_j \cos [4\pi (j-1)/5]}$$

$$q = \frac{\sum\limits_{j=1}^{5} z_j \sin [4\pi (j-1)/5]}{\cos \phi}.$$

There are two special conformations of the ring – the twist and envelope forms. For an equilateral five-membered ring the ideal twist form contains a twofold axis and the envelope form is mirror symmetric. The twist form is specified by phase angles ϕ of 18°, 54°, 90° ... and the envelope form by 0°, 36°, 72°, ... depending on how the atoms are numbered relative to the symmetry element. A phase change by 36° therefore corresponds to an apparent rotation of the ring by 72° (see Chapter 2, Section 2.4.4). The conformational flexibility of the five-membered rings was in fact the origin of the concept of pseudorotation. Kilpatrick, Pitzer and Spitzer [19b] realized that by changing the torsion angles in the ring all phases of ϕ can be obtained, equivalent to a rotation of the ring. In cyclopentane all conformations with a constant puckering amplitude have essentially the same energy. When heteroatoms or ring substituents are present, the pseudorotation path will no longer be equienergetic.

Murray-Rust and Motherwell [35] searched the CSD for all β-1'-aminofuranoside fragments **(12)** in nucleosides and nucleotides and made a principal component analysis of the conformations found for the five-membered ring and its direct substituents. A detailed description of this analysis is given in Chapter 4, Section 4.6.3.1. Here we merely summarize the results.

(12)

The analysis showed that essentially three factors contribute to the variation of the torsion angles. The first two could clearly be assigned as ring-puckering factors and the third as side chain $C(3')-C(4')-C(5')-O$ rotation. In a new analysis with the 1991 CSD version (only structures of unfused rings with a reported standard deviation of the $C-C$ bond lengths better than 0.01 Å without disorder were selected), taking only the five ring torsion angles, the first two factors included 99.9% of the variance! Since most of the molecules in the sample are chiral and of known absolute configuration, only the sample-point corresponding to the observed enantiomer was included (the other enantiomer would have torsion angles of opposite sign and phase angle $\phi + 180°$). Figure 9.12 shows a scattergram of the scores from the 181 fragments on the two factors. In Figure 9.13 the same data are expressed in terms of the puckering parameters q and ϕ. Most points lie close to a circle that describes the idealized pseudorotational path with the puckering amplitude $q = 0.394$ Å. Although principal component analysis gives only a phenomenological description of the data, it is striking that the distributions identify clearly two clusters with a high population of points. The region with $-18° < \phi < 18°$ includes all furanose rings with $C(2')$ *endo* and the range $126° < \phi < 162°$ contains rings with $C(3')$ in the *endo* position. But there is a non-negligible number of conformations outside those regions.

As mentioned above, the third factor of the principal component analysis describes rotation of the side chain. The factors are by definition mutually orthogonal

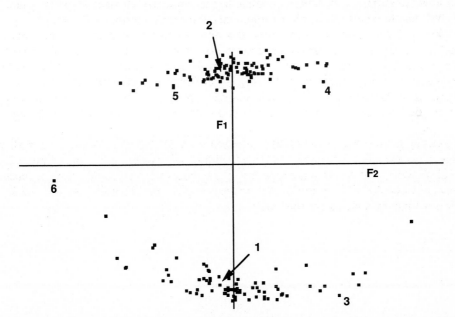

Fig. 9.12. Scattergram of the scores on the first two factors from a principal component analysis of the torsion angles of 181 furanose rings. The two factors include essentially 100% of the variance of the ring torsion angles. The numbers show the position of some selected ring conformations. 1: envelope (e) with mirror plane through $C(2')$ *endo*; 2: e $C(3')$ *endo*; 3: e $C(3')$ *exo*; 4: e $C(2')$ *exo*; 5: twist with C_2 axis through $C(1')$; 6: e $O(1')$ *endo*

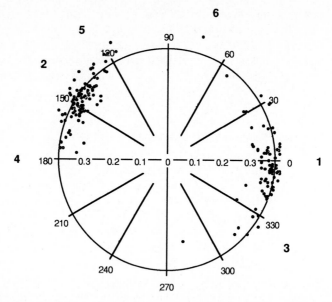

Fig. 9.13. Polar diagram of the puckering phase ϕ (°) vs. the puckering amplitude q (Å) of 181 non-fused furanose rings in fragments of type **11** found in the CSD release July 1991 (no disorder, no error flag, $R \leq 0.08$). The numbers show the position of some selected ring conformations; 1 (0°): envelope (e) with mirror plane through C(2') *endo*; 2(144°): e C(3') *endo*; 3(324°): e C(3') *exo*; 4(180°): e C(2') *exo*; 5(126°): twist with C_2 axis through C(1'); 6(72°): e O(1') *endo*

and hence uncorrelated. Moreover, the eigenvector corresponding to the third factor contains no significant contributions from the parameters that contribute heavily to the first two factors. Thus, the side-chain variation appears to be independent of the ring puckering.

Packing forces alone can change the ring puckering considerably. There are several examples of crystal structures with two symmetry independent molecules that show different ring conformations. For example, in oxanosine **(13)** [36] one molecule has the C(2') *exo* and the other the C(3') *exo* conformation, a phase change of about 150°.

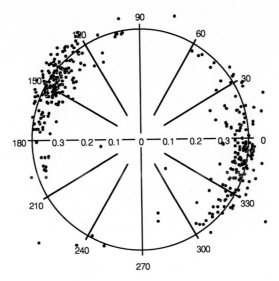

Fig. 9.14. Polar diagram of the puckering phase ϕ (°) vs. the puckering amplitude q(Å) of 376 acyclic substituted furan rings found in the CSD release July 1991 (no disorder, no error flag, $R \leq 0.08$)

It is instructive to compare the furanose sample with the complete set of furane rings. Figure 9.14 shows the ring puckering distribution for all furane fragments with the same selection criteria as was applied for the distribution in Figure 9.13. The sample size is more or less doubled (376) compared with the furanose compounds, which are, of course, still included in the diagram. The range covered by the phase angle ϕ is now extended from $-60°$ to $30°$ and from $120°$ to $200°$ with a few additional points in the more empty regions of Figure 9.13. This spread does not mean that there is a difference in the conformational preferences of the two sets. In the case of the furanose sample the molecules all have an N substituent on C(1') and a CH_2-O-R group at C(4'), which allows an unambiguous numbering of the five-membered ring. Secondly, most of the data for Figure 9.13 was obtained from nucleotides and nucleosides of known absolute configuration $\beta(R)$ at C(1'). In contrast, the furane sample contains also data from racemic crystals where the published coordinates refer to an arbitrarily chosen sense of chirality. Also, in the non-furanose rings the numbering system was not uniform with respect to interchange of C(1') and C(4') or of C(2') and C(3') and so interferes with the expression of any conformational preferences of similar non-symmetric ring fragments on the diagram. Changing the configuration at C(1') will for a specific phase angle ϕ change the *exo-endo* relationship which, of course, cannot be obtained from the puckering parameters; taking the enantiomeric structure will add 180° to the phase angle ϕ. Thus much of the wider scatter in Figure 9.14 can be explained by the non-unique definition of the ring fragment (see Chapter 2).

The asymmetric distribution in the diagram demonstrates one of the dangers associated with an automated analysis. Inattention to such details or careless handling of the analytical methods can lead to wrong interpretations if part of the data is biased by a certain chemical class as, in our case, by the nucleoside fragments.

9.3.5 Conformational Analysis of Cyclopentenones

The planar arrangement of the atoms in an equilateral five-membered ring with an angle of 108° close to the tetrahedral angle corresponds to minimum bond angle strain. Torsional strain and steric repulsion of substituents can deform the ring and lead to a smaller average bond angle. In unsaturated rings the resistance to twisting about a double bond is an extra force to keep the atoms in the plane. In addition, there is less strain from the exocyclic substituents of the sp^2 atoms. On the other hand, the deformation of the endocyclic bond angles has to be greater than in the saturated ring. A study has been done for cyclopentenone rings [37], where three of the five C-atoms are of the sp^2 type. The ring is therefore expected to be much flatter than cyclopentane and, indeed, as can be seen from Figure 9.15, there are many examples with an almost planar ring. Roughly 50% of the fragments have a puckering amplitude q smaller than 0.1 Å. The observed range of the phase angle ϕ is such that practically only the saturated C-atoms deviate from the plane: most conformations are close to the C(4), C(5) twist form ($\phi = -18°$) or to the envelope form with C(5) out of the plane ($\phi = 0°$). The two torsion angles C(3)– C(2)–C(1)–C(5) and C(4)–C(3)–C(2)–C(1) are essentially zero ($\phi = 18°$ to 90° and 270° to 306°). The bond angles in the ring (Figure 9.15 bottom) show that the angle deviation (circa 10°) from an unstrained system for the sp^2 atoms is about twice that of the saturated carbons (4 to 5°). The higher variation on C(5) ($\sigma = 2°$ vs. ca. 1° for the other angles) probably reflects the fact that this atom is the one most involved in ring puckering.

9.4 Space Groups as a Tool to Visualize Conformational Variation

As pointed out earlier in discussing the puckering amplitudes of five-membered rings (Figure 9.14) a particular selection of molecules in a sample can bias the apparent distribution of geometrical parameters in the population. In our case, the dominance of one enantiomer among a class of similar natural products caused an artificial unsymmetrical clustering. Even if symmetrically related conformations are distributed randomly over symmetry-equivalent forms an optical presentation of the data could give a wrong impression. One solution to that problem is to expand every observed

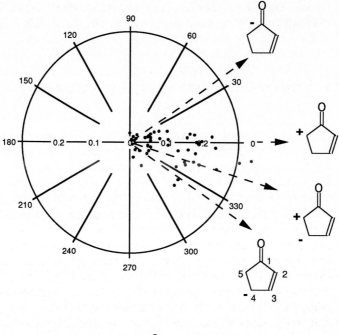

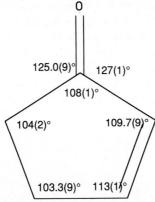

Fig. 9.15. Top: Polar diagram of the puckering phase $\phi(°)$ vs. the puckering amplitude $q(\text{Å})$ of 52 cyclopenten-2-one fragments with no cyclic substitution and $\sigma(C-C) \leq 0.01$ Å from the July 1987 release of the CSD. Since coordinates of chiral molecules from racemic crystals refer to an arbitrarily chosen enantiomer, all points were transferred into the range $\phi = 270-90°$ by adding 180° (corresponding to the enantiomeric structure) if necessary. Bottom: Mean bond angles with standard deviations from the 26 fragments from the above structures with $\sigma(C-C) \leq 0.005$ Å

conformation to give all symmetry-equivalent forms. If the parameters have a periodicity (like torsion angles) they can be described in a periodic configuration space with space-group like symmetries, as has been suggested by Murray-Rust, Bürgi and Dunitz [38] (see Chapter 2, Section 2.6). The method requires a configuration space whose dimension equals the number of periodic degrees of freedom. Obviously there is a practical limitation to this approach, since even if the complexity of space groups with more than three dimensions can be handled by mathematical methods, it is already very difficult to visualize the results in four dimensions. It is probably impossible in five- and higher-dimensional spaces unless one can use projections onto lower dimensional spaces without losing too much information.

9.4.1 Two Torsion Angles

An example of visualizing a conformational map as a two-dimensional periodic pattern is given for the ring flip process in 1,1-diarylvinyl systems by Kaftory et al. [39] and in benzophenones by Rappoport, Biali and Kaftory [40]. In the latter case, if one assumes a rigid frame built of the carbonyl group and the two carbon substituents, the torsion angles of the two rings (ϕ_1, ϕ_2) to the frame determine the conformation of the molecule. An idealized frame has C_{2v} symmetry. Because of the symmetry of the molecule, four energy-equivalent conformations exist: (ϕ_1, ϕ_2), (ϕ_2, ϕ_1), $(-\phi_1, -\phi_2)$ and $(-\phi_2, -\phi_1)$, which can be plotted in a two-dimensional coordinate system where the two axes describe the variation of the torsion angles (Figure 9.16). In both directions there is a periodicity of π, assuming ideal, unsub-

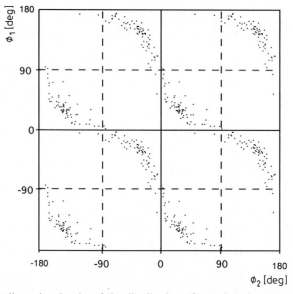

Fig. 9.16. Four two-dimensional units of the distribution of sample points (ϕ_1, ϕ_2) for the conformations of benzophenone. The unit translations are $180°$. The fragments were obtained from the January 1992 release of the CSD; 75 fragments included with no cyclic substituents

stituted phenyl rings. By choosing the appropriate translation vectors the conformational map represents a two-dimensional lattice of the plane group *cmm* (see Chapter 2, Section 2.6.3).

The calculated [40] low-energy form of benzophenone has a propeller conformation with $\phi_1 = \phi_2 = 30°$, which agrees well with the crystal structure (29.4°, 30.9°) [41]. Figure 9.17 shows the possible mechanisms for the interconversion of the molecule to the enantiomer with opposite helicity. Each phenyl ring can choose between two ways to attain its new position. One is rotation through the plane of the

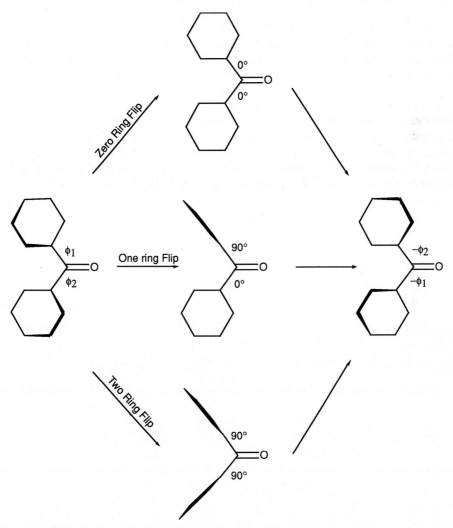

Fig. 9.17. Transition states for zero-, one-, and two-ring flips in benzophenone. A torsion angle of 0° indicates that the phenyl ring lies in the plane defined by the carbonyl group and the *ipso*-carbon atoms

carbonyl frame, and the other is rotation through a perpendicular position to the plane, which is called a ring flip. There are therefore different transition states for a zero-, one- or two-ring flip (for the one-ring flip there is, of course, the choice of either of the two rings). Because of the steric interactions the ring movements will not be independent.

In Figure 9.16 the symmetry-extended conformations of 75 benzophenone fragments found in the January 1992 release of the CSD are mapped. With a few exceptions the points lie in the region where both torsion angles have the same sign, which is the condition for a propeller conformation, whereas the region for torsion angles with opposite sign is practically empty. The points scatter along a curve close to the diagonal through (0°, 90°) and (90°, 0°). A denser population of points can be observed in the region of (30°, 30°), corresponding to the low-energy conformation of benzophenone.

What can we conclude from the conformational map for the stereoisomerization of benzophenone? A zero-ring flip mechanism would start from a point, for example, with torsion angles (30°, 30°) and follow a path along the diagonal through (0°, 0°) and end at the enantiomeric structure at (−30°, −30°). A two-ring flip would follow a path in the opposite direction through (90°, 90°) and end at (150°, 150°), which is equivalent to (−30°, −30°). Neither of those paths is supported by the distribution of points in the conformational map. For the one-ring flip, one ring (say ring 1) turns into the plane of the carbonyl skeleton and the other ring turns in the opposite direction to reach the transition state at (0°, 90°). Continuing these rotations leads to the point (−30°, 150°) = (−30°, −30°). This reaction path is very well represented by the experimental data. Thus the structure correlation method shows that the interconversion of benzophenone occurs preferably by the one-ring flip. At the beginning of the process both rings rotate by about the same amount. As one ring approaches the plane of the carbonyl frame (torsion angle 0°) the other ring turns ("flips") more rapidly through the perpendicular position, as can be seen from the larger variation in the torsion angle ϕ_2 for ϕ_1 close to 0° (or vice versa).

Conformational preferences have been studied for many other $C_6H_5-X-C_6H_5$ fragments (X = CH_2, CR_2, C=O, NH, NR, O, S, SO_2) [42], as well as for nitrobenzenes with $C(sp^2)$ or $N(sp^2)$ substituents in *ortho*-position [43]. If the terminal phenyl groups in the above examples are replaced by triptycyl-units, molecular analogs of bevel gear systems are obtained [44]. The conformational interconversion pathway mapped by bis(9-triptycyl)methane and related molecules follows almost exactly a gearing motion, as expected for two interlocked cogwheels [45]. (Note that the periodicity of the conformational map is now 120° instead of 180° as for the diphenylmethanes.)

A similar study was performed on systems containing two or more *t*-butyl groups in close proximity [46]. It has been complemented by comparative empirical force-field calculations for propane, di-*t*-butylmethane and bis(9-triptycyl)methane and by a group theoretical analysis [47] along the lines given in Chapter 2 (Section 2.6.3). Some of the detail, not obtainable from crystal structure data alone, could be extracted from NMR-measurements on appropriately substituted bis(9-triptycyl) methane derivatives [48].

9.4.2 Three Torsion Angles

The space-group method has been applied by Bye, Schweizer and Dunitz [49] to map the conformational distribution for triphenylphosphine fragments, in particular, triphenylphosphine oxides, as described in Chapter 2, Section 2.6.4. The phenyl rings on the phosphorus were taken as rigid rotors on a C_{3v} frame built by the P-atom and the bonds to the phenyl groups with O=P as the threefold axis. The conformation of a fragment is specified by the three torsion angles of the phenyl rings to the P=O group (Figure 9.18). Two conformations of the molecule show C_{3v} symmetry: one with all three torsion angles 0° and the other with all of them 90°. Adding a translational part for the periodicity of the torsion angles to the symmetry operations that give the isometric conformations leads to a distribution of points that can be described by the three-dimensional rhombohedral space group $R32$. The molecule with torsion angles of (0°, 0°, 0°) (C_{3v}) corresponds to a point at the origin

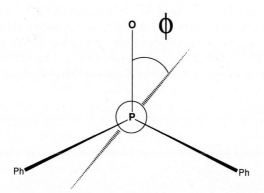

Fig. 9.18. Definition of the torsion angle for a phenyl ring on triphenylphosphine oxide. The figure shows a projection of the molecule down a P–C bond onto the corresponding phenyl group. Zero torsion angle means eclipsing of the O–P bond with the phenyl ring. The smaller of the two torsion angles is taken

of the cell. For better visualization, hexagonal axes were used (see Chapter 2 for the transformation from the rhombohedral to the hexagonal coordinate system).

Figure 9.19 top shows the distribution of the 62 points from the $Ph_3P=O$ fragments found in the CSD of the September 1981 release projected down the c_H axis. Figure 9.19 bottom is an orthogonal view of three clusters around the threefold axes, corresponding to molecules with torsion angles (ϕ, ϕ, ϕ), which seem to be the equilibrium conformation of a $Ph_3P=O$ fragment. There are two kinds of triad axes in the cell. In the range $0° \le z_H \le 60°$ the points on or close to the axis (0°, 0°, z_H) correspond to molecules with positive torsion angles, whereas the points close to the axis (120°, 60°, z_H) belong to the enantiomeric structures.

Between enantiomeric clusters one can recognize a pattern of sample points that connects the two dense regions. This represents a path for the stereoisomerization of a given molecule with threefold symmetry into its enantiomer. For example, a molecule with torsion angles (40°, 40°, 40°) ($x_H, y_H, z_H = 0°, 0°, 40°$) will interconvert to the corresponding enantiomeric molecule at $x_H, y_H, z_H = 120°, 60°, 20°$ with torsion angles (−40°, −40°, −40°). The stereoisomerization path starts out from one threefold axis in the direction to the other but avoids the midpoint at 60°, 30°, 30° which corresponds to a molecule with torsion angles (90°, 0°, 0°). The transition point is close to 60°, 40°, 30°, which represents a molecule with torsion angles (90°, 10°, −10°). When we follow the path from the triad axis at 0°, 0° 40° (corresponding to torsion angles (40°, 40°, 40°)) to the transition point we will see that at the beginning one torsion angle increases twice as fast as the other two decrease. The rotation of the two rings will not be synchronous any more as the torsion angle of the first ring approaches 90°. The third ring has to rotate a little faster to reach the mirror-symmetric transition state around 60°, 40°, 30° (torsion angles (90°, 10°, −10°)). So the first ring rotates by 50°, the second by −30° and the third by −50°. To reach the enantiomorphic structure with torsion angles (140°, −40°, −40°), which is equivalent to (−40°, −40°, −40°), the roles of the second and the third rings are interchanged.

There is an obvious thinning out of points in the region of the midpoint of the path, and this may be taken as an indication that the energy along the path reaches its maximum near this point (but see Chapter 5, Section 5.4). We therefore identify the transition state for the stereoisomerization as lying close to this point. Whether the transition state is actually mirror-symmetric (image of a point on the dyad axis of the space group) or only approximately so is impossible to decide from this kind of information. At any rate, the mirror-symmetric structure must correspond to a turning point in the energy, so it must be either an energy minimum between a pair of equivalent transition states or an energy maximum, but in the latter event it may not lie exactly on the reaction path. The passage of one ring through torsion angle 90° and of the other two rings through zero corresponds to what has been termed a "two-ring flip" mechanism for the stereoisomerization of the cognate molecule trimesitylmethane [50]. The structures of the two enantiomers and of the transition state are shown in Figure 9.20. Very similar conclusions were reached for other Ph_3P-X molecules. There does not appear to be any other experimental evidence so far on the mechanistic details of these stereoisomerization processes or on the

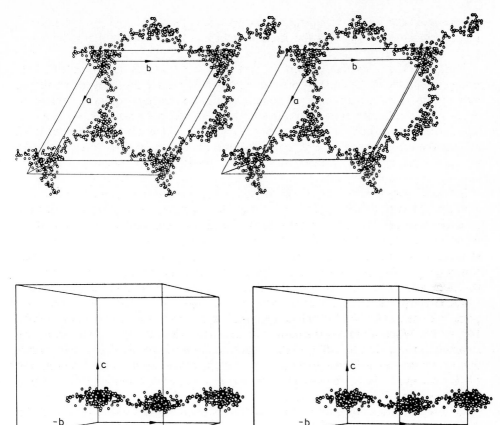

Fig. 9.19. Top: Stereoscopic view of the distribution of Ph_3PO sample points in a section of the hexagonal cell with $0° \leq z_H \leq 60°$. The unit translations are $180°$. Bottom: Sample points as in above but showing only clusters around the threefold axes $0°, 0°, z_H$; $120°, 60°, z_H$ and $180°, 0°, z_H$ centered at $0°, 0°, 40°$; $120°, 60°, 20°$ and $180°, 0°, 40°$

nature of the transition state, but for Ph_3PO itself force-field calculations [51] support the conclusions derived by the structure correlation method and suggest that the energy variation along the reaction path is only a few kcal mol^{-1}.

In square planar complexes of the type $XM(PR_3)_3$ the PR_3 groups can rotate about the $P-M$ bonds [52]. In this example, only two of three rotating groups are equivalent. This changes the relevant (space group) symmetry to $C2/c$. There are two pathways of conformational interconversion. Along one of them, the three groups move like interlocked cogwheels, along the other, two groups show a gearing, the other two a gear-clashing motion. This motion is accompanied by significant changes of the $M-P-R$ bond angles similar to the ones described in Section 9.3.2 for the $O-C-C$ angles in carboxylic esters and in Section 9.3.3 for the $C-O-C$ or $O-C-C$ angles in methoxyphenyl fragments.

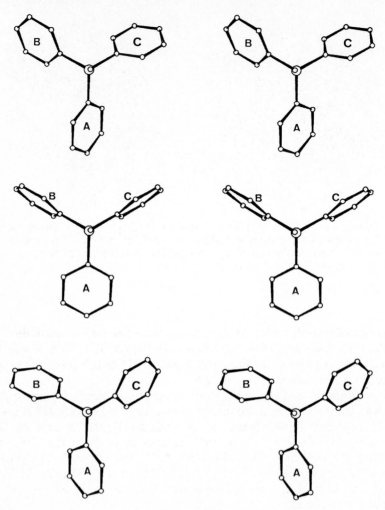

Fig. 9.20. Stereoscopic views of three structures along the stereoisomerization path of Ph$_3$PO. Top: C_3-symmetric conformation with torsion angles $40°, 40°, 40°$; middle: mirror-symmetric transition state with torsion angles $90°, 10°, -10°$; bottom: C_3-symmetric conformation with torsion angles $-40°, -40°, -40°$

9.4.3 Four and More Torsion Angles

When the same method is applied to conformational interconversion involving four torsion angles, the visualization problems become much more severe. One method, described in Chapter 2, Section 2.6.5, is to use a normal stereo plot for three of the four dimensions, and to represent the distance along the fourth coordinate by circles

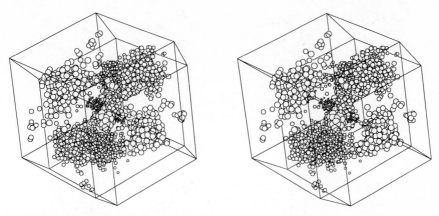

Fig. 9.21. Hyperstereoscopic view of the four-dimensional unit cell with axes length π of sample points for the conformation of tetraphenylborates. The projection from four into three dimensions is down the [1111] direction. Distance along this axis is indicated by the size of the spheres; they are small when they are far from the viewer

of increasing radius as the point approaches the viewer along the fourth dimension. This method has been applied to tetraphenyl derivatives XPh_4 with $X = B$, P and As. A detailed description of the symmetry aspects and the definitions of the torsion angles are given in Chapter 2, Section 2.6.5.

The distribution of the 24 symmetry-equivalent sample points for the conformations of 144 Ph_4B molecules in the primitive four-dimensional unit cell is shown in Figure 9.21. A preferred conformation of the tetraphenylborate anion shows "open" $^oD_{2d}$ symmetry [53] (Figure 9.22 top left) represented by the points $(30°, 30°, 30°, 30°)$, $(90°, 90°, 90°, 90°)$ and $(150°, 150°, 150°, 150°)$. The "closed" $^cD_{2d}$ conformation (Figure 9.22 top right) would be located at $(0°, 0°, 0°, 0°)$, $(60°, 60°, 60°, 60°)$ and $(120°, 120°, 120°, 120°)$, but is not found in the distribution, suggesting that this is a high-energy conformation.

Figure 9.21 shows two sets of symmetry-equivalent clusters. One set, in the middle of the scatter plot, consists of three clusters, each centered at a $^oD_{2d}$ special position and extending along a line that preserves S_4 molecular symmetry. There is also a second set of six clusters. These correspond to conformations with approximate S_4 symmetry (Figures 9.22 bottom) with torsion angles around $(-30°, 30°, -30°, 30°)$, etc. In Figure 9.21 the first set of clusters are heavily superimposed and are to be distinguished in the stereo-view by their radii, which correspond to "height" in the fourth dimension. The sausage-like shapes of these three clusters are turned by 120° to one another. One of these sausage-like clusters (centered at $(90°, 90°, 90°, 90°)$ and marked **1**) is shown in Figure 9.23, extending along opposite S_4 directions towards two clusters of the second set. This suggests a low-energy distortion of the $^oD_{2d}$ conformation that preserves S_4 symmetry, leading to a second group of low-energy conformations with this symmetry and centered around $(-30°, 30°, -30°, 30°)$ (or $(150°, 30°, 150°, 30°)$, marked **2** in Figure 9.23). The closest symmetry-equivalent

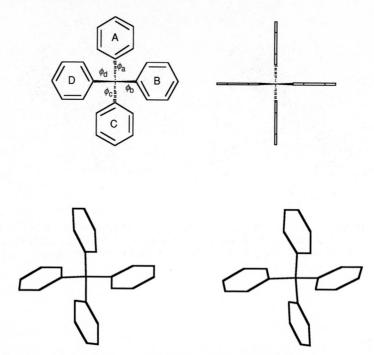

Fig. 9.22. Top: The two conformations of Ph_4X molecules with highest symmetry D_{2d}. The left side shows the "open" $^oD_{2d}$ form with torsion angles $\phi_a = \phi_b = \phi_c = \phi_d = 90\,°$; on the right the "closed" $^cD_{2d}$ form with $\phi_a = \phi_b = \phi_c = \phi_d = 0\,°$. Bottom: Stereoview of a conformation with S_4 symmetry (torsion angles $= -30\,°$, $30\,°$, $-30\,°$, $30\,°$)

cluster would be centered at $(210°, 30°, 90°, 90°)$, outside the limits of the primitive cell, but there is no well defined path between the two.

The Ph_4P and Ph_4As distributions are very similar, except that the first set of clusters has practically disappeared (see Figure 2.22 in Chapter 2). The steric interaction of the phenyl rings is certainly strongest in the Ph_4B fragments. With the elongation of the bond from the phenyl rings to the central atom, the non-bonding interactions become less severe. Energy calculations with any of the common molecular mechanics force fields predict a relatively flat energy profile for the ring rotation in Ph_4As compounds, assuming the same torsion potentials for the $As-C$ bond as for $B-C$ (in fact the $C-C$ potential was used). The non-bonding forces can therefore not be responsible for the avoidance of the $^oD_{2d}$ region. It suggests that there must be some electronic interactions between the central atom and the phenyl rings that favor the conformations of cluster type 2.

The most complicated problem of this kind analyzed so far is the fragment $M(PPh_3)_2$ [54]. It shows eight conformational degrees of freedom, two about the $M-P$ bonds and six about the $P-Ph$ bonds. The study showed gearing motion of the two PPh_3 groups alternating with helicity inversion of one and then the other PPh_3 group (see Section 9.4.2).